AF606535

Gene Therapy Protocols

METHODS IN MOLECULAR BIOLOGY™

John M. Walker, SERIES EDITOR

459. **Prion Protein Protocols**, edited by *Andrew F. Hill, 2008*
458. **Artificial Neural Networks:** *Methods and Applications*, edited by *David S. Livingstone, 2008*
457. **Membrane Trafficking**, edited by *Ales Vancura, 2008*
456. **Adipose Tissue Protocols, Second Edition**, edited by *Kaiping Yang, 2008*
455. **Osteoporosis**, edited by *Jennifer J. Westendorf, 2008*
454. **SARS- and Other Coronaviruses:** *Laboratory Protocols*, edited by *Dave Cavanagh, 2008*
453. **Bioinformatics, Volume 2:** *Structure, Function, and Applications*, edited by *Jonathan M. Keith, 2008*
452. **Bioinformatics, Volume 1:** *Data, Sequence Analysis, and Evolution*, edited by *Jonathan M. Keith, 2008*
451. **Plant Virology Protocols:** *From Viral Sequence to Protein Function*, edited by *Gary Foster, Elisabeth Johansen, Yiguo Hong, and Peter Nagy, 2008*
450. **Germline Stem Cells**, edited by *Steven X. Hou and Shree Ram Singh, 2008*
449. **Mesenchymal Stem Cells:** *Methods and Protocols*, edited by *Darwin J. Prockop, Douglas G. Phinney, and Bruce A. Brunnell, 2008*
448. **Pharmacogenomics in Drug Discovery and Development**, edited by *Qing Yan, 2008*
447. **Alcohol:** *Methods and Protocols*, edited by *Laura E. Nagy, 2008*
446. **Post-translational Modification of Proteins:** *Tools for Functional Proteomics, Second Edition*, edited by *Christoph Kannicht, 2008*
445. **Autophagosome and Phagosome**, edited by *Vojo Deretic, 2008*
444. **Prenatal Diagnosis**, edited by *Sinhue Hahn and Laird G. Jackson, 2008*
443. **Molecular Modeling of Proteins**, edited by *Andreas Kukol, 2008.*
439. **Genomics Protocols:** *Second Edition*, edited by *Mike Starkey and Ramnanth Elaswarapu, 2008*
438. **Neural Stem Cells: Methods and Protocols**, *Second Edition*, edited by *Leslie P. Weiner, 2008*
437. **Drug Delivery Systems**, edited by *Kewal K. Jain, 2008*
436. **Avian Influenza Virus**, edited by *Erica Spackman, 2008*
435. **Chromosomal Mutagenesis**, edited by *Greg Davis and Kevin J. Kayser, 2008*
434. **Gene Therapy Protocols: Volume 2:** *Design and Characterization of Gene Transfer Vectors*, edited by *Joseph M. Le Doux, 2008*
433. **Gene Therapy Protocols: Volume 1:** *Production and In Vivo Applications of Gene Transfer Vectors*, edited by *Joseph M. Le Doux, 2007*
432. **Organelle Proteomics**, edited by *Delphine Pflieger and Jean Rossier, 2008*
431. **Bacterial Pathogenesis:** *Methods and Protocols*, edited by *Frank DeLeo and Michael Otto, 2008*
430. **Hematopoietic Stem Cell Protocols**, edited by *Kevin D. Bunting, 2008*
429. **Molecular Beacons:** *Signalling Nucleic Acid Probes, Methods and Protocols*, edited by *Andreas Marx and Oliver Seitz, 2008*
428. **Clinical Proteomics:** *Methods and Protocols*, edited by *Antonio Vlahou, 2008*
427. **Plant Embryogenesis**, edited by *Maria Fernanda Suarez and Peter Bozhkov, 2008*
426. **Structural Proteomics:** *High-Throughput Methods*, edited by *Bostjan Kobe, Mitchell Guss, and Huber Thomas, 2008*
425. **2D PAGE: Volume 2:** *Applications and Protocols*, edited by *Anton Posch, 2008*
424. **2D PAGE: Volume 1:**, *Sample Preparation and Pre-Fractionation*, edited by *Anton Posch, 2008*
423. **Electroporation Protocols**, edited by *Shulin Li, 2008*
422. **Phylogenomics**, edited by *William J. Murphy, 2008*
421. **Affinity Chromatography:** *Methods and Protocols, Second Edition*, edited by *Michael Zachariou, 2008*
420. **Drosophila:** *Methods and Protocols*, edited by *Christian Dahmann, 2008*
419. **Post-Transcriptional Gene Regulation**, edited by *Jeffrey Wilusz, 2008*
418. **Avidin-Biotin Interactions:** *Methods and Applications*, edited by *Robert J. McMahon, 2008*
417. **Tissue Engineering**, *Second Edition*, edited by *Hannsjörg Hauser and Martin Fussenegger, 2007*
416. **Gene Essentiality:** *Protocols and Bioinformatics*, edited by *Svetlana Gerdes and Andrei L. Osterman, 2008*
415. **Innate Immunity**, edited by *Jonathan Ewbank and Eric Vivier, 2007*
414. **Apoptosis in Cancer:** *Methods and Protocols*, edited by *Gil Mor and Ayesha Alvero, 2008*
413. **Protein Structure Prediction**, *Second Edition*, edited by *Mohammed Zaki and Chris Bystroff, 2008*
412. **Neutrophil Methods and Protocols**, edited by *Mark T. Quinn, Frank R. DeLeo, and Gary M. Bokoch, 2007*
411. **Reporter Genes for Mammalian Systems**, edited by *Don Anson, 2007*
410. **Environmental Genomics**, edited by *Cristofre C. Martin, 2007*
409. **Immunoinformatics:** *Predicting Immunogenicity In Silico*, edited by *Darren R. Flower, 2007*
408. **Gene Function Analysis**, edited by *Michael Ochs, 2007*
407. **Stem Cell Assays**, edited by *Vemuri C. Mohan, 2007*
406. **Plant Bioinformatics:** *Methods and Protocols*, edited by *David Edwards, 2007*
405. **Telomerase Inhibition:** *Strategies and Protocols*, edited by *Lucy Andrews and Trygve O. Tollefsbol, 2007*
404. **Topics in Biostatistics**, edited by *Walter T. Ambrosius, 2007*
403. **Patch-Clamp Methods and Protocols**, edited by *Peter Molnar and James J. Hickman 2007*
402. **PCR Primer Design**, edited by *Anton Yuryev, 2007*

METHODS IN MOLECULAR BIOLOGY™

Gene Therapy Protocols

Volume 2: Design and Characterization of Gene Transfer Vectors

Third Edition

Edited by

Joseph M. Le Doux

Department of Biomedical Engineering, Georgia Institute of Technology and Emory University, Atlanta, GA

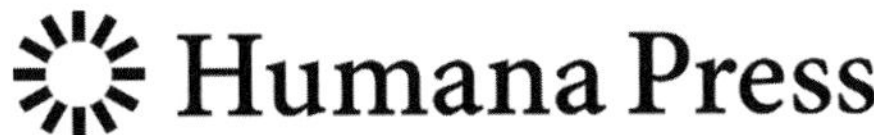

Editor
Joseph M. Le Doux
Department of Biomedical Engineering
Georgia Institute of Technology and Emory University
Atlanta, GA 30322-0355
joe.ledoux@bme.gatech.edu

Series Editor
John M. Walker
School of Life Sciences
University of Hertfordshire
College Lane
Hatfield Campus
Hatfield, Herts., UK AL10 9AB
john.walker543@ntlworld.com

ISBN: 978-1-60327-247-6 e-ISBN: 978-1-60327-248-3
ISSN: 1064-3745 e-ISSN: 1940-6029

Library of Congress Control Number: 2007941274

Cover illustration: Background image: Photo provided courtesy of Dongsheng Duan. Modified by Nancy Fallatt with permission. Cover image: Figure 5, Chapter 15, "Regulated Expression of Adenoviral Vectors-Based Gene Therapies," by James F. Curtin, Marianela Candolfi, Mariana Puntel, Weidong Xiong, A.K.M. Muhammad, Kurt Kroeger, Sonali Mondkar, Chunyan Liu, Niyati Bondale, Pedro R. Lowenstein, and Maria G. Castro.

Printed on acid-free paper

9 8 7 6 5 4 3 2 1

springer.com

Preface

Gene therapy has the potential to significantly impact human healthcare in the twenty-first century. The idea behind gene therapy is simple: to deliver genetic material to cells that will slow down or halt the progression of disease, or to help repair or regenerate damaged or lost tissues. To successfully implement this simple idea, however, we must first address a number of technological challenges. One such challenge is that of vector design. Methods are needed to design vectors that exhibit specific attributes that will help them achieve safe and effective gene transfer. For example, targeted vectors are needed that transfer their genetic material to the diseased cell type and no others. Often, strict control of the level and timing of gene expression is required. The first gene transfer vectors were designed to augment expression of the delivered gene. However, vectors are also needed to reduce or eliminate gene expression or to replace or repair damaged genetic material within diseased cells. Frequently, in order to treat chronic diseases, it is necessary to achieve stable gene expression, either by integrating the genetic material into the chromosomal DNA of the host cell, or by maintaining the transferred gene as an episome. Another critical issue is the manufacture of gene transfer vectors. Methods must be developed that will allow gene transfer vectors to be produced on a large scale and in highly concentrated and pure form, and the outcome of using these vectors to genetically modify cells should be predictable and reproducible.

The field of gene therapy is rapidly advancing on all of these fronts. Significant improvements have been made to the core gene transfer technologies in critical areas such as gene transfer efficiency, targeted gene delivery, regulated gene expression, and vector processing and characterization. In this new and entirely revised third edition, *Gene Therapy Protocols, Volumes 1 and 2* present a comprehensive collection of detailed methods and protocols used by the leaders in the field of gene therapy. The first volume covers current and emerging methods for the processing and characterization of major viral and nonviral gene transfer vectors. The second volume presents some of the most promising approaches available to design vectors for efficient, targeted, and regulated gene delivery and expression. This compilation of protocols is expected to serve as a valuable resource for graduate students and postdoctoral fellows, as well as for basic and clinical researchers in the industry and academia.

Joseph M. Le Doux

Contents

Contributors

PATRICK AEBISCHER • *Neurosciences Institute, Swiss Federal Institute of Technology Lausanne, Lausanne, Switzerland*

SALVADOR F. AUSAR • *Department of Pharmaceutical Chemistry, University of Kansas, Lawrence, KS*

BRUCE J. BAUM • *Gene Therapy and Therapeutics Branch, NIDCR, NIH, DHHS, Bethesda, MD*

LALITHA R. BELUR • *Arnold and Mabel Beckman Center for Transposon Research, Gene Therapy Program, Institute of Human Genetics, and Department of Genetics, Cell Biology and Development, University of Minnesota, Minneapolis, MN*

KRISTIAN BERG • *Department of Radiation Biology, Institute for Cancer Research, Rikshospitalet-Radiumhospitalet Medical Center, Montebello, Oslo, Norway*

NIYATI BONDALE • *The Board of Governors' Gene Therapeutics Research Institute, Cedars-Sinai Medical Center and Department of Molecular and Medical Pharmacology, David Geffen School of Medicine, University of California, Los Angeles, Los Angeles, CA*

ANETTE BONSTED • *Department of Radiation Biology, Institute for Cancer Research, Rikshospitalet-Radiumhospitalet Medical Center, Montebello, Oslo, Norway*

VINCENT BRONDANI • *Charité Medical School, Institute of Virology, Berlin, Germany*

MARIANELA CANDOLFI • *The Board of Governors' Gene Therapeutics Research Institute, Cedars-Sinai Medical Center and Department of Molecular and Medical Pharmacology, David Geffen School of Medicine, University of California, Los Angeles, Los Angeles, CA*

MARIA G. CASTRO • *The Board of Governors' Gene Therapeutics Research Institute, Cedars-Sinai Medical Center and Department of Molecular and Medical Pharmacology, David Geffen School of Medicine, University of California, Los Angeles, Los Angeles, CA*

TONI CATHOMEN • *Charité Medical School, Institute of Virology, Berlin, Germany*

F. L. COSSET • *Ecole Normale Supérieure de Lyon, Lyon, France*

DAVID T. CURIEL • *Division of Human Gene Therapy, Departments of Medicine, Pathology, Surgery, and Obstetrics and Gynecology, and the Gene Therapy Center, University of Alabama at Birmingham, Birmingham, AL*

JAMES F. CURTIN • *The Board of Governors' Gene Therapeutics Research Institute, Cedars-Sinai Medical Center and Department of Molecular and Medical Pharmacology, David Geffen School of Medicine, University of California, Los Angeles, Los Angeles, CA*

MARTIN FUSSENEGGER • *Institute for Chemical and Bioengineering, ETH Zurich, Zurich, Switzerland*

ALAIN GARNIER • *Department of Chemical Engineering, Centre de Recherche sur la Fonction, la Structure et l'Ingénierie des Protéines, Université Laval, Québec, Québec, Canada*

VALERIA GONZALEZ-NICOLINI • *Institute for Chemical and Bioengineering, ETH Zurich, Zurich, Switzerland*

JUSTIN HANES • *Department of Chemical and Biomolecular Engineering, Johns Hopkins University, Baltimore, MD*

ROBERT G. HAWLEY • *Department of Anatomy and Regenerative Biology, The George Washington University Medical Center, Washington, DC*

TERESA S. HAWLEY • *Flow Cytometry Core Facility, The George Washington University Medical Center, Washington, DC*

ANDERS HØGSET • *PCI Biotech AS, Oslo, Norway*

SANGEETA B. JOSHI • *Department of Pharmaceutical Chemistry, University of Kansas, Lawrence, KS*

AMINE KAMEN • *Biotechnology Research Institute, NRC, Montreal, Québec, Canada*

JAMES T. KOERBER • *The Department of Chemical Engineering and The Helen Wills Neuroscience Institute, The University of California, Berkeley, CA*

JOHN O. KONZ • *Bioprocess R&D, Merck & Co., Inc., West Point, PA*

ROBERT M. KOTIN • *Laboratory of Biochemical Genetics, National Heart, Lung, and Blood Institute, Bethesda, MD*

KURT KROEGER • *The Board of Governors' Gene Therapeutics Research Institute, Cedars-Sinai Medical Center and Department of Molecular and Medical Pharmacology, David Geffen School of Medicine, University of California, Los Angeles, Los Angeles, CA*

SAMUEL K. LAI • *Department of Chemical and Biomolecular Engineering, Johns Hopkins University, Baltimore, MD*

JOSEPH M. LE DOUX • *Department of Biomedical Engineering, Georgia Institute of Technology and Emory University, Atlanta, GA*

JOHN A. LEWIS • *Merck Research Laboratories, Merck & Co., Inc., West Point, PA*

CHUNYAN LIU • *The Board of Governors' Gene Therapeutics Research Institute, Cedars-Sinai Medical Center and Department of Molecular and Medical Pharmacology, David Geffen School of Medicine, University of California, Los Angeles, Los Angeles, CA*

PEDRO R. LOWENSTEIN • *The Board of Governors' Gene Therapeutics Research Institute, Cedars-Sinai Medical Center and Department of Molecular and Medical Pharmacology, David Geffen School of Medicine, University of California, Los Angeles, Los Angeles, CA*

BILL C. MATHIS • *Vaccine and Biologics Research, Merck Research Laboratories, Merck & Co., Inc., West Point, PA*

R. SCOTT MCIVOR • *Arnold and Mabel Beckman Center for Transposon Research, Gene Therapy Program, Institute of Human Genetics, and Department of Genetics, Cell Biology and Development, University of Minnesota, Minneapolis, MN*

C. RUSSELL MIDDAUGH • *Department of Pharmaceutical Chemistry, University of Kansas, Lawrence, KS*

SONALI MONDKAR • *The Board of Governors' Gene Therapeutics Research Institute, Cedars-Sinai Medical Center and Department of Molecular and Medical Pharmacology, David Geffen School of Medicine, University of California, Los Angeles, Los Angeles, CA*

ALLISON MONTALVO • *Merck Research Laboratories, Merck & Co., Inc., West Point, PA*

A. K. M. MUHAMMAD • *The Board of Governors' Gene Therapeutics Research Institute, Cedars-Sinai Medical Center and Department of Molecular and Medical Pharmacology, David Geffen School of Medicine, University of California, Los Angeles, Los Angeles, CA*

FELIX MÜLLER-LERCH • *Charité Medical School, Institute of Virology – CBF, Berlin, Germany*

D. NÈGRE • *Ecole Normale Supérieure de Lyon, Lyon, France, and BioSciences Lyon-Gerland, Lyon, France*

CATHERINE R. O'RIORDAN • *Genzyme Corporation, Framingham, MA*

LEE R. PITTS • *Bioprocess R&D, Merck & Co., Inc., West Point, PA*

LINA PRASMICKAITE • *Department of Tumor Biology, Institute for Cancer Research, Rikshospitalet-Radiumhospitalet Medical Center, Montebello, Oslo, Norway*

MARIANA PUNTEL • *The Board of Governors' Gene Therapeutics Research Institute, Cedars-Sinai Medical Center and Department of Molecular and Medical Pharmacology, David Geffen School of Medicine, University of California, Los Angeles, Los Angeles, CA*

ALI RAMEZANI • *Department of Anatomy and Regenerative Biology, The George Washington University Medical Center, Washington, DC*

SANGEETHA L. SAGAR • *Bioprocess R&D, Merck & Co., Inc., West Point, PA*
DAVID V. SCHAFFER • *The Department of Chemical Engineering and A137, The Helen Wills Neuroscience Institute, The University of California, Berkeley, CA*
DAVID J. SEGAL • *Charité Medical School, Institute of Virology – CBF, Berlin, Germany*
MARÍA DE LAS MERCEDES SEGURA • *Department of Chemical Engineering, Centre de Recherche sur la Fonction, la Structure et l'Ingénierie des Protéines, Université Laval, Québec, Québec, Canada*
RICHARD H. SMITH • *Laboratory of Biochemical Genetics, National Heart, Lung, and Blood Institute, Bethesda, MD*
ANTONIUS SONG• *Genzyme Corporation, Framingham, MA*
JOLANTA SZULC • *Neurosciences Institute, Swiss Federal Institute of Technology Lausanne, Lausanne, Switzerland*
E. VERHOEYEN • *Ecole Normale Supérieure de Lyon, Lyon, France, and BioSciences Lyon-Gerland, Lyon, France*
ERNST WAGNER • *Pharmaceutical Biology-Biotechnology, Department of Pharmacy, Ludwig-Maximilians-Universitaet, Munich, Germany*
FUBAO WANG • *Vaccine and Biologics Research, Merck Research Laboratories, Merck & Co., Inc., West Point, PA*
ANDREW WILBER • *Arnold and Mabel Beckman Center for Transposon Research, Gene Therapy Program, Institute of Human Genetics, and Department of Genetics, Cell Biology and Development, University of Minnesota, Minneapolis, MN*
JAYANTHI J.WOLF • *Merck Research Laboratories, Merck & Co., Inc., West Point, PA*
HONGJU WU • *Division of Human Gene Therapy, Department of Obstetrics and Gynecology, and the Gene Therapy Center, University of Alabama at Birmingham, Birmingham, AL*
WEIDONG XIONG • *The Board of Governors' Gene Therapeutics Research Institute, Cedars-Sinai Medical Center and Department of Molecular and Medical Pharmacology, David Geffen School of Medicine, University of California, Los Angeles, Los Angeles, CA*
LINDA YANG • *Laboratory of Biochemical Genetics, National Heart, Lung, and Blood Institute, Bethesda, MD*
CHANGYU ZHENG • *Gene Therapy and Therapeutics Branch, NIDCR, NIH, DHHS, Bethesda, MD*

1

Purification of Retrovirus Particles Using Heparin Affinity Chromatography

María de las Mercedes Segura, Amine Kamen, and Alain Garnier

Summary

Retroviral vectors have been used as gene delivery vehicles for more than two decades and continue to be the best available tool for stable and efficient transfer of therapeutic genes into various cell types. Although most gene therapy preclinical studies presently use crude or concentrated retroviral vector supernatants, purification to eliminate serum and host-derived impurities contained in these stocks will be a necessary requirement for clinical applications. Chromatography is deemed the most promising technology for large-scale purification of viral vectors. Heparin affinity chromatography offers the possibility to selectively and efficiently purify retroviruses. This chapter gives a simple, reproducible, and scaleable protocol for the purification of bioactive VSV-G pseudotyped retroviral vectors that employs membrane and chromatography technologies. The protocol can be easily adapted for the purification of different retroviral vector pseudotypes and lentiviral vectors. The purification techniques described here represent a significant improvement over the conventional sucrose density gradient methodology used for retrovirus purification and will hopefully contribute to the technological progress in the field of gene therapy.

Key Words: Recombinant retrovirus vector; gene therapy; purification; scaleable processes; membrane filtration; chromatography; heparin affinity chromatography.

1. Introduction

Retroviral vectors have attracted the attention of gene therapy researchers for their ability to stably integrate the transgene of interest into the target cell genome providing the possibility of long-term gene expression and ultimately long-term therapeutic effect. Being one of the most popular viral vectors used

From: *Methods in Molecular Biology, vol. 434: Volume 2: Design and Characterization of Gene Transfer Vectors*
Edited by: J. M. Le Doux © Humana Press, Totowa, NJ

in clinical trials, vectors derived from the oncovirus Moloney murine leukemia virus (MoMLV) have demonstrated great potential as gene delivery vehicles, both *ex vivo* ***(1–3)*** and *in vivo* ***(4,5)***. Major obstacles associated with retroviral vectors include the production of low-titer viral stocks and the instability of the viral particles produced. Concentration of vector stocks is required for most gene therapy applications to improve transduction efficiencies. Moreover, concentrated or not, viral stocks still contain contaminants that need to be removed to increase the potency and safety of the final product. Non-purified vector preparations contain contaminating molecules that are toxic to cells, reduce transduction efficiencies ex vivo ***(6)*** and can induce a systemic immune response and inflammation when injected in vivo ***(7,8)***.

Although ultracentrifugation has traditionally been used for isolating retroviruses in small quantities, the increasing demand for large volumes of high-quality retroviral vector stocks for pre-clinical and clinical trials has more recently motivated the exploration of scaleable purification technologies, mainly membrane filtration and chromatography (for review see **ref.** ***9***). As a result, various filtration devices and chromatography supports have been adapted for the purification of retroviral vectors in the past few years ***(10–17)***. The selectivity of affinity chromatography is particularly interesting because it offers the potential to reduce the number of purification steps increasing product yields and decreasing process costs. In principle, retroviral vectors could be purified by immunoaffinity chromatography by relying on the specific interaction between immobilized antibodies and the viral Env-protein. However, the high costs associated with antibody purification and immobilization, the low stability of these ligands toward sanitizing agents and the harsh conditions usually required to break antibody–antigen interactions do not favor the use of this method for large-scale purification of retroviral vectors ***(18)***. Another possibility is to engineer vectors to contain affinity tags inserted on the surface of the virus to facilitate their purification ***(14,15)***. However, engineering vectors by inserting tags or chemically modifying the viral Env-protein without reducing or eliminating the viruses' ability to transduce cells has proved to be difficult ***(19–21)***. This has been partly attributed to the inability of the Env-protein to provide fusogenic functions for viral entry once its structure has been modified.

An attractive alternative is to explore the natural ability of these viruses to bind commercially available affinity ligands or immobilized viral receptors. In this case, no alteration of the viral Env-protein is required. Heparin, a relatively inexpensive and stable affinity chromatography ligand used to purify various biomolecules and viruses ***(22–25)*** has shown to be useful for the purification of VSV-G pseudotyped retroviral vectors derived from 293 producer cells giving excellent results in terms of yield, selectivity, and reproducibility ***(11)***.

Although the method was described for the purification of VSV-G pseudotyped particles, we have found that other MoMLV-derived vector pseudotypes, as well as lentivirus vectors also show affinity for heparin ligands and can be efficiently purified using the purification protocol described herein.

This chapter describes in detail a scaleable multi-step purification scheme from crude retrovirus supernatant to clinical-grade virus based on heparin affinity chromatography (*see* **Fig. 1**). Clarification, concentration, buffer exchange, and partial purification of retroviral particles from crude supernatants are achieved by membrane filtration with essentially no loss in vector infectivity. Concentrated viral particles, suspended in the appropriate binding buffer, are then selectively captured by immobilized heparin ligands. Elution of retrovirus particles from the heparin affinity column is achieved under mild conditions (neutral pH and 0.35 M NaCl) resulting in high recoveries of infective particles (61%) while purifying the viral particles from the majority of serum proteins. A final polishing step can be introduced to achieve higher purity by loading the semi-purified concentrated material onto a size-exclusion chromatography column. However, this step is not required in the case of vector stocks produced in serum-free media. The overall yield of infective particles for the complete purification strategy is approximately 38%.

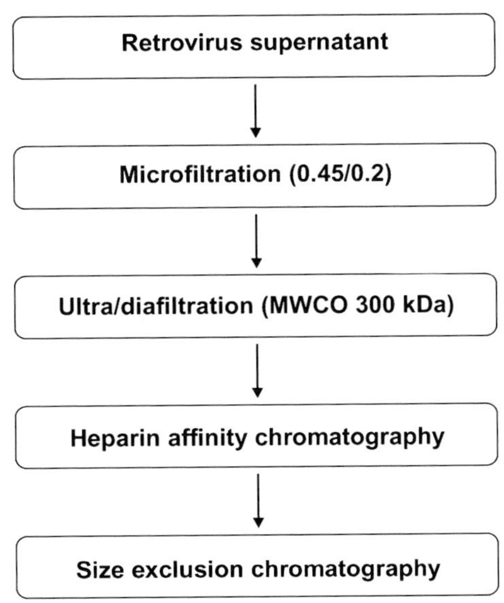

Fig. 1. Scheme for the purification process of retroviral particles.

2. Materials

2.1. Microfiltration

1. Dual HT Tuffryn® polysulfone membrane (0.45/0.2 μm) capsule filtration device (Pall Corporation).

2.2. Ultrafiltration

1. Amicon Stirred cell unit (Millipore).
2. Omega™ polyethersulfone membrane disc filters with a molecular weight cut off of 300,000 (Pall Corporation).
3. Dulbecco's Modified Eagle's Medium (DMEM, phenol red-free) supplemented with 10% heat-inactivated fetal bovine serum (FBS) (*see* **Note 1**).
4. Filter sterilized diafiltration buffer: 20 mM Tris–HCl, 150 mM NaCl, pH 7.5 (*see* **Note 2**).

2.3. Heparin Affinity Chromatography

1. Low-pressure liquid chromatography system (GradiFrac™; GE Healthcare).
2. Fractogel® EMD heparin (S) gel (Merck) packed into an HR 5/5 column (GE Healthcare) to a final volume of 1 ml (*see* **Note 3**).
3. 0.45-μm pore size Acrodisc syringe-mounted filters with HT Tuffryn® polysulfone membrane (Pall Corporation).
4. Buffer A: 20 mM Tris–HCl, pH 7.5 filtered and degassed (*see* **Note 4**).
5. Buffer B: 20 mM Tris–HCl, 2 M NaCl, pH 7.5 filtered and degassed.
6. Storage buffer: 150 mM NaCl in Milli-Q H_2O, 20% Ethanol filtered and degassed.
7. Regeneration/sanitization buffer: 2 M NaCl in Milli-Q H_2O, 0.1 M NaOH filtered and degassed (*see* **Note 5**).

2.4. Size Exclusion Chromatography

1. Low-pressure liquid chromatography system (GradiFrac™; GE Healthcare).
2. Sepharose CL-4B gel (GE Healthcare) packed into a XK 16/70 (GE Healthcare) to a final volume of approximately 100–120 ml (*see* **Note 6**).
3. 0.45-μm pore size Acrodisc syringe-mounted filters with HT Tuffryn® polysulfone membrane (Pall Corporation).
4. Running buffer: Phosphate-buffered saline (PBS), pH 7.4 filtered and degassed.
5. Storage buffer: 150 mM NaCl in Milli-Q H_2O, 20% ethanol filtered and degassed.
6. Regeneration/sanitization: 1 M NaCl in Milli-Q H_2O, 0.5 M NaOH filtered and degassed.

3. Methods

3.1. Microfiltration

1. Thaw retrovirus supernatants using water bath at 37 °C.
2. Pellet cells and cell debris by centrifugation at 10,000 × *g* for 10 min. Decant the supernatant into a sterile bottle.
3. Aliquot starting material samples for analyses and measure the starting volume. Keep the supernatant at 4 °C until further use.
4. Set up the filtration system in the biological safety cabinet. Flush the microcapsule membrane with 500 ml of Milli-Q H_2O and empty the tubing (*see* **Note 7**).
5. Filter supernatants at constant flow rate (start at 10 ml/min and gradually increase to 50 ml/min). Collect the permeate in sterile bottles.
6. Aliquot clarified supernatant samples for analyses.
7. Keep the clarified supernatant at 4 °C until the next downstream processing step or store at –80 °C along with the aliquoted samples (*see* **Notes 8** and **9**).

3.2. Ultrafiltration

1. Use **Table 1** to select the stirred cell model and membrane size according to the application.
2. Soak the membrane in a container with Milli-Q H_2O to remove trace amounts of glycerine humectants and sodium azide bacteriostat. The membrane should be kept wet at all times from this point on.
3. Clean and assemble the stirred cell unit. The glossy side of the membrane should be facing up.
4. Fill the stirred cell unit with Milli-Q H_2O. Adjust the nitrogen pressure to 10 psig. Pre-wash the membrane with Milli-Q H_2O and check the water flux to ensure the membrane is not broken or displaced. A water flux of 1–3 ml/(min cm^2) indicates the 300,000 MWCO membrane is intact.

Table 1
Selection of the Appropriate Stirred Cell Unit (Amicon®, Millipore) and Membrane Size

Unit model	8050	8400	2000
Cell capacity (ml)	50	400	2000
Stirred minimum volume (ml)	2.5	10	50
Stirring rate (rpm)	200	100	50
Membrane diameter (mm)	43	76	150
Effective filtration area (cm^2)	13.4	41.8	162

The stirring rates shown correspond to a tip speed of 33.5 cm/s.

5. Pre-foul the membrane with approximately 0.25 ml/cm^2 of phenol red-free DMEM, 10% FBS medium. Adjust the tip speed to 33.5 cm/s and increase the nitrogen pressure to 30 psig. Be careful not to let the membrane dry. Add PBS if necessary.
6. Aliquot the clarified retrovirus samples for analyses and measure the starting volume. Considering this volume, calculate the final volume desired in order to attain a predetermined concentration factor at the end of the process.
7. Add the clarified supernatant into the stirred cell unit and carry out ultrafiltration under constant nitrogen pressure (30 psig) and tip speed (33.5 cm/s). Collect permeate in sterile bottles. Monitor the process at different times by determining the flow rate and volume of permeate (*see* **Note 10**).
8. Once the desired volume of retentate is reached, add cold diafiltration buffer into the stirred cell unit (~1 ml/cm^2) and carry out diafiltration under constant nitrogen pressure (30 psig) and tip speed (33.5 cm/s). Repeat diafiltration three times in discontinuous mode. The permeates obtained during ultrafiltration and diafiltration can be collected separately for subsequent analyses.
9. The ultra/diafiltration process is stopped once the desired volume of retentate is reached. Slowly resuspend the virus and gently wash the membrane with diafiltration buffer to reach the desired final volume. Aliquot the concentrated supernatant samples for analyses.
10. Keep the concentrated supernatant at 4 °C until the next downstream processing step or store at –80 °C along with the previously aliquoted samples.

3.3. Heparin Affinity Chromatography

1. Turn on the UV lamp and rinse the GradiFrac® system thoroughly with Milli-Q H_2O.
2. Install the 1-ml Fractogel® heparin column and remove the storage buffer with 10 column volumes (CV) of Milli-Q H_2O at 0.3 ml/min.
3. Rinse lines A (for buffer A) and B (for buffer B) with the corresponding buffers and equilibrate the column with 10 CV of binding/wash buffer containing 150 mM NaCl (7.5% buffer B) at 0.5 ml/min (or 153 cm/h linear flow rate) (*see* **Note 11**).
4. Filter the sample using an Acrodisc syringe-mounted filter with a pore size of 0.45 μm. Aliquot starting material samples for analyses.
5. Set the sensitivity to 2.0 absorbance units full scale (AUFS) and monitor the UV absorbance at 280 nm. When a stable baseline is achieved, load 3 ml of the concentrated virus sample. Apply a step-wise gradient elution strategy that includes a wash step at 150 mM NaCl (7.5% buffer B, 15 CV) to remove the bulk of serum contaminating proteins, followed by virus elution at 350 mM NaCl (17.5% buffer B, 13 CV) and a final high-stringency wash step at 1200 mM NaCl (60% buffer B, 6.5 CV) to remove tightly bound contaminants (*see* **Table 2, Fig. 2,** and **Note 12**). The process is carried out at room temperature and 0.5 ml/min.
6. The virus particles elute in a defined peak at 350 mM NaCl. Pool virus-containing fractions (# 16–19) and aliquot for analyses.

Table 2
Heparin Affinity Chromatography Step Elution Method for a 1-ml Column

Total volume(ml)	Conc. B (%)	Flow (ml/min)	Fraction volume (ml)	Notes
0	8	0.5	1.5	Start run
1.5	8	0.5	1.5	Inject position at 1.5 ml
4.5	8	0.5	1.5	Load position at 4.5 ml
19.5	8	0.5	1.5	Return to baseline
20	17	0.5	1.5	Gradient change
33	17	0.5	1.5	Virus elution at 350 mM NaCl
34	60	0.5	1.5	Gradient change
40.5	60	0.5	1.5	1.2 M NaCl stringent wash
41.5	8	0.5	1.5	Re-equilibration
49.5	8	0.5	1.5	End run

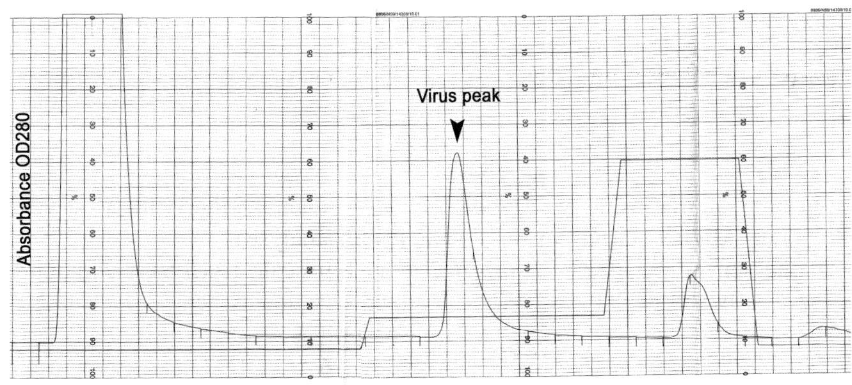

Fig. 2. Typical heparin affinity chromatography elution profile. 3 ml of a 20-fold concentrated virus were loaded onto a 1-ml Fractogel® EMD Heparin (S) column. The virus was eluted by addition of 350 mM NaCl into the mobile phase. Retroviral particles were recovered in a sharp peak (6 ml). The sensitivity is set to 2 absorbance units full scale (AUFS). Chart speed: 5 mm/min.

7. After each run, re-equilibrate the column with binding buffer (10 CV) at 0.5 ml/min or store the column in storage buffer (10 CV) at 0.3 ml/min.
8. Store semi-purified samples at –80 °C along with the previously aliquoted samples.

3.4. Size Exclusion Chromatography

1. Install the Sepharose CL-4B column and remove the storage buffer with 10 column volumes (CV) of Milli-Q H_2O at 0.5 ml/min.
2. Rinse line A with running buffer and equilibrate the column with 5 CV of this buffer.
3. Filter the sample using an Acrodisc syringe-mounted filter with a pore size of 0.45 µm. This step can be omitted when loading semi-purified retrovirus samples that have not been frozen in between chromatography steps. Aliquot starting material samples for analyses.
4. Monitor UV absorbance at 280 nm. When a stable baseline is achieved, load 7.5 ml of the semi-purified virus sample and perform an isocratic elution with PBS running buffer (1.2 CV). The process is carried out at room temperature and 0.5 ml/min.
5. The virus particles elute in the void volume of the column. Pool virus-containing fractions and aliquot for analyses.
6. After each run, re-equilibrate the column with running buffer (5 CV) at 0.5 ml/min or store the column in storage buffer (5 CV) at 0.5 ml/min (or 15 cm/h linear flow rate).
7. Store purified samples at –80 °C along with the previously aliquoted samples.

4. Notes

1. Phenol red is present in most culture media as pH indicator. It is very important that the retrovirus supernatants to be purified by chromatography do not contain phenol red to avoid UV signal interference. The virus should be produced in phenol-red-free DMEM culture medium, and this medium should also be used to pre-foul the ultrafiltration membrane.
2. All solutions should be prepared using ultrapure water (organic content < 5 ppb and resistivity > 18.2 MΩ cm) that can be obtained using a Milli-Q purification system (Millipore) referred here as Milli-Q H_2O.
3. It should be noted that chromatographic columns, such as the HR5/5, are provided with top and bottom filters designed for protein purification purposes. To avoid virus losses on the filter, replace the original filters with 10-µm mesh filter.
4. It is essential that all buffers used for chromatography are freshly filtered (0.45 µm) and degassed to protect the column. Samples should also be filtered through 0.45-µm membranes before being loaded into the column.
5. It is recommended that after 3–4 purification runs the columns are regenerated/sanitized. This is normally performed by washing with 2–3 CV of regeneration/sanitization solution followed by an extensive wash with Milli-Q H_2O to bring back to neutral pH. Monitor the column effluent pH before next run.

6. For best peak resolution, consider that the longer the column, the greater the separation between different size components. Additionally, always load sample volumes <10% of the total bed volume for better separation by size-exclusion chromatography.
7. Install a pressure gauge to monitor pressure at all times during clarification. Using a capsule filtration device with an effective filtration area of 500 cm^2, 1–2 L of retrovirus vector supernatants can be filtered with low back pressure (below 0.01 MPa).
8. Because of retrovirus temperature sensitivity ***(9)***, it is recommended to carry out the complete purification procedure in 1 or 2 working days. In this case, samples in between downstream processing operations can be safely kept at 4°C without significant loss in viral titer. Alternatively, samples can also be stored at –80°C to protect the virus from thermal inactivation during longer periods. At this temperature, retrovirus stocks were shown to maintain their potency over months ***(26,27)***. Keep the number of freeze-and-thaw cycles to a minimum.
9. To better evaluate the performance and/or estimate the recovery of infective particles for each purification step, take the following into consideration prior to titration. First, it is suggested to titer all samples immediately after purification, in this way avoiding unnecessary freeze-and-thaw cycles that might affect samples at different purification stages differently. Second, it is very important to titer all samples to be compared within the same titration assay to avoid inter-assay variability. Also, all samples should be processed in duplicate or triplicate to minimize intra-assay variability. Using the GFP expression assay and FACS analysis typically results in low intra-assay variability (RSD <10%). However, the inter-assay variability could be very high (RSD values up to 100%) depending on the assay conditions and operator. Third, only virus dilutions that result in percentage of GFP+ cell values ranging from 3 to 20% should be selected for titer determination. As such, several virus dilutions should be assayed.
10. The flow rate during ultrafiltration typically decreases gradually over time because of membrane fouling, particularly with samples containing 10% FBS media supplement. A constant flow rate throughout the process could be an indication of membrane damage.
11. It is recommended to perform an acetone test to verify the packing efficiency and column integrity. A blank test before the run with running buffer and/or phenol red-free DMEM, 10% FBS medium is also advisable.
12. In case the gradient programmer does not allow decimal point adjustments, program the elution gradient method as in **Table 2**.

Acknowledgments

The authors thank Normand Arcand and Alice Bernier for helpful discussions and Gavin Whissell for careful review of this manuscript. This work was supported by a NSERC Strategic Project grant and the NCE Canadian Stem Cell Network.

References

1. Aiuti, A., Slavin, S., Aker, M., Ficara, F., Deola, S., Mortellaro, A., Morecki, S., Andolfi, G., Tabucchi, A., Carlucci, F., Marinello, E., Cattaneo, F., Vai, S., Servida, P., Miniero, R., Roncarolo, M. G., and Bordignon, C. (2002) Correction of ADA-SCID by stem cell gene therapy combined with nonmyeloablative conditioning. *Science* **296,** 2410–3.
2. Cavazzana-Calvo, M., Hacein-Bey, S., de Saint Basile, G., Gross, F., Yvon, E., Nusbaum, P., Selz, F., Hue, C., Certain, S., Casanova, J. L., Bousso, P., Deist, F. L., and Fischer, A. (2000) Gene therapy of human severe combined immunodeficiency (SCID)-X1 disease. *Science* **288,** 669–72.
3. Gaspar, H. B., Parsley, K. L., Howe, S., King, D., Gilmour, K. C., Sinclair, J., Brouns, G., Schmidt, M., Von Kalle, C., Barington, T., Jakobsen, M. A., Christensen, H. O., Al Ghonaium, A., White, H. N., Smith, J. L., Levinsky, R. J., Ali, R. R., Kinnon, C., and Thrasher, A. J. (2004) Gene therapy of X-linked severe combined immunodeficiency by use of a pseudotyped gammaretroviral vector. *Lancet* **364,** 2181–7.
4. Gomez-Navarro, J., Curiel, D. T., and Douglas, J. T. (1999) Gene therapy for cancer. *Eur J Cancer* **35,** 2039–57.
5. Rainov, N. G., and Ren, H. (2003) Clinical trials with retrovirus mediated gene therapy–what have we learned? *J Neurooncol* **65,** 227–36.
6. Yamada, K., McCarty, D. M., Madden, V. J., and Walsh, C. E. (2003) Lentivirus vector purification using anion exchange HPLC leads to improved gene transfer. *Biotechniques* **34,** 1074–8, 80.
7. Baekelandt, V., Eggermont, K., Michiels, M., Nuttin, B., and Debyser, Z. (2003) Optimized lentiviral vector production and purification procedure prevents immune response after transduction of mouse brain. *Gene Ther* **10,** 1933–40.
8. Scherr, M., Battmer, K., Eder, M., Schule, S., Hohenberg, H., Ganser, A., Grez, M., and Blomer, U. (2002) Efficient gene transfer into the CNS by lentiviral vectors purified by anion exchange chromatography. *Gene Ther* **9,** 1708–14.
9. Segura, M. M., Kamen, A., and Garnier, A. (2006) Downstream processing of oncoretroviral and lentiviral gene therapy vectors. *Biotechnol Adv* **24,** 321–37.
10. Kuiper, M., Sanches, R. M., Walford, J. A., and Slater, N. K. (2002) Purification of a functional gene therapy vector derived from Moloney murine leukaemia virus using membrane filtration and ceramic hydroxyapatite chromatography. *Biotechnol Bioeng* **80,** 445–53.
11. Segura, M., Kamen, A., Trudel, P., and Garnier, A. (2005) A novel purification strategy for retrovirus gene therapy vectors using heparin affinity chromatography. *Biotechnol Bioeng* **90,** 391–404.
12. Transfiguracion, J., Jaalouk, D. E., Ghani, K., Galipeau, J., and Kamen, A. (2003) Size-exclusion chromatography purification of high-titer vesicular stomatitis virus G glycoprotein-pseudotyped retrovectors for cell and gene therapy applications. *Hum Gene Ther* **14,** 1139–53.
13. Williams, S. L., Eccleston, M. E., and Slater, N. K. (2005) Affinity capture of a biotinylated retrovirus on macroporous monolithic adsorbents: towards a rapid single-step purification process. *Biotechnol Bioeng* **89,** 783–7.

14. Williams, S. L., Nesbeth, D., Darling, D. C., Farzaneh, F., and Slater, N. K. (2005) Affinity recovery of Moloney Murine Leukaemia Virus. *J Chromatogr B Analyt Technol Biomed Life Sci* **820,** 111–9.
15. Ye, K., Jin, S., Ataai, M. M., Schultz, J. S., and Ibeh, J. (2004) Tagging retrovirus vectors with a metal binding peptide and one-step purification by immobilized metal affinity chromatography. *J Virol* **78,** 9820–7.
16. Reeves, L., and Cornetta, K. (2000) Clinical retroviral vector production: step filtration using clinically approved filters improves titers. *Gene Ther* **7,** 1993–8.
17. Cruz, P. E., Goncalves, D., Almeida, J., Moreira, J. L., and Carrondo, M. J. (2000) Modeling retrovirus production for gene therapy. 2. Integrated optimization of bioreaction and downstream processing. *Biotechnol Prog* **16,** 350–7.
18. Andreadis, S. T., Roth, C. M., Le Doux, J. M., Morgan, J. R., and Yarmush, M. L. (1999) Large-scale processing of recombinant retroviruses for gene therapy. *Biotechnol Prog* **15,** 1–11.
19. Katane, M., Takao, E., Kubo, Y., Fujita, R., and Amanuma, H. (2002) Factors affecting the direct targeting of murine leukemia virus vectors containing peptide ligands in the envelope protein. *EMBO Rep* **3,** 899–904.
20. Palù, G., Parolin, C., Takeuchi, Y., and Pizzato, M. (2000) Progress with retroviral gene vectors. *Rev Med Virol* **10,** 185–202.
21. Tai, C. K., Logg, C. R., Park, J. M., Anderson, W. F., Press, M. F., and Kasahara, N. (2003) Antibody-mediated targeting of replication-competent retroviral vectors. *Hum Gene Ther* **14,** 789–802.
22. Navarro del Cañizo, A. A., Mazza, M., Bellinzoni, R., and Cascone, O. (1996) Foot and mouth disease virus concentration and purification by affinity chromatography. *Appl Biochem Biotechnol* **61,** 399–409.
23. O'Keeffe, R. S., Johnston, M. D., and Slater, N. K. (1999) The affinity adsorptive recovery of an infectious herpes simplex virus vaccine. *Biotechnol Bioeng* **62,** 537–45.
24. Zolotukhin, S., Byrne, B. J., Mason, E., Zolotukhin, I., Potter, M., Chesnut, K., Summerford, C., Samulski, R. J., and Muzyczka, N. (1999) Recombinant adeno-associated virus purification using novel methods improves infectious titer and yield. *Gene Ther* **6,** 973–85.
25. Burnouf, T., and Radosevich, M. (2001) Affinity chromatography in the industrial purification of plasma proteins for therapeutic use. *J Biochem Biophys Methods* **49,** 575–86.
26. McTaggart, S., and Al-Rubeai, M. (2000) Effects of culture parameters on the production of retroviral vectors by a human packaging cell line. *Biotechnol Prog* **16,** 859–65.
27. Wikstrom, K., Blomberg, P., and Islam, K. B. (2004) Clinical grade vector production: analysis of yield, stability, and storage of gmp-produced retroviral vectors for gene therapy. *Biotechnol Prog* **20,** 1198–203.

2

Scaleable Purification of Adenovirus Vectors

John O. Konz, Lee R. Pitts, and Sangeetha L. Sagar

Summary

Adenovirus vectors currently are being evaluated in gene delivery studies ranging from prophylactic vaccination to therapeutic gene therapy. The quantity of purified virus required for these studies necessitate that purification methods must shift from classical density gradient ultracentrifugation to scaleable approaches. A methodology is described herein using batch centrifugation, tangential flow ultrafiltration, and chromatography to purify adenovirus particles at a scale of approximately 10^{13} viral particles. This method has been demonstrated to easily scale an additional 40-fold. While purification of human adenovirus type 5 is exemplified, modifications are suggested for the purification of other serotypes.

Key Words: Adenovirus; virus purification; anion exchange chromatography; ultrafiltration; nuclease.

1. Introduction

In the period from 1989 to 2004, over 240 clinical trials were initiated with adenovirus vectors, predominantly adenovirus type 5 (Ad5) *(1)*. Interest in clinical applications, as well as basic research aimed at improving vector function, has created a need for purification methodologies that are scaleable and accessible to both academic and industrial labs. Historically, production of adenovirus for research use has utilized cesium chloride density gradient ultracentrifugation *(2,3)*. While density gradient ultracentrifugation is useful for the purification of very small quantities of virus, i.e., approximately 10^{12} viral particles (vp), it becomes cumbersome for quantities of 10^{13} vp or greater.

Purification of adenovirus using anion exchange chromatography dates back to the late 1950s *(4–6)*, but has become common practice in the past

From: *Methods in Molecular Biology, vol. 434: Volume 2: Design and Characterization of Gene Transfer Vectors*
Edited by: J. M. Le Doux © Humana Press, Totowa, NJ

decade. Anion exchange chromatography is particularly selective for adenovirus because of the high net surface charge of the virus *(7)*, which leads to unusually tight binding to the resin. Ultrafiltration also can be highly selective for adenovirus purification, because of the large size differential between adenovirus particles and most host cell impurities. Several groups have now published on scaleable methods for adenovirus purification. These processes and additional considerations for commercial-scale process development have recently been reviewed ***(8)***.

The methodology described herein, which relies on batch centrifugation, nuclease treatment, tangential flow ultrafiltration, and anion exchange chromatography, was developed to achieve high purity without requiring the use of non-routine or technically demanding unit operations. For Ad5, it has been shown to generate a highly pure and potent product, as measured by protein purity, residual host cell DNA content, and tissue-culture infectivity assays ***(9)***. The specific procedure described here accommodates 1 l of adenovirus culture corresponding to infection of roughly 10^9 cells, with an expected productivity of 10^{13} vp, although it can easily be scaled up at least 40-fold. Considerations for the purification of other adenovirus serotypes also are provided in the notes section.

Although we have attempted to highlight some areas of concern, appropriate biosafety precautions should be taken for infectious agents, as outlined in guidelines as well as institutional requirements. Appropriate inactivation procedures for equipment and waste also should be developed before initiating work ***(10–12)***. *See* **Note 1** for more information.

2. Materials

2.1. Cell Harvest and Lysis

1. Resuspension buffer: 5 mM Tris–HCl, 75 mM NaCl, 1 mM $MgCl_2$, 5% sucrose, 1% Polysorbate-80 (PS-80), pH 8.0.
2. 250-ml polypropylene centrifuge tube with conical bottom.
3. 50-ml polypropylene conical centrifuge tubes (*see* **Note 2**).
4. Centrifuge and rotor capable of handling 250-ml conical tubes.
5. Water bath.

2.2. Clarification

1. Centrifuge and rotor capable of handling 50-ml conical tubes at 9000 × *g* (e.g., Beckman Avanti J-20 centrifuge with Beckman JA-10 rotor).
2. Rotor inserts for 50-ml conical tubes.
3. 125-ml sterile polyethylene terephthalate copolyester (PETG) bottle (Nalgene, Rochester, New York, USA).
4. 1-inch magnetic stir bar.

2.3. Nuclease Treatment

1. 7500 U. Benzonase® nuclease (Merck KGaA, Darmstadt, Germany).
2. Magnetic stir plate.

2.4. Ultrafiltration

1. 50 cm^2, 1000 kDa regenerated cellulose tangential flow ultrafiltration cartridge (Pellicon XL w/ C screen, Millipore, Billerica, Massachusetts, USA).
2. Pellicon XL stand.
3. Nalgene "top-works" three-port bottle top cap.
4. Nalgene PETG bottles (2 × 125 ml, 1 × 500 ml).
5. Pressure gauges *(3)*.
6. 0.2-micron disposable air filter.
7. Size 16 platinum-cured silicone tubing (Masterflex, Vernon Hills, Illinois, USA).
8. Tie wraps and gun.
9. Size 16 male-male connectors and "Y" connector.
10. Three peristaltic pumps with size 16 heads (e.g., Masterflex low speed pump).
11. Magnetic stir plate.
12. Electronic balance.
13. Ring stand and clamp.
14. Diafiltration buffer: 50 mM HEPES, 2 mM $MgCl_2$, 1 M NaCl, pH 7.5.
15. 1-inch magnetic stir bar.

2.5. Anion Exchange Chromatography

1. 1 cm D × 10 cm L high-pressure chromatography column (e.g., Waters AP-1) packed with Source 15Q resin (GE Healthcare, Piscataway, New Jersey, USA) as per manufacturer specifications (*see* **Note 3**).
2. Dual pump high-pressure chromatography system [e.g., standard HPLC or Akta (GE Healthcare)] with dual wavelength UV detector, capable of delivering 1.2 ml/min and delivering linear gradients (*see* **Note 4**). Ensure that the system does *not* have an in-line filter installed (*see* **Note 5**).
3. Fraction collector or containers for collection.
4. Adjustment buffer: 50 mM HEPES, 2 mM $MgCl_2$, 0.1% PS-80, pH 7.5.
5. Chromatography buffer A: 50 mM HEPES, 2 mM $MgCl_2$, 0.39 M NaCl, 0.1% PS-80, pH 7.5 (*see* **Note 6**).
6. Chromatography buffer B: 50 mM HEPES, 2 mM $MgCl_2$, 0.47 M NaCl, 0.1% PS-80, pH 7.5 (*see* **Note 6**).
7. 4 C "deli-box" style refrigerator or cold room.
8. Conductivity meter and probe, e.g., Orion 130 A conductivity meter (Thermo Fisher, Waltham, Massachusetts, USA) and associated conductivity probe (Orion 013010A conductivity cell).
9. 0.1 N NaOH.

2.6. Buffer Exchange

1. Desalting columns (GE Healthcare PD-10).
2. Stabilizing buffer: 5 mM Tris–HCl, 75 mM NaCl, 1 mM $MgCl_2$, 5% sucrose, 0.005% PS-80, pH 8.0 (*see* **Note 7**).

2.7. Sterile Filtration

1. Polypropylene syringe with luer lock fitting.
2. 0.22 μm polyvinylidene fluoride (PVDF) sterile filter (Millex GV25, Millipore).

3. Methods

3.1. Cell Harvest and Lysis

1. Cells should be harvested by centrifugation (e.g. 6000 × *g*, 15 min) in five 200-ml aliquots in 250-ml conical tubes when the intracellular concentration of virus is at a maximum. In general, this will occur at roughly 48 h post-infection *(8)* (*see* **Note 8**).
2. Resuspend each cell pellet in 10 ml of resuspension buffer.
3. Transfer into two 50-ml conical tubes (~25 ml each).
4. Freeze the tubes at –70 °C for at least 1 h. The frozen harvested cells can be held indefinitely at this point.
5. Thaw the tubes in a 37 °C water bath.
6. Repeat **steps 3** and **4** two times. (Step potentially not needed, *see* **Note 9.**)

3.2. Clarification

1. Centrifuge the aliquots for 10 min at 9000 × *g*.
2. Tare a new 125-ml bottle with stir bar.
3. Carefully collect the supernatant and transfer to the 125-ml bottle with stir bar.

3.3. Nuclease Treatment

1. Add Benzonase® to a final concentration of 150 U/ml, mixing thoroughly (*see* **Note 10**).
2. Incubate 1 h at room temperature with gentle mixing on a stir plate.
3. The batch can optionally be held overnight at 4 °C.

3.4. Ultrafiltration (see Note 11)

3.4.1. Equipment Preparation

These steps can be completed in advance to streamline processing.

1. Cut the tubes on the 3-port cap to allow the end of one tube to reach the bottom of the 125-ml retentate container and the ends of the other two tubes to remain

above the liquid level. This arrangement is shown in **Fig. 1** where the retentate container is indicated as "Product Bottle."

2. Using size 16 silicone tubing, male-to-male polypropylene connectors, and tie wraps, set up the ultrafiltration system as depicted in **Fig. 1** and described below.
3. Place the Pellicon XL filter on a Pellicon XL filter stand.
4. Install a pressure gauge at the retentate inlet (feed line) of the ultrafiltration cartridge. Connect a below-surface tube to the pressure gauge inlet, after threading tube through a low speed peristaltic pump head.
5. Install a pressure gauge to the retentate outlet of the ultrafiltration cartridge.
6. Attach a plastic "Y" connector to an above-surface tube and connect the pressure gauge on the retentate outlet to one side of the "Y."
7. Feed a line from an empty bottle through a peristaltic pump to the open port of the "Y" connected to the retentate outlet. This is the "buffer feed line."
8. Attach a vent filter to the third inlet (above-surface line).
9. Attach a pressure gauge to the top permeate outlet connection on the ultrafiltration cartridge.
10. Run a line from the pressure gauge on the permeate outlet through a low-speed peristaltic pump to a 500-ml bottle.
11. Flush, sanitize, and integrity test the membrane as per manufacturer's specification. Drain the system. These steps are important to remove storage chemicals and ensure that the membrane is integral (i.e., that the adenovirus will be retained).
12. Pump approximately 30 ml of diafiltration buffer through the membrane at a rate of 30 ml/min while directing the permeate to waste. The permeate pump can be set to the target of 1.5 ml/min flow rate during the flush. This step ensures that the sanitizing agent is adequately removed from the system. Drain the system.

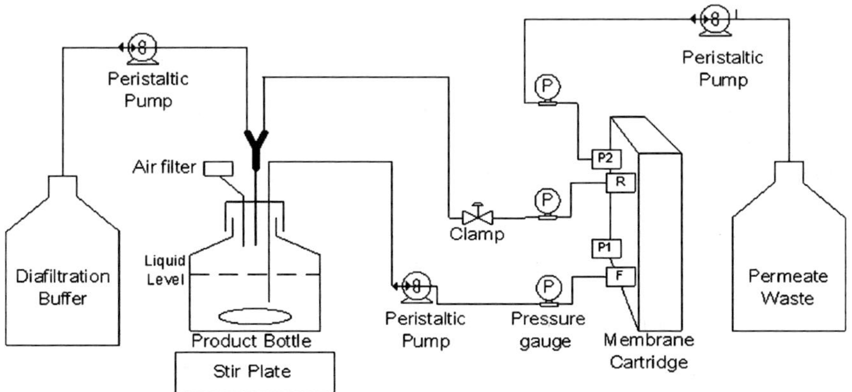

Fig. 1. Ultrafiltration set-up.

3.4.2. Step Operation

If possible, step operation should occur at 4 °C (*see* **Note 11**).

1. Place the buffer feed line into a container with at least 250 ml of diafiltation buffer.
2. In a biosafety cabinet, careful replace the current retentate bottle with the nuclease-treated batch. Carefully tighten the multi-port cap.
3. Place the retentate bottle on a stir plate. Clamp the retentate bottle in place using a ring stand. Begin gentle stirring.
4. Increase the recirculation rate, permeation rate and buffer feed rate to 30, 1.5, and 1.5 ml/min, respectively. *See* **Note 12** for comments on flow rates. Mark the current batch level in the bottle.
5. Operate the diafiltration until 250 ml of buffer has permeated. It may be necessary to adjust the flow rate of either the buffer feed or permeate pump to maintain a constant volume in the system. Ensure that the inlet pressure remains at or below 20 psi. If necessary, reduce the recirculation flow rate of the system to ensure that the tubing and connections remain integral.
6. Stop all three pumps. Reverse the feed pump to drain the membrane contents back into the retentate container. In a biosafety cabinet, carefully remove the 3-port cap and replace with the original bottle cap.
7. Weigh the batch and calculate the current volume.
8. If desired, the batch can be held overnight at 4 °C.

3.5. Anion Exchange Chromatography

3.5.1. Preparation

Preparation steps can be completed in advance (e.g., during operation of the ultrafiltration step) to save time.

1. Pack the column with Source 15Q resin as per column and resin manufacturers' instructions.
2. Sanitize the column by pumping 0.1 N NaOH though the column at 1.2 ml/min until the pH is elevated in the effluent. Disconnect the column and hold in NaOH for 30 min.
3. Flush pump B thoroughly with chromatography buffer B (volume required is system dependent).
4. Verify that the effluent is at the same conductivity and pH as the buffer.
5. Repeat above steps with pump A and buffer A.
6. Reconnect the column and equilibrate the column with five-column volumes (40 ml) of buffer A by pumping at 1.2 ml/min.

3.5.2. Operation

The chromatography step should be performed at 4 °C to minimize risk of product aggregation.

1. Place the batch on a stir plate with moderate mixing (a vortex should be present, but not causing extensive air entrainment, as air/liquid interfaces can promote aggregation).
2. Slowly reduce the conductivity of the batch by adding 1.7 times the current batch volume of cold dilution buffer (*see* **Note 13** for serotypes other than Ad5). The batch volume should now be roughly 135 ml.
3. If preparation steps occurred in advance of the day of processing, re-equilibrate the column with an additional 2–5-column volumes of buffer A at 1.2 ml/min.
4. Carefully transfer the pump A feed line into the batch, and seal with parafilm. Using a ring-stand, clamp the bottle in place to eliminate the possibility of spills.
5. Load the batch on to the column by pumping at 1.2 ml/min. Be extremely careful not to allow any air into the feed line by carefully monitoring the liquid level and position of the feed line. Air in the column can lead to pressure increases and/or poor performance.
6. When approximately 5–10 ml of batch remain in the feed bottle, carefully add 20–30 ml of buffer A to the feed bottle to flush the feed line.
7. When the feed bottle is nearly empty, stop the pump and transfer the pump A feed line back into the buffer A bottle.
8. Wash the column with five-column volumes (40 ml) of buffer A. Ensure that the absorbance has returned to baseline. If not, continue washing.
9. Elute the virus with a linear gradient to 100% buffer B over four-column volumes. Fraction collection can be done manually or with a fraction collector, provided that the collector can be placed in a contained area that can be decontaminated. The virus peak will have an A_{260}/A_{280} ratio of approximately 1.25, and the product should elute in approximately 1–1.5-column volumes. **Fig. 2** shows an example of a typical chromatogram.
10. Provided an input of 10^{13} vp to the process, approximately 10 ml of product at approximately 5×10^{11} vp/ml should have been collected. Proceed immediately to the next step (*see* **Note 14**).

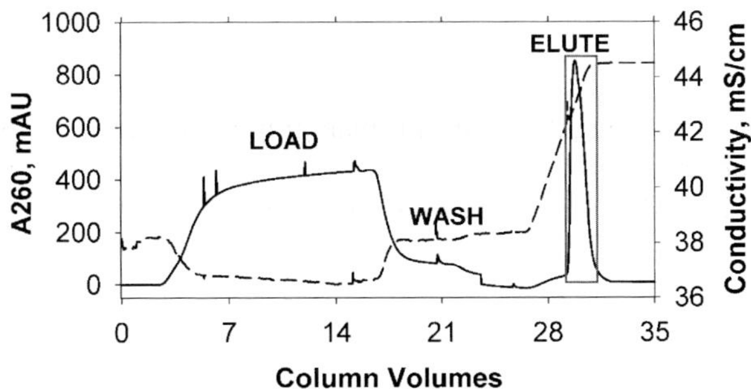

Fig. 2. Chromatogram from preparative anion exchange chromatography. Absorbance at 260 nm indicated by solid line, conductivity by dashed line. The collected product fraction is indicated by the gray box.

3.6. Buffer Exchange with Desalting Columns (see Note 15)

1. Pre-equilibrate 1 PD-10 desalting column for each 2.5 ml of batch volume, using 25 ml of stabilizing buffer.
2. Allow column to drain completely. If the product will not be applied to the column within about 5 min, recap so that the column will not dry out.
3. Within a biosafety cabinet, pipette 2.5 ml of product on each column, and allow to drain.
4. Add 3.5 ml of stabilizing buffer and collect this fraction as product.
5. Pool the product fractions.

3.7. Sterile Filtration (see Note 16)

1. Inside a biosafety cabinet, carefully draw the product into a luer-lock syringe.
2. Attach a Millex GV-25 syringe filter.
3. Slowly push the product through the filter. If resistance increases, replace the filter and continue.

4. Notes

1. As adenovirus is an infectious agent and vector preparations may contain wild-type virus, careful planning should precede any attempt to follow this protocol. Planning should include an evaluation of required personal protective equipment as well as decontamination procedures for all product-contact equipment. Institutional biosafety officers should be involved in this process. Pertinent guidelines in the USA that will help in this process include (1) NIH Guidelines for Research Involving Recombinant DNA molecules (http://www4.od.nih.gov/oba/rac/guidelines/guidelines.html), and (2) the CDC Manual, "Biosafety in Microbiological and Biomedical Laboratories" (http://www.cdc.gov/od/ohs/biosfty/bmbl4/bmbl4toc.htm). Examples of methodologies that may be appropriate for inactivation of adenovirus can be found in the literature ***(10–12)***.
2. If alternative centrifuge tubes are substituted, ensure that they are rated for the specified centrifugal force. Most 50-ml conical tubes are not and may break during the operation.
3. The scale of the anion exchange column should be dictated by both the size and the productivity of the cell culture process. In general, columns sized to accommodate purification of 10^{12} vp per ml of bed volume should reproducibly result in good product yields. In practice, it may be useful to measure the concentration of adenovirus in the Benzonase®-treated supernatant before selecting the column size, and then adjusting the anion exchange chromatography protocol accordingly. A number of vendors manufacture columns rated for the moderate pressures required for this application; the user should also verify that the frit size is less than 15 micron. For small-scale purification, pre-packed columns are available from the vendor (GE Healthcare).

4. Chromatography system should have a pressure cutoff that can be set to the minimum pressure limit for any of the system components (column, tubing, fittings, etc.). Other system requirements will be scale-dependent. For larger-scale processing, the flow rate required for the system can be calculated using a linear velocity of 1.5 cm/min (e.g., for 5-cm diameter column, flow rate = $3.14 \times 5^2/4$ [cm^2] $\times$ 1.5 [cm/min] = 29 [ml/min]).
5. Many commercial chromatography systems come installed with a small in-line filter to remove particulates from the buffers. In this application, where the system pump is used to load the virus on to the column, an in-line filter will be rapidly fouled causing product loss and a large pressure increase.
6. Because of differences in surface charge, serotypes other than Ad5 will elute at different salt concentrations ***(13,14)***. It is recommended that the salt concentration required for elution be checked by comparing the elution time in an analytical anion exchange HPLC run with the time for elution of Ad5, as described ***(13)***. Benzonase®-treated clarified lysate or ultrafiltration retentate can be used for this determination if CsCl-ultracentfugation-purified virus is not available. Following this test, the sodium chloride content of both buffer A and B should be shifted based on the elution differential versus Ad5.
7. For considerations in the selection of a stabilizing buffer, see Altaras et al. and Evans et al. ***(8,15)***. Users interested in commercial applications should consider intellectual property.
8. Harvest time may be dependent on serotype and/or specifics of vector construction. Experiments may be required to optimize the time of harvest. These experiments can be performed by harvesting small aliquots during a timecourse and analyzing adenovirus content using standard anion exchange HPLC methods ***(14,16)***.
9. In general, more than one freeze/thaw cycle will not be necessary to achieve complete cell lysis. It may be desirable to confirm this by standard anion exchange HPLC methods.
10. The intent of this step is to digest both DNA and RNA. Other nucleases that have DNase and RNase activities (or nuclease cocktails with both activities) may potentially be substituted for Benzonase®.
11. To facilitate both cold processing and use of a biosafety cabinet, it may be convenient to set up the system on a cart, allowing the user to roll the cart from a cold room to a biosafety cabinet for open handling. Alternatively, the system can be configured to use sterile-welding technology to avoid open handling.
12. These flow rates were carefully selected to minimize membrane fouling caused by gel-layer formation, while remaining in a range-of-operation consistent with the equipment specified. While precise control of these flow rates is not critical, large deviations (>30%) are not recommended.
13. In **Note 6**, a procedure for identifying the required sodium chloride concentration for a serotype other than Ad5 is described. In addition to changing the chromatography buffers, the batch must be diluted sufficiently to ensure virus binding. In general, it is adequate to dilute the batch to a conductivity less than that of buffer A.

14. Adenovirus has a propensity to aggregate, especially at high concentrations such as those achieved at this step. As a result, high concentration solutions should not be stored for extended periods.
15. If this process has been scaled up and batch volume exceeds 30 ml, PD-10 desalting columns become impractical. In that case, diafiltration against five volumes of stabilizing buffer using a similar setup as used in **Subheading 3.4** is recommended. A membrane with a MWCO of 300 kDa or less is recommended to ensure virus retention. The product of this ultrafiltration is likely to be of slightly higher purity than product that is only desalted.
16. Sterile filtration of the product may not be necessary for all applications, but is recommended. If the product volume ranges from 30 to 300 ml, the product should be sterile filtered through a Sterivex GV (Millipore) filter at a pump rate of 10 ml/min. It is recommended that pressure be monitored to ensure that filter fouling does not occur.

References

1. Edelstein, M. L., Abedi, M. R., Wixon, J., and Edelstein, R. M. (2004) Gene therapy clinical trials worldwide 1989–2004-an overview. *J. Gene Med.* **6**, 597–602.
2. Green, M., and Pina, M. (1963) Biochemical studies on adenovirus multiplication. IV. Isolation, purification, and chemical analysis of adenovirus. *Virology* **20**, 199–207.
3. Maizel, J. V. J., White, D. O., and Scharff, M. D. (1968) The polypeptides of adenovirus: I. Evidence for multiple protein components in the virion and a comparison of types 2, 7A, and 12. *Virology* **36**, 115–125.
4. Klemperer, H. G., and Pereira, H. G. (1959) Study of adenovirus antigens fractionated by chromatography on DEAE-cellulose. *Virology* **9**, 536–545.
5. Haruna, I., Yaoi, H., Kono, R., and Watanabe, I. (1961) Separation of adenovirus by chromatography on DEAE-cellulose. *Virology* **13**, 264–267.
6. Philipson, L. (1960) Separation on DEAE cellulose of components associated with adenovirus reproduction. *Virology* **10**, 459–465.
7. Goerke, A. R., To, B. C. S., Lee, A. L., Sagar, S. L., and Konz, J. O. (2005) Development of a novel adenovirus purification process utilizing selective precipitation of cellular DNA. *Biotechnol. Bioeng.* **91**, 12–21.
8. Altaras, N. E., Aunins, J. G., Evans, R. K., Kamen, A., Konz, J. O., and Wolf, J. J. (2005) Production and formulation of adenovirus vectors. *Adv. Biochem. Eng. Biotechnol.* **99**, 193–260.
9. Konz, J. O., Lee, A. L., Lewis, J. A., and Sagar, S. L. (2005) Development of a purification process for adenovirus: controlling virus aggregation to improve the clearance of host cell DNA. *Biotechnol. Prog.* **21**, 466–472.
10. Jannat, R., Hsu, D., and Maheshwari, G. (2005) Inactivation of adenovirus type 5 by caustics. *Biotechnol. Prog.* **21**, 446–450.
11. McCormick, L., and Maheshwari, G. (2004) Inactivation of adenovirus types 5 and 6 by Virkon S. *Antiviral Res.* **64**, 27–33.

12. Maheshwari, G., Jannat, R., McCormick, L., and Hsu, D. (2004) Thermal inactivation of adenovirus type 5. *J. Virol. Methods* **118**, 141–146.
13. Konz, J. O., Livingood, R. C., Bett, A. J., Goerke, A. R., Laska, M. E., and Sagar, S. L. (2005) Serotype specificity of adenovirus purification using anion-exchange chromatography. *Hum. Gene Ther.* **16**, 1346–1353.
14. Blanche, F., Cameron, B., Barbot, A., Ferrero, L., Guillemin, T., Guyot, S., Somarriba, S., and Bisch, D. (2000) An improved anion-exchange HPLC method for the detection and purification of adenoviral particles. *Gene Ther.* **7**, 1055–1062.
15. Evans, R. K., Nawrocki, D. K., Isopi, L. A., Williams, D. M., Casimiro, D. R., Chin, S., Chen, M., Zhu, D. M., Shiver, J. W., and Volkin, D. B. (2004) Development of stable liquid formulations for adenovirus-based vaccines. *J. Pharm. Sci.* **93**, 2458–2475.
16. Shabram, P. W., Giroux, D. D., Goudreau, A. M., Gregory, R. J., Horn, M. T., Huyghe, B. G., Liu, X. D., Nunnally, M. H., Sugarman, B. J., and Sutjipto, S. (1997) Analytical anion-exchange HPLC of recombinant type-5 adenoviral particles. *Hum. Gene Ther.* **8**, 453–465.

3

Quantifying the Titer and Quality of Adenovirus Stocks

Fubao Wang, Bill C. Mathis, Allison Montalvo, Jayanthi J. Wolf, and John A. Lewis

Summary

The broad application of recombinant adenoviruses to the development of vaccines and gene therapy vectors has encouraged the development of molecular assays for the facile quantitation of adenoviral particles and the assignment of their infectious potency. The Genome Quantitation Assay (GQA) and the QPCR-Based Potency Assay (QPA) developed for adenoviruses offer the attributes of precision, rapidity, and high throughput either performed manually or facilitated by simple automated liquid handling systems. These assay attributes allow for accelerated process development support and product characterization and release. The assays for adenovirus could offer the additional advantage in that their quantitation is based on viral replication independent of cytopathology permitting quantitation of serotypes that cause minimal cytopathic effect (CPE) in 293 cells and specificity that allows the components of multivalent vaccines to be discriminated and quantitated for release.

Key Words: Adenovirus; GQA; QPA; adenoviral genome; infectious potency; QPCR.

1. Introduction

Adenoviruses were independently isolated by Rowe et al. *(1)* and Hilleman and Werner *(2)* and subsequently shown to be widespread in human populations responsible for a wide range of relatively mild pathologies of respiratory, gastrointestinal, and conjunctival disease *(3)*. The family *Adenoviridae* includes 47 human serotypes organized into Groups A–E of which the Adenoviruses 1, 2, 5, and 6 of Group C have been most widely studied. The development of adenovirus vaccine of Group B to protect military recruits advanced by

From: *Methods in Molecular Biology, vol. 434: Volume 2: Design and Characterization of Gene Transfer Vectors*
Edited by: J. M. Le Doux © Humana Press, Totowa, NJ

Hilleman et al. ***(4)*** ushered in a half-century biological, genetic, and clinical study of the *Adenoviridae* viruses, leading to an understanding that invited the development of the Group C viruses as recombinant vectors for vaccine development ***(5,6)*** and as vectors for gene therapy ***(7–9)***. More recently, attention has been focused on the isolation and manipulation of genomes of Group B and Group D adenoviruses as a strategy to circumvent the seroprevalence of the more common Adenovirus serotypes of Group C ***(10)*** that complicate vaccine efficacy.

The development of adenovirus vectors for various clinical objectives has been classically supported by the propagation of replication defective E1 deleted viruses in the human embryonic kidney cell line 293 ***(11)*** and more recently the human embryonic retinoblast PER.C6® cell line ***(12)***. The PER.C6® line precludes the generation of replication competent adenovirus (RCA) by rational design of the transgene cassette to exclude nucleotide sequences contained in most commonly used adenovirus vectors. The 293 cell line has been most commonly used for the quantitation of adenoviral infectious potency, using either plaque or $TCID_{50}$ (tissue culture infectious dose 50%) endpoint dilution methods. These traditional assays suffer the disadvantages of labor intensity, a 12–14-day term for the development of cytopathic effect (CPE), throughput restriction and an inherent variability that especially compromises their suitability for the support of bioprocess development. Their dependence on CPE to reveal the presence of infectious virus furthermore restricts the utility of these methods to serotypes that do not form obvious CPEs in 293 or PER.C6®.

The quantitation of adenovirus for bioprocess and vaccine release has been most commonly assigned by physico-chemical methods ***(13,14)*** and more recently by high-performance liquid chromatography (HPLC) ***(15)*** and real-time quantitative polymerase chain reaction (PCR) ***(16,17)***. We used the real-time quantitative PCR technology ***(18)*** to develop assays for quantitating adenoviral genomes and infectious potency ***(19)***. In the following sections, we present the basic protocols for quantifying the titer and quality of adenovirus stocks and encourage their adoption.

2. Materials

2.1. Cell Culture and Virus Infection

1. Alpha Minimum Essential Medium (Invitrogen, Carlsbad, CA, USA) supplemented with 10% fetal bovine serum (FBS, HyClone, Logon, UT, USA) and 1% penicillin/streptomycin.
2. Trypsin/EDTA solution (0.05% Trypsin) is used to harvest 293 cells. Cells are placed in 96-well, flat-bottomed plates.

2.2. DNA Extraction

1. Qiagen 96-well Centrifugation System (Qiagen, Valencia, CA, USA).
2. 96-well Blood DNA Purification Kits (Qiagen).

2.3. Cell Lysis

1. 96/384™ Multimek (Beckman Coulter, Fullerton, CA, USA). Other liquid handling robotic systems can be used for the lysis step.
2. Triton X-100 (1.44%): Dispense 8.56 ml of molecular biology grade water and 1.44 ml of 10% Triton X-100. This solution is prepared immediately before use.
3. Foil Seals are used to seal plates after lysis and before storage at –70 °C.

2.4. PCR Quantitation

1. Primer Express Software (Applied Biosystems, Foster city, CA, USA) is used to design the primers and probes (*see* **Note 1**).
2. TaqMan Universal PCR Master Mix (Applied Biosystems).
3. Molecular biology grade water.
4. PCR primers and probe.
5. PCR workstation.
6. 10% sodium dodecyl sulfate (SDS).
7. MicroAmp optical adhesive covers (Applied Biosystems).
8. MicroAmp optical 96-well or 384-well reaction plates (Applied Biosystems).
9. DNA/RNA decontamination solution (DNAZAP, Ambion, Austin, TX)
10. ABI Prism 7900 Sequence Detection Systems with a 96-well or 384-well heat block and detector (Applied Biosystems). It should be noted that other 96-well Sequence Detection Systems (e.g., ABI 7300, 7500, or 7700) could also be used.
11. Sequence Detection Systems Software (Applied Biosystems).

3. Methods

3.1. Genome Quantitation Assay

Conventional methods utilize the extinction coefficient and OD_{260nm} absorption of lysed adenovirus to calculate the particle quantities. In the Genome Quantitation Assay (GQA), viral particles are lysed by incubation at 56 °C in the presence of SDS. Viral DNAs released upon lysis are then quantitated by TaqMan PCR using a standard curve (SC) that is calibrated by OD_{260nm} absorption. The genome quantity derived from the SC represents the genome containing viral particles in the test article (TA).

3.1.1. Preparation of Primers and Probes

Primer and probe sequences are designed to target the genome backbone or a specific transgene by using Applied Biosystems' Primer Express software.

HPLC-purified primers are lyophilized and should be dissolved in 1× TE to a concentration of 100 μM. The probe concentration should be adjusted to 100 μM by adding 1× TE. The primers and probe can then be aliquoted in increments of 100 μl (master stock) and 10 μl (working stock). Storage of the primers and probe should be at –20 °C.

3.1.2. Preparation of Standards

1. Prepare three independent dilutions of a stock adenovirus plasmid preparation by diluting 4 μl of plasmid with 396 μl of molecular grade water (*see* **Note 2**).
2. Determine the absorbance at 260 nm using 100 μl of diluted plasmids. Each independent dilution should be read in triplicate using the spectrophotometer.
3. The DNA concentration can be determined by multiplying the OD at 260 nm (average of three readings) by 50 (1.0 OD equals 50 μg/ml) and by 100 (dilution factor). Average the three independent DNA concentration results to achieve a final plasmid DNA concentration.
4. Adjust the DNA quantity to molecule copy number per 10 μl based on molecular weight (base pair × 660 g/mole) and concentration.
5. Dilute an aliquot further to 10^7molecules per 10 μl.
6. Prepare a SC by diluting the 10^7molecules per 10 μl stock serially by adding 100 μl of the previous dilution to 900 μl of water.
7. Pool three independent serial dilutions at the same concentrations.
8. Aliquot 350 μl of each into 2-ml Sarstedt tubes as Master Stock for SC and store at –20 °C.
9. Aliquot 35 μl of each Master Stock concentration into 2 ml Sarstedt tubes as Working Stock for SC and store at –20 °C.
10. Test pooled serially diluted SC DNAs by PCR. The materials are qualified when the $R^2$of the SC is >0.98.

3.1.3. Procedure

1. Use 2-ml Sarstedt tubes with Sarstedt racks for easy manipulations. Prepare a set of tubes, three tubes for no template control (NTC), three for each SC concentration, and three for each test sample. Label each tube.
2. Prepare tubes for serial dilution of samples. Three for each sample (for 1 × 10^{11} viral particles/ml concentration), two for each subset of master reaction mix, and three for water (aliquot from a full bottle).
3. Equilibrate temperature of H_2O from 4 °C to room temperature, distribute to respective tubes as indicated in **Table 1**, aliquot 3 × 2 ml of H_2O from a 50-ml bottle into three Sarstedt tubes, distribute to respective tubes as indicated in **Table 1**, distribute H_2O to tubes for master reaction mix, and distribute H_2O to tubes for sample dilution.
4. To prepare the Master Reaction Mix, add 2× TaqMan PCR universal reaction mix each to the master reaction mix tube (*see* **Note 3**). Universal master reaction

Table 1
TaqMan PCR Compositions and Well Positions

Well position	Template	Forward primer	Reverse primer	TaqMan probe	DNA	MBG water	MBG water	Master mix
A1-3	NTC	0.1	0.1	0.1	0	20	4.7	25
A4-6	1 x 10^6	0.1	0.1	0.1	10	10	4.7	25
A7-9	5 x 10^5	0.1	0.1	0.1	10	10	4.7	25
A10-12	1 x 10^5	0.1	0.1	0.1	10	10	4.7	25
B1-3	5 x 10^4	0.1	0.1	0.1	10	10	4.7	25
B4-6	1 x 10^4	0.1	0.1	0.1	10	10	4.7	25
B7-9	5 x 10^3	0.1	0.1	0.1	10	10	4.7	25
B10-12	1 x 10^3	0.1	0.1	0.1	10	10	4.7	25
C1—	Test articles	0.1	0.1	0.1	10	10	4.7	25

mix is stored at 4 °C and should be warmed up by holding in hands until room temperature. Mix well before aliquoting. Add forward and reverse primers and probe for each subset (4 µl each). Vortex gently but thoroughly (the probe color is a good indication). Distribute 33 µl of master mix to every reaction tube.

5. Lyse, dilute, and distribute adenoviral samples as follows (*see* **Note 4**). Prepare 0.2% SDS by adding 10 µl of 20% SDS stock to 990 µl of H_2O. Mix by vortexing. Transfer 50 µl of 0.2% SDS to each Sarstedt tube (three per sample or viral standards). Add 50 µl of virus sample each tube. Mix by vortexing. Spin briefly in a Pico-centrifuge (RPI, model: HF-120) at full speed for 5 s. Incubate at 56 °C for 10 min then cool to room temperature. Dilute lysed samples by 100-fold (10 µl lysed virus in 990 µl of H_2O). Mix well by vortexing after each dilution, distribute 3 × 11 µl of the 100-fold diluted materials to each respective sample tubes (*see* **Note 5**).
6. Distribute SC material. Thaw a frozen set of SC serial dilution.
7. Spin briefly in a Pico-centrifuge at full speed for 5 s. Add 3 × 11 µl of each concentration into SC tubes.
8. Transfer complete mixtures to 96-well plate (*see* **Note 6**). The positions of each reaction are specified as shown in **Table 1**. Vortex all the tubes gently and thoroughly. Place a MicroAmp optical 96-well reaction plate on a plate base. Load 50 µl of complete reaction mixture in wells in a 96-well reaction plate. Change tips after every loading.
9. Handle the loaded plate. The loaded plate is covered by 12 optical 8-well strips. Change the plate position 90 ° counterclockwise, so that the first column is close to you. Use the cap installing tool to close each cap loosely (halfway closed), start from the column 1 (the closest column to you), and close one column a time onwards. Switch the plate back to the original orientation when all 12 columns are covered. Use the other end of the cap-installing tool to roll over all eight wells in one strip to tighten the caps. Hold the covered plate in hand halfway

upside down. Adhesive optical covers can also be used instead of the caps. Pull the plate quickly to drive air bubbles out of the liquid. (It is fine to have some air bubbles on the top surface of the reaction mix.) Put the plate onto the PCR chamber in a 7700 or 7900 sequence detector. Press the whole plate gently for close contact between the wells and the metal base.

10. Run the PCR according to the 7700 or 7900 instruction manual. The whole reaction process is operated by a computer with the program Sequence Detection Systems. The default PCR condition is Stage I, 50 °C, 2 min, 1 cycle; Stage II, 95 °C, 10 min, 1 cycle; Stage III, 95 °C, 15 s, 40 cycles 60 °C, 60 s with data collection at all stages. Click on *run* to start the PCR. The reaction takes about 2 h using the default condition.

3.1.4. Data Analysis

1. Analyze PCR results in the Sequence Detection Systems program. Assign corresponding wells as NTC and *SC*. Input the exact amount for SC points. Click on *analysis/Analyze* to analyze the results. When the analysis is done, a window with amplification plot default in log scale will pop up, choose cycle 3 (start) to cycle 11 (stop) for base line calculation. Set the threshold at 0.05 by manually input this value into the box "Use Threshold." Once the threshold line is set, click OK. Highlight all the wells, click on *analysis/amplification plot* to show (1) amplification plot ((Rn vs. cycle numbers), (2) amplification plot (Cycle threshold Ct vs. wells). Click on *analysis/SC*, a SC window will pop up. In the window, black dots represent the SC points and red dots on the curve represent all other samples. Three parameters are given regarding the SC: slope, intercept, and R square. Assess the quality of a SC by comparing these results with the previously compiled results.
2. Analyze the PCR results in a Microsoft Excel spreadsheet. Copy and paste the data into Excel to calculate averages and standard deviations of triplicates.

3.2. QPCR-Based Potency Assay

The QPCR-based potency assay (QPA) uses QPCR to quantitate the mass of adenovirus genomes replicated 24 h after the inoculation of a TA on 293 cell monolayers ***(19)***. The mass of replicated adenovirus is then related to potency by interpolation to a SC of replicated adenovirus genomes constructed with a reference adenovirus standard to which potency has been assigned by an end-point dilution method. The QPA assay was originally developed with a step whereby DNA from infected cells was purified using a Qiagen kit before QPCR (DNA purification QPA). An alternative QPA method was subsequently developed where DNA from infected cells is released using a triton lysis step before PCR (Lysate QPA). The final assay protocol can be chosen based on the availability of the instruments and both of them perform similarly and are amenable for automation for a high-throughput laboratory. The Lysate QPA further decreases assay cost, time, and assay steps. QPCR can be carried out using 96-well or 384-well plates, and Cycle Threshold (Ct) values for test samples can be linearly interpolated against a SC created by the reference adenovirus standard.

3.2.1. Cell Culture and Infection of 293 Cells

1. The 293 cells are established in T150 flask cultures in cMEM and passaged when approaching confluence with trypsin/EDTA. The cells are incubated at 37 °C in an atmosphere of 95% air, 5% CO_2, and 90% humidity.
2. 293 cells are plated in 96-well plates at a density of 2×10^4cells/well (~6×10^4 cells/cm^2) in a 100 μl volume of cMEM for 24 h before infection.
3. A well-characterized virus preparation to which infectious potency has been previously assigned using an end-point titer assay ***(20)*** is serially diluted fourfold to concentrations ranging from 2×10^5 infectious units (IU)/10 μl to 3.1×10^3 IU/10 μl. These dilutions can be frozen at –70 °C for long-term storage in 50 μl aliquots and used as the SC in the QPA assay. A single aliquot of each concentration can be removed from a –70 °C freezer and thawed at room temperature before each infection.
4. At the time of infection, virus TAs are removed from a –70 °C freezer and thawed at room temperature. TAs are diluted to target the SC's mid-point by obtaining the viral particle concentration by the GQA assay before potency assay testing. Using approximate physical to infectious particle ratios, undiluted sample potencies can be estimated and used to calculate the necessary sample dilutions.
5. The 96-well plate, containing 293 cells that were plated for 24 h, is removed from the incubator. Ten microliters of the appropriate diluted TA, SC, or positive control (PC) is added to each well as shown in the plate layout (*see* **Table 2**). Columns 1 and 12 in the plate are not used. The plate is placed back in the 37 °C incubator for 24 h. The infected plates can then be treated in one

Table 2
QPA Infection and PCR Quantitation Plate Layout

	1	2	3	4	5	6	7	8	9	10	11	12
A		SC1	SC3	TA1	TA3	TA5	TA7	TA9	TA11	TA13	PC	
B		SC1	SC3	TA1	TA3	TA5	TA7	TA9	TA11	TA13	PC	
C		SC1	SC3	TA1	TA3	TA5	TA7	TA9	TA11	TA13	PC	
D		SC1	SC3	TA1	TA3	TA5	TA7	TA9	TA11	TA13	PC	
E		SC2	SC4	TA2	TA4	TA6	TA8	TA10	TA12	TA14	TA15	
F		SC2	SC4	TA2	TA4	TA6	TA8	TA10	TA12	TA14	TA15	
G		SC2	SC4	TA2	TA4	TA6	TA8	TA10	TA12	TA14	TA15	
H		SC2	SC4	TA2	TA4	TA6	TA8	TA10	TA12	TA14	TA15	

of two ways: DNA purification (*see* **Subheading 3.2.2.**) or Triton Lysis (*see* **Subheading 3.2.3.**).

3.2.2. DNA Purification QPA

The procedure below follows the manual recommended by Qiagen for the QIAamp 96 Spin Blood Kits with slight modification.

1. Prepare buffers needed for DNA extraction.
    a. Prepare protease by adding 7 ml of molecular biology grade water to 140 mg of lyophilized Protease. Mix 2.5 ml of suspended protease solution with 20 ml of PBS.
    b. Prepare Qiagen buffer AL by decanting reagent AL1 into buffer AL. Mix thoroughly by shaking.
    c. Prepare buffer AW by adding 160 ml of ethanol. Mix thoroughly by shaking.
2. Remove the plate from the 37 °C incubator and aspirate the supernatant.
3. Add 50 µl of Protease/PBS mixture.
4. Add 50 µl of buffer AL and incubate at 70 °C for 10 min.
5. After the 10-min incubation, add 50 µl of Ethanol. Mix the contents in each well and transfer all of the contents to the 96-well spin kit column plate.
6. Remove liquid by centrifuging the spin plate at 6000 × *g* for 4 min in the Qiagen centrifuge.
7. Wash the columns twice with 500 µl of buffer AW centrifuging between each wash at 6000 × *g* for 4 min in the Qiagen centrifuge.
8. After washing, the spin plate is then transferred to the collection tubes and incubated at 70 °C for 10 min.
9. After this incubation period, begin elution of DNA by adding 200 µl of buffer AE that has been pre-heated to 70 °C. Centrifuge the plate at 6000 × *g* for 4 min in the Qiagen centrifuge.

3.2.3. Lysate QPA

1. The infected 96-well plate is removed from the 37 °C incubator and observed under the microscope. For lytic adenoviruses, the cell monolayer may appear to be disrupted.
2. A solution of Triton X-100 (1.44%) is prepared and placed on the Multimek deck.
3. Using the Multimek and a V-groove reservoir, 50 µl of 1.44% Triton is added to all wells of a 96-well plate (yielding a volume of 160 µl of lysate at a final triton concentration of 0.45%). The wells are mixed using the Multimek with five 90-µl mixes at 100% speed.
4. The plates are sealed with foil seals, labeled with assay and plate identifiers, and stored at –70 °C (for a minimum of 1 h).
5. Before QPCR, the 96-well plates are thawed at room temperature (takes approximately 1 h).

6. After thawing, 1:20 dilutions of the lysates are made using the Multimek by adding 95-μl molecular biology grade water to all wells of 96-well plates and then adding 5-μl of the lysates. The lysates are mixed into the water using ten 40-μl mixes at 100% speed.

3.2.4. QPCR in 96-Well Plates

1. Prepare work area and pipetman by cleaning with DNA ZAP solutions 1 and 2 to prevent DNA contamination.
2. Prepare the sample plate by aliquoting 10 μl of each sample into a well in a 96-well optical plate.
3. Prepare the QPCR master mix by adding 12.5 μl of each primer and probe, 1.9 ml of molecular biology grade water and 3.2 ml of 2× universal master mix to a 15-ml tube. Mix gently by vortexing.
4. Transfer 40 μl of this reaction mix to each of the sample wells in the 96-well plate. Cap the plate wells and centrifuge at 1500 × *g* for 1 min in the Qiagen centrifuge.
5. Place reaction plate in the ABI Prism 7700 or 7900 Sequence Detector.

3.2.5. QPCR in 384-Well Plates

1. QPCR master mix is prepared using either an adenovirus backbone-specific or transgene-specific primer and probe set (*see* **Note 7**). The concentration of primer and probes needs to be determined empirically, and usually ranges between 200 and 400 nM for each component.
2. Taqman Universal QPCR master mix is diluted to a final concentration of 1× (0.5× has also been used successfully) in a final volume of 10 μl per well (8 μl master mix and 2 μl sample).
3. Using the Multimek, 2 μl of purified DNA or 1:20-diluted triton lysates can be added to the appropriate quadrant of a 384-well optical QPCR plate (*see* **Note 8**). Eight microliters of the QPCR master mix containing primer and probes is then added to each well using either the Multimek or a manual multichannel pipette.
4. After reagent addition, the 384-well optical plate is sealed with a transparent optical seal, centrifuged at 1500 × *g* for 30 s, and analyzed on an ABI 7900 sequence detection system.
5. The QPCR data can be analyzed with a spreadsheet designed to arrange the exported Ct values and perform the linear interpolation of Ct values to assigned potencies.

4. Notes

1. It should be noted that Minor Groove Binding (MGB) probes with non-florescent quenchers must be used for QPCR in the Lysate QPA to avoid inhibition in the QPCR assay.

2. Caution must be taken to avoid inadvertent contamination of TAs with SC reagents or other identified or unidentified materials within the general laboratory area.
3. Master Reaction Mixtures must be prepared using a PCR workstation using pipettes and tips dedicated exclusively for that purpose. No DNA can be handled in this restricted area.
4. TA DNAs should ideally be manipulated in a contained working space with a working surface, readily subject to disinfection by UV irradiation. In the absence of these ideal facilities, special caution should be taken to work with DNAs on working surfaces covered with disposable pads removed on a daily basis. The development of a controlled system for the changing of gloves and gowns is recommended. Use only aerosol-resistant disposable tips during operations involved in the manipulation of DNA samples. Eliminate any residual DNA on pipettes and racks from previous experiments using reagents such as DNAZap before each experiment.
5. Ten microliters are taken for each reaction to pipette accurately. In case of more diluted samples are required, the dilution volume can be proportionally increased, e.g., 20 μl DNA plus 180 μl of H_2O.
6. Caution should be given while moving liquid-containing pipette and tips above the plate. No liquid drop is allowed.
7. In experiments directly comparing the performance of adenovirus-specific and transgene-specific primer and probe sets on the same triton lysates, no significant differences have been observed.
8. It should be noted that four 96-well plates can therefore be compressed into one 384-well plate.

Acknowledgments

The authors thank our colleagues who provided strong support throughout this work (alphabetically): Andrew Bett, Barry Buckland, Carrie Harper, Anthonise Louis, Jennifer McMackin, Allen Puddy, Volker Sandig, Timothy Schofield, Robert Sitrin, Charles Tan, Jenny Xu, and Yuhua Zhang.

References

1. Rowe, W.P., Huebner, R.J., Gilmore, L.K., Parrott, R.H., and Ward T.G. (1953) Isolation of a cytopathogenic agent from human adenoids undergoing spontaneous degeneration in tissue culture. *Proc. Soc. Exp. Biol. Med.* **84**, 570–573.
2. Hilleman, M.R. and Werner, J.H. (1954) Recovery of new agents from patients with acute respiratory illness. *Proc. Soc. Exp. Biol. Med.* **85**, 183–188.
3. Horwitz, M.S. (1996) *Adenoviruses in Virology* by Fileds, B.N., Knipe D.M., and Howley P.M. eds. Lippincott Williams & Wilkins **Vol. 2**, pp. 22149–22171.
4. Hilleman, M.R. (1958) Efficacy of and indications for use of adenovirus vaccine. *Am. J. Public Health* **48**, 153–158.

5. Shiver, J.W., Fu, T.M., Chen, L., Casimiro, D.R., Davies, M.E., Evans, R.K., Zhang, Z.Q., Simon, A.J., Trigona, W.L., Dubey, S.A., Huang, L., Harris, V.A., Long, R.S., Liang, X., Handt, L., Schleif, W.A., Zhu, L., Freed, D.C., Persaud, N.V., Guan, L., Punt, K.S., Tang, A., Chen, M., Wilson, K.A., Collins, K.B., Heidecker, G.J., Fernandez, V.R., Perry, H.C., Joyce, J.G., Grimm, K.M., Cook, J.C., Keller, P.M., Kresock, D.S., Mach, H., Troutman, R.D., Isopi, L.A., Williams, D.M., Xu, Z., Bohannon, K.E., Volkin, D.B., Montefiori, D.C., Miura, A., Krivulka, G.R., Lifton, M.A., Kuroda, M.J., Schmitz, J.E., Letvin, N.L., Caulfield, M.J., Bett, A.J., Youil, R., Kaslow, D.C., Emini, E.A. (2002) Replication-incompetent adenoviral vaccine vector elicits effective anti-immunodeficiency-virus immunity. *Nature* **415**, 331–335.
6. Shiver, J.W. and Emini, E.A. (2004) Recent advances in the development of HIV-1 vaccines using replication-incompetent adenovirus vectors. *Ann. Rev. Med.* **55**, 355–372.
7. Engelhardt, J.F., Yang, Y., Stratford-Perricaudet, L.D., Allen, E.D., Kozarsky, K., Perricaudet, M., Yankaskas, J.R., and Wilson, J.M. (1993) Direct gene transfer of human CFTR into human bronchial epithelia of xenografts with E1-deleted adenoviruses. *Nat. Genet.* **4**, 27–34.
8. Zabner, J., Couture, L.A., Gregory, R.J., Graham, S.M., Smith, A.E., Welsh, M.J. (1993) Adenovirus-mediated gene transfer transiently corrects the chloride transport defect in nasal epithelia of patients with cystic fibrosis. *Cell* **75**, 207–216.
9. Crystal, R.G., McElvaney, N.G., Rosenfeld, M.A., Chu, C.S., Mastrangeli, A., Hay, J.G., Brody, S.L., Jaffe, H.A., Eissa, N.T., and Danel, C. (1994) Administration of an adenovirus containing the human CFTR cDNA to the respiratory tract of individuals with cystic fibrosis. *Nat. Genet.* **8**, 42–51.
10. Lemckert, A.A., Sumida, S.M., Holterman, L., Vogels, R., Truitt, DM., Lynch, D.M., Nanda, A., Ewald, B.A., Gorgone, D.A., Lifton, M.A., Goudsmit, J., Havenga, M.J., and Barouch, D.H. (2005) Immunogenicity of heterologous prime-boost regimens involving recombinant adenovirus serotype 11 (Ad11) and Ad35 vaccine vectors in the presence of anti-ad5 immunity. *J. Virol.* **79**, 9694–9701.
11. Graham, F.L., Smiley, J., Russell, W.C., and Nairn, R. (1977) Characteristics of a human cell line transformed by DNA from human adenovirus type 5. *J. Gen. Virol.* **36**, 59–74.
12. Fallaux, F.J., Bout, A., van der Velde, I., van den Wollenberg, D.J., Hehir, K.M., Keegan, J., Auger, C., Cramer, S.J., van Ormondt, H., van der Eb, A.J., Valerio, D., and Hoeben, R.C. (1998) New helper cells and matched early region 1-deleted adenovirus vectors prevent generation of replication-competent adenoviruses. *Hum. Gene Ther.* **9**, 1909–1917.
13. Maizel, J.V., White D.O., and Scharff, M.D. (1968) The polypeptides of adenovirus: evidence for multiple protein components in the virions and a comparison of types 2, 7A and 12. *Virology* **36**, 115–125.
14. Sweeney, J.A. and Hennessey, J.P. (2002) Evaluation of accuracy and precision of adenovirus absorptivity at 260 nm under conditions of complete DNA disruption. *Virology* **295**, 284–288.

15. Shabram, P.W., Giroux, D.D., Goudreau, A.M., Gregory, R.J., Horn, M.T., Huyghe, B.G., Liu, X., Nunnally, M.H., Sugarman, B.J., and Sutjipto, S. (1997) Analytical anion-exchange HPLC of recombinant type-5 adenoviral particles. *Hum Gene Ther*. **8**, 453–465.
16. Ma, L., Bluyssen, H.A., De Raeymaeker, M., Laurysens, V., van der Beek, N., Pavliska, H., van Zonneveld, A.J., Tomme, P., and van Es, H.H. (2001) Rapid determination of adenoviral vector titers by quantitative real-time PCR. *J. Virol. Methods* **93**, 181–188.
17. Wang, L., Wang, C.J., Tan, C.Y., Hsu, D., and Hennessey, J.P. Jr. (2006) A robust approach for the quantitation of viral concentration in an adenoviral vector-based human immunodeficiency virus vaccine by real-time quantitative polymerase chain reaction. *Hum. Gene Ther*. **17**, 1–13.
18. Heid, C.A., Stevens, J., Livak, K.J., and Williams, P.J. (1996) Real time quantitative PCR. *Genome Res*. **6**, 986–994.
19. Wang, F., Puddy, A.C., Mathis, B.C., Hager, A., Louis, A., McMackin J., Xu, J., Zhang, Y., Tan, C., Schofield, T.S., Wolf, J.J. and Lewis J.A. (2005) Using QPCR to assign infectious potencies to adenovirus based vaccines and vectors for gene therapy. *Vaccine* **23**, 4500–4508.
20. Wigund, R. and Kumel, G. (1977) The kinetics of adenovirus infection and spread in cell cultures infected with low multiplicity. *Arch. Virol.* **54,** 177–187.

4

Chromatography-Based Purification of Adeno-Associated Virus

Richard H. Smith, Linda Yang, and Robert M. Kotin

Summary

Virus-mediated gene transfer shows great potential as a therapeutic strategy for the management of various inherited and acquired human diseases. Among the current viral vectors, adeno-associated virus (AAV) has become the vector of choice for numerous gene therapy applications. As AAV-based vectors approach the clinic, the need for scalable methods of production and purification is steadily increasing. In this chapter, we present a column chromatography-based protocol for the purification of recombinant AAV type 1 (AAV-1) to near homogeneity. The protocol, which can be completed within one working day, employs three major purification steps: (1) polyethylene glycol-mediated vector precipitation, (2) anion-exchange chromatography, and (3) gel filtration chromatography. This method provides a basic strategy, or "platform," that can be adapted to the purification of other recombinant AAV vector serotypes.

Key Words: AAV-1; adeno-associated virus; PEG precipitation; chromatography; purification.

1. Introduction

Adeno-associated viruses (AAVs) are small, DNA viruses of the *Parvoviridae* family (reviewed in **refs. *1–4***). The non-enveloped AAV virion has a diameter of 20–25 nm and exhibits icosahedral symmetry ***(5,6)***. The viral capsid harbors a single-stranded, linear DNA genome, typically just under 5 kb in length. The AAV genome contains two protein-coding genes designated *rep* and *cap*. The *rep* gene encodes a family of related polypeptides (the Rep proteins) essential for viral DNA replication. The *cap* gene encodes three overlapping viral capsid proteins, known as VP1, VP2, and VP3. The

From: *Methods in Molecular Biology, vol. 434: Volume 2: Design and Characterization of Gene Transfer Vectors*
Edited by: J. M. Le Doux © Humana Press, Totowa, NJ

capsid proteins occur in 60 copies per virion *(7)*. VP3 is the major constituent of the viral capsid, whereas VP1 and VP2 occur in only a few copies per infectious unit. The viral capsid proteins determine the physical and biological properties of the virion, such as tissue tropism and stability within various physical environments. AAV particles are particularly robust and have been reported to be heat stable, as well as resistant to organic solvents and extremes of pH.

Classically, purification of AAV is performed by multiple rounds of isopycnic banding in cesium chloride (CsCl) density gradients. Although effective, repeated banding in CsCl gradients is time consuming and recalcitrant to scale up. Much recent work has focused on the development of liquid chromatography-based strategies for the purification of recombinant AAV *(8–20)*. To this end, we present a chromatography-based protocol for the purification of recombinant AAV serotype 1. The process described in this chapter (outlined in **Fig.** 1) contains three major purification steps. The first is polyethylene glycol (PEG)-mediated precipitation of nuclease-resistant vector from the crude cellular lysate. This step is amenable to relatively large volumes of lysate and provides both concentration and partial purification of the vector. PEG precipitation utilizes differences in the solubility of various biological molecules in the presence of a given concentration of PEG, a nonionic polymer of ethylene oxide. Protein solubility is affected by both the concentration and

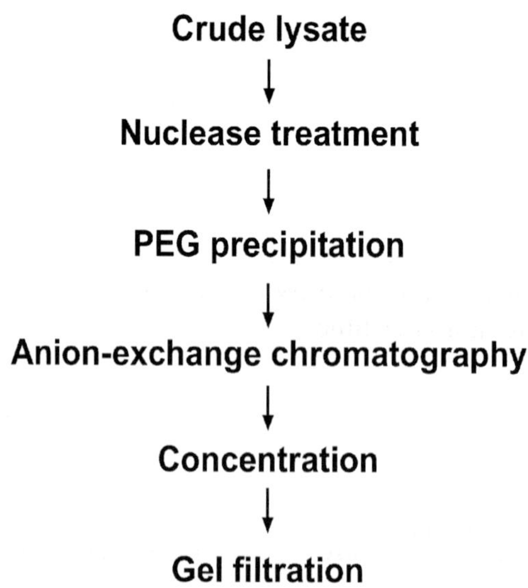

Fig. 1. Purification scheme.

the length of the PEG polymer *(21)*. This protocol uses PEG with an average molecule weight of 8000 Da at an approximately 4% (w/v) final concentration.

After PEG precipitation, recombinant vector is purified further by ion-exchange chromatography. Ion-exchange chromatography separates charged molecules based on electrostatic interaction between the molecules and the ionized groups incorporated into the column matrix material. Bound substances are eluted from the column by increasing the ionic strength of the fluid phase through an increase in salt concentration. To date, all AAV serotypes that we have examined were found to bind positively charged (anion-exchange) resins under the proper conditions of ionic strength and pH. Various anion-exchange resins are available and differ in a number of parameters, such as cost, binding strength and capacity, maximum flow rate, and resolving capacity. An important determination of the resolving capacity of ion-exchange chromatography medium is the size and uniformity of the beads composing the exchange resin. Generally, the smaller and more uniform the bead, the greater the resolving capacity. The method described here utilizes a high resolution, anion-exchange bulk medium, SOURCE15Q, composed of monodisperse, 15 μm beads. This resin was chosen for its resolving power, rapid flow rate, relatively large protein binding capacity, and scale-up capability.

Following anion-exchange chromatography, the peak vector-containing column fractions are identified by quantitative, real-time PCR, pooled, and prepared for subsequent gel filtration chromatography by concentration with a centrifugal filtration device. Gel filtration chromatography separates biological molecules according to size through differential diffusion of the molecules within the pores of the column matrix material. Although AAVs are among the smallest of the DNA animal viruses, the viral particles are still quite large compared with the majority of soluble proteinaceous cellular components. This fact has been exploited by choosing a gel filtration medium, Superdex 200, in which the exclusion limit of the beads (1.3×10^6 Da) is near the estimated molecular weight of the assembled AAV capsid proteins (~3.6×10^6 Da). Recombinant AAV is excluded from the gel filtration medium and elutes in the void volume, whereas the majority of cellular proteins are retarded within the column and elute at a later time.

2. Materials

2.1. Preparation of the Cellular Lysate

1. Benzonase® nuclease (*see* **Note 1**).
2. Phosphate-buffered saline (PBS): 0.9% NaCl, 0.0210% KH_2PO_4, 0.0726% $Na_2HPO_4 \cdot 7H_2O$ (all w/v), pH 7.2.
3. 250-ml polypropylene centrifuge tubes.

4. 50-ml polypropylene conical tubes.
5. Dry ice (pelleted).

2.2. PEG-Mediated Precipitation of Recombinant Vector

1. Fifty percent (w/v) PEG stock solution (average PEG molecular weight: 8000) (*see* **Note 2**).
2. Syringe-driven (preferably large diameter; e.g., 33 mm), 0.2-μm pore-size filter units with polyethersulfone (PES) filtration membrane.
3. Oak Ridge polypropylene centrifuge tubes, 50 ml (nominal).

2.3. Anion-Exchange Chromatography

1. Tricorn 10/100 column packed with SOURCE 15Q ion exchange medium (both from GE Healthcare, Piscataway, NJ, USA). Pre-packed 1- and 6-ml columns (RESOURCE columns) are available from GE.
2. Buffer A: 20 mM bis-Tris–HCl, pH 6.0, 10 mM NaCl (*see* **Note 3**).
3. Buffer B: 20 mM bis-Tris–HCl, pH 6.0, 1 M NaCl.

2.4. Quantitative, Real-Time PCR

1. QuantiTect SYBR Green PCR Kit (QIAGEN, Inc., Chatsworth, CA, USA).
2. Optical-grade, real-time PCR tubes and caps.
3. iCycler iQ Real-Time PCR Detection System (BioRad, Hercules, CA, USA) or equivalent.
4. Vector-specific primer set (*see* **Note 4**).

2.5. Vector Concentration and Gel Filtration Chromatography

1. Centricon Plus-20 centrifugal filter device, PES membrane, 100,000 molecular weight cut-off (Millipore, Corp., Bedford, MA, USA).
2. Superdex 200 10/300 GL gel filtration column (GE Healthcare).
3. Buffer C: PBS, pH 7.2 (*see* **Subheading 2.1.**), 2 mM $MgCl_2$, 2% glycerol (*see* **Note 5**).

2.6. Sodium Dodecyl Sulfate–Polyacrylamide Gel Electrophoresis

The following sodium dodecyl sulfate–polyacrylamide gel electrophoresis (SDS–PAGE) reagents are from Invitrogen (Carlsbad, CA, USA). Other equivalent SDS–PAGE systems and/or reagents may be used (including traditional Tris–glycine-based gel reagents).

1. XCell Surelock Mini-Cell electrophoresis unit.
2. NuPAGE pre-cast 4–12% polyacrylamide gels, 1.5 mm.
3. NuPAGE SDS sample buffer (4×), NuPAGE antioxidant, 10× NuPAGE reducing agent.

4. 20× MOPS running buffer.
5. Protein molecular weight markers.

2.7. Silver Staining

1. SilverXpress® Silver Staining Kit (Invitrogen).
2. Anhydrous methanol.
3. Glacial acetic acid.
4. Ultrapure (18 MΩ/cm) water.

2.8. Western Blotting

1. XCell II Western blotting module (Invitrogen).
2. NuPAGE Transfer Buffer (20×) (Invitrogen).
3. Non-fat dry milk.
4. PBST: PBS (*see* **Subheading 2.1.**) with 0.05% (v/v) Tween 20, pH 7.4.
5. Pre-cut PVDF filter sandwiches (Invitrogen).
6. Anti-AAV capsid, rabbit antiserum (commercially available from American Research Products, Inc., Belmont, MA, USA; Research Diagnostics, Inc., Concord, MA, USA; and also Progen, GmbH, Heidelberg, Germany).
7. Goat anti-rabbit horseradish peroxidase-conjugated secondary antibody.
8. Horseradish peroxidase-compatible chemiluminescent detection reagent.
9. X-ray film.

3. Methods

The purification method described in this chapter was developed using recombinant AAV serotype 1 (AAV-1), bis-Tris-based ion exchange chromatography buffers adjusted to pH 6.0 (*see* **Note 6**), and an AKTA-FPLC liquid chromatography system (GE Healthcare). Other equivalent liquid chromatography systems capable of gradient formation may be used. Recombinant AAV-1 was produced by triple transfection of human embryonic kidney cells (HEK 293F) grown in serum-free medium ***(18)***. The method provides an average 26% yield with a purity of ≥95%. Routine starting material consisted of approximately 6.5×10^8 cells at the time of harvest. All procedures involving the manipulation and handling of cells of human origin and rAAV-containing fluids should be performed in a biological safety cabinet. Proper laboratory attire, including protective eyewear, should be worn at all times.

3.1. Preparation of the Crude Cellular Lysate

1. Harvest adherent cells into serum-free medium by gentle scraping (*see* **Note 7**). Most laboratory-scale preps (<10^9 cells) can be harvested into a total pooled volume of 200 ml or less. Cells grown in suspension are harvested directly by low-speed centrifugation (*see* **step 2**).

2. Pellet the AAV-containing cells in a 250-ml centrifuge tube at 200 × *g* for 10 min and remove the supernatant (*see* **Note 8**).
3. Rinse the cell pellet by suspension in 10 ml of ice-cold PBS and centrifuge as above.
4. Resuspend the washed cell pellet in 8 ml of serum-free culture medium and transfer to a 50-ml polypropylene conical tube.
5. Plunge the conical tube directly into pelleted dry ice and incubate until the medium is frozen solid. Thermal gloves and protective eyewear should be worn when coming into contact with dry ice.
6. Place the 50-ml conical tube in a 37 °C water bath until the contents are completely thawed.
7. Vortex briefly and repeat the freeze-thaw cycle above three times.
8. Pellet the cellular debris by centrifugation at 3000 × *g* for 10 min.
9. Using a pipette, carefully transfer the vector-containing supernatant to a fresh 50-ml conical tube. Take care not to disturb the pelleted cellular debris as transfer of debris may hamper subsequent filtration.
10. To digest non-encapsidated DNA, add Benzonase® nuclease to a final concentration of 150 units nuclease per milliliter of lysate and incubate the lysate in a 37 °C water bath for 1 h.

3.2. PEG-Mediated Vector Precipitation

1. Filter the nuclease-treated lysate through a 0.2-μm syringe-driven filter unit (*see* **Note 9**) connected to a 12-ml luer-lock syringe. Place the filtrate in a sterile Oak Ridge tube.
2. Add 1/12th volume of the 50% PEG stock solution to the Oak Ridge tube (for a final concentration of approximately 4% PEG).
3. Place the tube on ice for at least 1 h (*see* **Note 10**).
4. Collect the PEG precipitate by centrifugation for 30 min at 12,000 × *g*.
5. Discard the supernatant.
6. Spin the Oak Ridge tube briefly at 120 × *g* and remove residual PEG solution.
7. Resuspend the precipitate in 8 ml of buffer A (*see* **subheading 2.3**) by vigorous, prolonged pipetting. The suspended precipitate (which will remain opaque) should be mixed until particulates are no longer visible (*see* **Note 11**).

3.3. Anion-Exchange Chromatography

1. Attach the SOURCE 15Q ion-exchange column (Tricorn 10/100 column, ~8 ml bed volume) to the liquid chromatography apparatus. Previously unused columns should be conditioned by passage of at least 3 column volumes (CV) of low-salt buffer (buffer A) followed by 2–3 CV of high-salt buffer (buffer B), and then a final equilibration with at least 3 CV of low-salt buffer. Column equilibration may be performed at a volumetric flow rate of 4 ml/min (*see* **Note 12**).
2. Once the column is equilibrated with buffer A, reduce the flow rate to 2 ml/min.

3. Filter the resuspended PEG precipitate using a large diameter (33 mm) 0.2-μm pore-size syringe filter (*see* **Note 9**) and load into the injection loop of the chromatography apparatus. For the AKTA-FPLC system, we have found the 10 ml "Superloop" (GE Healthcare) to be very convenient (*see* **Note 13**).
4. Inject the sample. If desired, the flow-through may be collected as a single fraction of 10–20 ml (depending on the amount of non-binding material).
5. Wash the column with buffer A until the absorbance at 280 nm (OD_{280}) is "flat-lined" (i.e., no change in absorbance).
6. To elute the bound protein, initiate a linear salt gradient by ramping from 0 to 50% buffer B over a course of 20 ml (10 min at 2 ml/min) (*see* **Note 14**).
7. Beginning 3 ml after the start of the salt gradient, collect the eluate as 1-ml fractions over the course of the gradient. A representative anion-exchange chromatography elution profile is shown in **Fig. 2A**.
8. To identify the AAV-positive peak fractions, analyze gradient fraction samples by quantitative, real-time PCR (*see* **Subheading 3.4** and **Note 15**) using vector-specific primers.
9. Regenerate the SOURCE 15Q column by passage of at least one CV of 100% buffer B, followed by re-equilibration in buffer A (*see* **Note 16**).

3.4. Quantitative, Real-Time PCR (Q-PCR)

1. Determine the number of individual PCR tubes necessary to perform triplicate assays of each column fraction, six quantitative standards, and a template-minus control.
2. On the basis of 50 μl per reaction, prepare a sufficient quantity of 1× Q-PCR mix containing a 0.4 μM concentration of each primer.
3. Place 50 μl of the Q-PCR mix in each PCR tube.
4. Prepare a quantitative Q-PCR standard by making a 10-fold serial dilution of AAV vector (i.e., transgene-containing) plasmid starting with a concentration of 10^8 copies/μl and ending with a concentration of 10^3 copies/μl (*see* **Note 17**).
5. Dilute 2 μl of each chromatography fraction 1:100 in PCR-grade distilled water and mix thoroughly.
6. Transfer 2 μl of each diluted fraction sample and 2 μl of each standard into separate Q-PCR tubes (prepared in **step 3**). Use 2 μl of PCR-grade water for the template-minus control.
7. Start the real-time PCR thermal cycler in data collection mode. For the QuantiTect reagents used here, a typical PCR program is 15 min at 95 °C, followed by 40 cycles of 95 °C for 30 s, 55 °C for 30 s, and 72 °C for 30 s.
8. To identify the peak AAV-containing fractions, plot vector copy-number versus fraction number. Examples are shown in **Figs. 2B and 3B**.

3.5. Vector Concentration

To reduce the sample volume before gel filtration chromatography, the peak AAV-containing anion-exchange fractions are concentrated using a Centricon Plus-20 centrifugal filtration device. The volume of the column fractions is

(a)

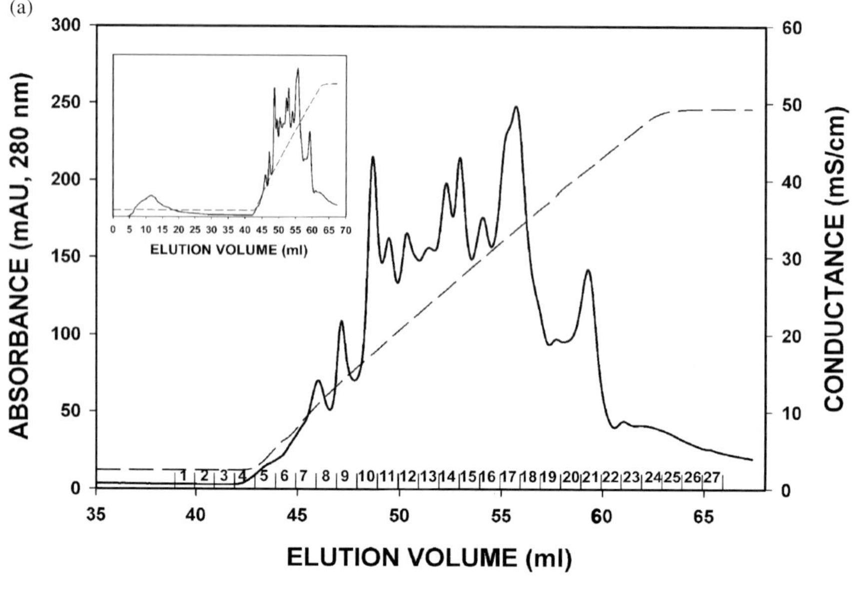

(b)

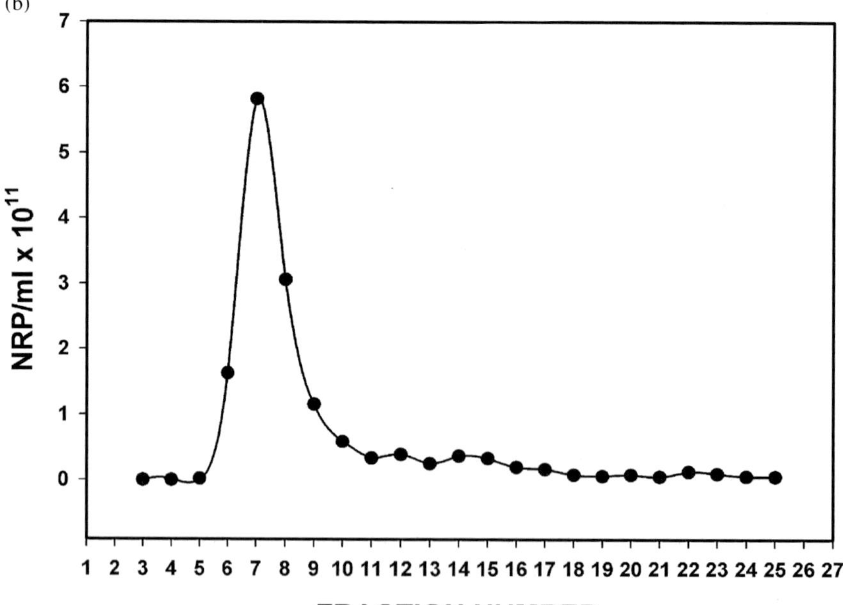

Fig. 2. Anion-exchange chromatography. **(A)** PEG-precipitated protein was suspended in anion-exchange column buffer, filtered, and loaded onto a Tricorn 10/100 column packed with SOURCE 15Q anion-exchange medium (8 ml bed volume). After washing, bound protein was eluted from the column with a linear 10–500 mM NaCl gradient (represented by the buffer conductance in mS/cm; dashed line) over a course of

reduced to the final concentrate volume (or "stop" volume) of the filtration device (approximately 150 µl) and then adjusted to a 0.5 ml loading volume.

1. Pre-wet a Centricon Plus-20 centrifugal filtration unit with the addition of 10 ml of gel filtration buffer (buffer C).
2. Centrifuge the filtration unit at 3500 × *g* for 4 min. If the minimum retention volume (~150 µl) is not reached, repeat the centrifugation step.
3. Discard the buffer from the lower chamber of the filtration unit.
4. Place another 10 ml of buffer C in the filtration unit and add the peak AAV-containing anion-exchange fractions (e.g., fractions 6–9 in **Fig. 2B**) to the buffer.
5. Centrifuge at 3500 × *g* for 4 min and discard the filtrate. If the minimum retention volume (~150 µl) is not reached, repeat the centrifugation step.
6. Insert the recovery cup in the top portion of the filtration unit.
7. Recover the concentrated retentate by detaching the upper portion of the filtration unit (now containing the recovery cup), inverting, and centrifuging at 1000 × g for 1 min.
8. Transfer the viral concentrate (approximately 150 µl total volume) to a sterile microfuge tube and bring to the total volume to 0.5 ml with buffer C.

3.6. Gel Filtration Chromatography

1. Attach the Superdex 200 column (24 ml bed volume) to the liquid chromatography apparatus.
2. Equilibrate the column with at least 2 CVs of buffer C at a volumetric flow rate of 1 ml/min. The maximum pressure should not exceed 1.5 MPa (*see* **Note 18**).
3. Decrease the flow rate to 0.8 ml/min.
4. Load the sample injection loop with the concentrated vector and inject the sample.
5. Beginning 2 ml after sample injection, collect the eluate as 1 ml fractions for a duration of approximately 1 CV (*see* **Note 19**). An example chromatogram is shown in **Fig. 3A**.
6. AAV peak fractions may be identified, and the vector titered, by quantitative, real-time PCR using vector-specific primers (*see* **Subheading 3.4.**). Further characterization of the purified material may include silver staining (*see* **Note 20**) and Western blot analysis as described below.
7. AAV-containing fractions may be filter-sterilized (0.2 µm membrane) and stored at –80 °C.

Fig. 2. 20 ml. The solid line (absorbance at 280 nm) represents the protein elution profile. One-milliliter fractions were collected over the course of the gradient. The fraction number is indicated at the bottom of the graph. The elution profile of the entire chromatography run is shown in the panel insert. Note that a minimal amount of protein occurs in the flow-through fraction. (**B**) The number of nuclease-resistant particles (NRP) per 1-ml fraction collected from the anion-exchange column was determined by quantitative, real-time PCR using vector-specific primers.

(a)

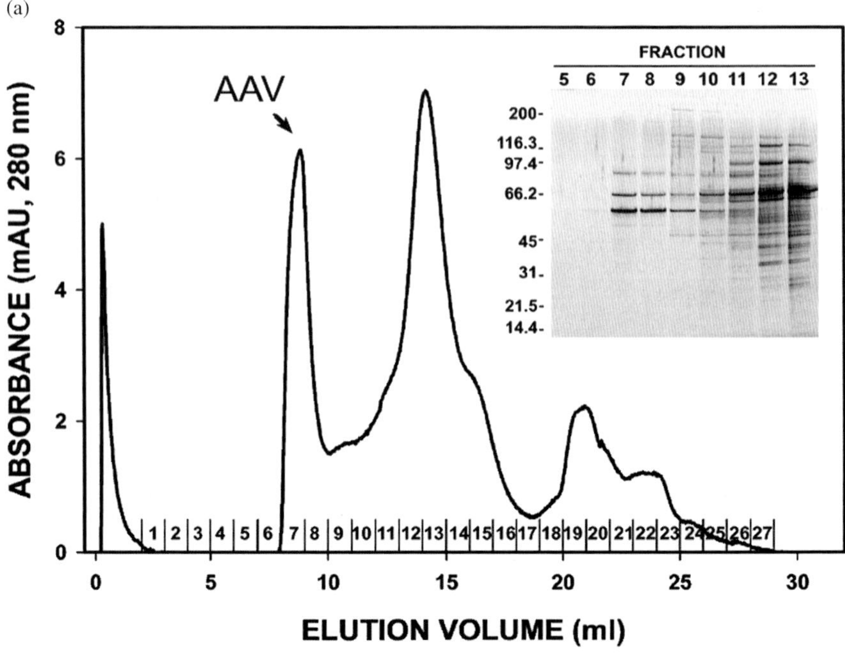

(b)

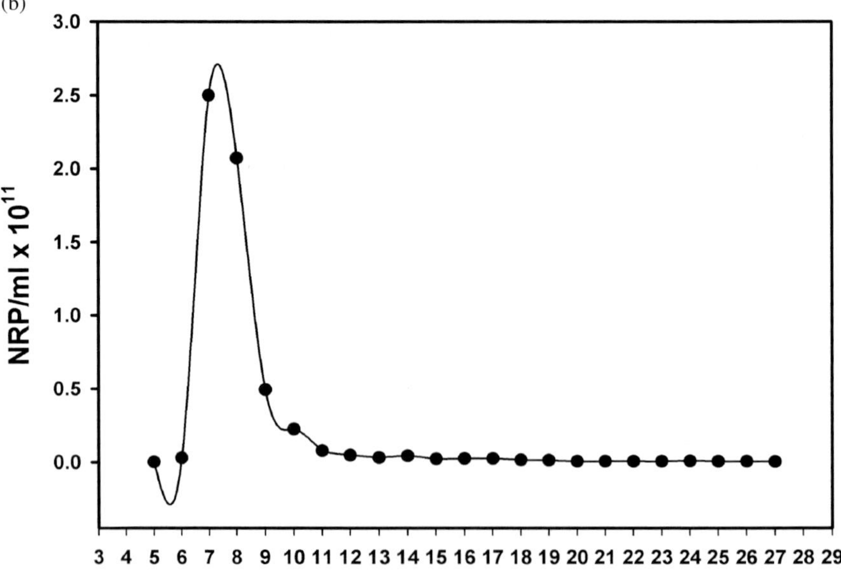

Fig. 3. Gel filtration chromatography. (**A**) Anion-exchange column fractions 6–9 were pooled and concentrated using a 100-kDa molecular weight cut-off centrifugal filtration device. The concentrated sample was loaded onto a Superdex 200 10/300 GL

3.7. Characterization of Purified Recombinant AAV Particles

3.7.1. Vector Titration by Quantitative, Real-Time PCR

Same as **Subheading 3.4**.

3.7.2. SDS–PAGE of Fraction Samples

1. Before PAGE analysis, prepare the following by diluting the concentrated stock reagents in distilled water: 800 ml of 1× MOPS running buffer, 1 ml of 1× reducing NuPAGE sample buffer, and 1 ml of 2× reducing NuPAGE sample buffer.
2. Remove a 4–12%, 1.5-mm NuPAGE gel from its pouch, rinse with distilled water, and remove the gel comb and tape.
3. Place in the gel apparatus and fasten securely.
4. Fill the inner chamber of the gel apparatus with 1× MOPS buffer and add 0.5 ml of NuPAGE antioxidant reagent.
5. Fill the outer portion of the gel unit with the remaining running buffer.
6. If the gel is to be silver-stained, dilute the protein molecular weight markers to a final concentration of approximately 5 ng per band per µl and mix 4 µl of the diluted marker with 38 µl 1× sample buffer in a thermal cycler tube.
7. In separate thermal cycler tubes, mix 20 µl of each column fraction to be analyzed with 20 µl of 2× sample loading buffer.
8. Heat the molecular weight standard and fraction samples in a thermal cycler at 98 °C for 15 min, followed by a 25 °C "hold" cycle.
9. Using a 10-ml syringe fitted with a 200-µl pipette tip, rinse the wells of the NuPAGE gel with 1× running buffer and load the heated samples.
10. Apply voltage (200 V constant) until the tracking dye reaches the bottom of the gel. The gel is now ready for silver staining or western blotting.

3.7.3. Silver Staining using SilverXpress® Reagents

The following is based on the manufacturer's suggested protocol. All solutions should be prepared in ultrapure (18 MΩ/cm) water.

Fig. 3. column. The solid line (absorbance at 280 nm) represents the protein elution profile. One-milliliter fractions were collected. Fraction numbers are indicated at the bottom of the graph. The insert shows a silver-stained polyacrylamide gel in which each lane received 20 µl of the indicated gel filtration chromatography fraction. Note that the three AAV serotype 1 capsid proteins, VP1 (81.4 kDa), VP2 (66.2 kDa), and VP3 (59.6 kDa), are observed in fractions 7 and 8. (**B**) The number of nuclease-resistant particles (NRP) per 1-ml fraction collected from the gel filtration column was determined by quantitative, real-time PCR using vector-specific primers.

1. Remove the gel cassette from the electrophoresis apparatus and carefully pry the cassette apart.
2. Soak the polyacrylamide gel in 200 ml of a 50% methanol, 10% acetic acid solution (both v/v) for 20 min with shaking.
3. Prepare 200 ml of sensitizer solution (50% methanol, 2.5% SilverXpress Sensitizer, both v/v) and soak the fixed gel in 100 ml of this solution for 30 min with gently shaking. Replace with 100 ml fresh sensitizer solution and repeat for an additional 30 min.
4. Wash the gel twice with 200 ml of ultrapure water for 10 min each time.
5. Stain the gel for 15 min with 100 ml of 5% SilverXpress Stainer A, 5% SilverXpress Stainer B (both v/v). Shake gently.
6. Wash the gel twice (5 min each time) with 200 ml ultrapure water.
7. To visualize the protein bands, incubate the gel in 100 ml volume of a 5% (v/v) SilverXpress Developer solution. When the protein bands are readily visible (usually between 2 and 10 min), stop the staining reaction with the addition of 5 ml of kit-supplied stop solution and incubate for 10 min.
8. Rinse the gel with three 200-ml volumes of ultrapure water. The gel is now ready for photography and/or drying. Representative silver-stained gels are shown in **Fig. 3A** (panel insert) and **Fig. 4A**.

3.7.4. Western Blotting

1. Prepare 1 l of 1× transfer buffer containing 10% methanol (v/v) by adding 50 ml of 20× transfer buffer concentrate to 850 ml distilled water and bringing to a total volume of 1 l with the addition of ~100 ml of anhydrous methanol.
2. Briefly soak the pads of the blot module in transfer buffer. Squeeze-out any air bubbles.
3. Place two of the pads in the cathode portion of the blot module.
4. Wet a pre-cut PVDF filter membrane in methanol for 30 s, rinse briefly in distilled water, and then soak in transfer buffer for 30 s.
5. Wet a piece of pre-cut filter paper in transfer buffer and place on top of the pads on the cathode plate.
6. Place the following items in order on top of the filter paper: the polyacrylamide gel to be blotted, a pre-soaked PVDF membrane, a second piece of wetted filter paper, and two pre-soaked pads.
7. Close the blot module with the anode plate and carefully place the module within the transfer tank.
8. Fill the transfer tank and blot module with 1× transfer buffer.
9. To transfer the separated polypeptides to the PVDF membrane, apply 30 V (constant) for 1.5 h.
10. Prepare blocking solution by adding 10 g of non-fat dry milk to 200 ml of PBST. Mix thoroughly.
11. After blotting is complete, disassemble the transfer module and place the PVDF membrane in a small container with a sufficient volume of blocking solution to

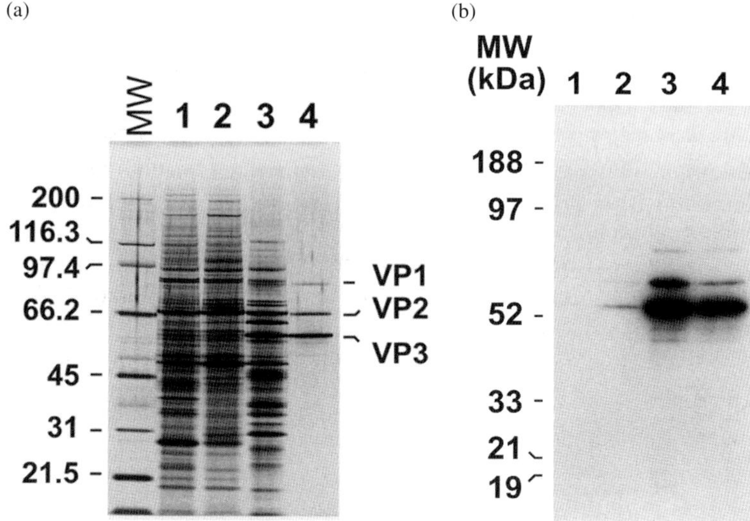

Fig. 4. Silver staining and Western blot analysis of process samples. (**A**) Samples of the crude cellular lysate and PEG-precipitated protein pellet, as well as peak fraction samples from the anion-exchange and gel filtration columns were separated on a 4–12% polyacrylamide gradient gel and visualized by silver staining. Lane 1 was loaded with one microgram of total cellular protein from the crude lysate. Lane 2 was loaded with 1 μg of PEG-precipitated protein. Lane 3 was loaded with a sample containing ~5 × 10^9 nuclease-resistant particles from the peak anion-exchange column fraction (fraction 7). Lane 4 was loaded with a sample containing ~5 × 10^9 nuclease-resistant particles from the peak gel filtration column fraction (fraction 7). Lane marked "MW" received protein molecular weight markers (size in kilodaltons). Note that the amount of protein in lanes 1 and 2 was normalized for comparison. The protein concentration of the total cellular lysate was approximately 10 times that of an equivalent volume of the PEG-precipitated material. (**B**) A 4–12% polyacrylamide gradient gel was loaded with protein samples as described in "A." Separated polypeptides were blotted onto a PVDF membrane and visualized by Western blot analysis using a polyclonal anti-AAV capsid protein rabbit antiserum, goat anti-rabbit horseradish peroxidase (HRP)-conjugated secondary antibody, and a chemiluminescent HRP substrate. The three AAV serotype 1 capsid proteins, VP1 (81.4 kDa), VP2 (66.2 kDa), and VP3 (59.6 kDa), are observed.

completely cover the membrane. Incubate at room temperature for at least 1 h, with shaking (*see* **Note 21**).

12. Add the primary anti-AAV capsid antibody at the appropriate dilution (*see* **Note 22**) and incubate for 1.5 h at room temperature with shaking.
13. Remove the primary antibody solution. Wash the blot three times by agitation in the presence of a sufficient volume of PBST to float the membrane (5 min each wash).

14. Place the blot in fresh blocking solution containing the appropriate dilution of the secondary detection antibody and incubate for 1 h with shaking.
15. Remove the secondary antibody solution. Wash the blot three times with PBST (5 min each wash).
16. Remove excess PBST by holding the PVDF membrane vertically and touching one corner to a piece of filter paper. Cover the blot with chemiluminescent detection reagent and incubate for 3–5 min.
17. Remove excess detection reagent as above, cover the blot in plastic wrap, and expose to X-ray film. A representative Western blot is shown in **Fig. 4B**.

4. Notes

1. Benzonase® is a non-specific nuclease capable of degrading both double-stranded and single-stranded nucleic acids (RNA or DNA). The enzyme functions best within a pH range of 6 to 10 and requires magnesium cations for activity (low mM range). High concentration preparations ($\geq$250 U/μl) are most convenient for the cell lysis protocol because of the relatively large volume of cell lysate.
2. To make a 50% (w/v) PEG stock solution, slowly add 50 g of PEG to 50 ml of distilled water with constant stirring. Once the PEG has begun to dissolve, adjust to a final volume of 100 ml and stir until the PEG is completely in solution. Filter through a 0.2-μm pore-size membrane and store at room temperature. The PEG stock solution should be made in advance as both dissolving the PEG and filtering the viscous PEG solution take an extended amount of time. A 50% PEG stock solution helps to reduce the total volume of the sample when performing PEG precipitation. This may be significant if large volumes (such as the medium of an AAV-producing culture) are subjected to PEG precipitation. However, if relatively small volumes of sample are processed, then a less concentrated PEG stock solution, classically 20% (w/v), may be more convenient as it should dissolve and filter more quickly.
3. Chromatography buffers should be filtered through a 0.2-μm pore-size membrane and "degassed" by leaving the fluid under vacuum for at least 10 min, with occasional tapping to dislodge any gas bubbles that have developed.
4. Primers should be specific to the transgene being packaged. Ideally, primers should be 20–21 nt length and matched as closely as possible in terms of melting temperature (T_M). Primers should be chosen such that the amplified region is in the order of 100–150 bp in length.
5. Glycerol is optional. It increases the viscosity (and therefore the back-pressure) of the buffer but may ameliorate aggregation of AAV at high particle concentrations ***(22)***.
6. The hydrogen ion concentration (pH) of the binding buffer is an important parameter in purification of rAAV through ion-exchange chromatography as it can influence the overall net charge (and hence the binding characteristics) of the recombinant virion. This protocol utilizes a bis-Tris buffering system

at pH 6.0 for the ion exchange chromatography step. We have found this pH to be efficacious in the purification of rAAV serotypes 1 and 5. Other AAV serotypes may require a different pH for optimal binding. For an AAV serotype for which information concerning ion-exchange chromatography parameters is lacking, one should "scout" rAAV resin-binding characteristics in buffers varied by 0.5 pH unit increments in the range of pH 6.0–8.5.

7. We have found "cell lifters" (sterile plastic paddles with a wide beveled edge; Costar/Corning, Corning, NY, USA) rather than traditional cell scrapers or rubber policeman, to be particularly suitable for this purpose.
8. Significant amounts of cell-free vector have been reported for some serotypes. Cell-free vector may be recovered from the tissue culture medium through precipitation with 4% PEG as outlined in **Subheading 3.2**. For larger volumes, incubate the PEG-containing medium at 4 °C for 2–3 h on an orbital platform. Collect the PEG precipitate by centrifugation in a GSA rotor (Sorvall) at 10,000 × *g* for 1 h.
9. Use a luer-lock syringe and filter. Do not apply excessive force as this may result in mechanical failure of the filtration device and/or uncontrolled discharge of the syringe contents. Work in a biological safety cabinet as this serves as a splash guard and provides protection from aerosols. If the filter is clogged, simply retract the syringe plunger to recover as much sample as possible from the filtration unit and apply a new filter.
10. Overnight incubation may improve recovery.
11. It is important that the PEG precipitate be resuspended to a homogenous mixture capable of filtration. If insoluble material persists, it can be pelleted by a brief (5–10 min) centrifugation at 120 × *g*.
12. The volumetric flow rate employed here is appropriate for the separation medium and column dimensions (10 × 100 mm) used in the development of the protocol. To determine an equivalent flow rate for a column with different dimensions, calculate the linear flow rate using the equation $L = V(60)/\pi r^2$ and then convert this value to the appropriate volumetric flow rate using the equation $V = L\pi r^2/60$, where V is the volumetric flow rate in milliliters per minute (ml/min), L is the linear flow rate in centimeters per hour (cm/h), r is the radius of the inside diameter of the column (in cm), and π represents the constant pi (~3.14).
13. If a large capacity injection loop is not available, the sample can be injected sequentially in smaller volumes. Alternately, the sample may be applied to the column with a peristaltic pump, after which the column is reattached to the liquid chromatography system.
14. When using a large-volume loading loop (such as a 10 ml Superloop), it is important that the loop be in the load position (i.e., "off-line") before the gradient is initiated to prevent distortion of the gradient.
15. We have found quantitative, real-time PCR (Q-PCR) to be the fastest, most sensitive method for detecting peak AAV-containing fractions. Compared to other detection methods, Q-PCR also has the advantage of identifying "full" (i.e., nucleic acid-containing) recombinant particles. Alternatively, column

fractions may be analyzed for AAV content by Western blotting, cellular transduction (if an assayable transgene is packaged), or by an AAV-specific enzyme-linked immunosorbent assay (ELISA). AAV-specific ELISA kits are commercially available (American Research Products, Research Diagnostics, Inc., and also Progen, GmbH).

16. The column may be placed at 4 ºC for short-term storage. For long-term storage, equilibrate the column with 20% ethanol and store at 4 ºC.
17. The real-time PCR standards may be stored at –20 °C for repeated use.
18. To shorten overall process time, equilibration of the gel filtration column may be performed during the data collection interval of the Q-PCR assay used to identify the peak anion-exchange fractions.
19. Once the elution prosperities of a given AAV serotype are established (under standardized conditions), fraction collection can be limited to peak AAV-containing fractions, rather than collecting an entire CV.
20. Upon SDS–PAGE analysis, purified AAV particles present as three protein bands representing the AAV capsid proteins: VP1, VP2, and VP3. The computed molecular weights of AAV-1 VP1, -2, and -3 are 81.4, 66.2, and 59.6 kDa, respectively. These bands are observed in fractions 7 and 8 of the silver-stained gel depicted the panel insert in **Fig. 3A** (see also **Fig. 4A**). The detection of AAV particles by SDS–PAGE and protein staining is limited by the sensitivity of the protein stain (for typical silver-staining reagents, this is approximately 1 ng per band). Empirically, we have found the limit of AAV detection by silver staining to be roughly 10^8 recombinant particles loaded per lane of the gel.
21. Blots may be blocked overnight at 4 °C, thus providing a convenient stopping point.
22. Optimal dilutions of primary and secondary antibodies must be determined empirically. Good starting points are 1:100 to 1:1000 for the primary antibody and 1:1000 to 1:10,000 for the secondary antibody.

Acknowledgments

This research was supported by the Intramural Research Program of the National Heart, Lung, and Blood Institute, National Institutes of Health. We thank Sylvain Cecchini for critical review of the draft manuscript.

References

1. Hoggan, M. D. (1970) Adenovirus associated viruses. *Prog. Med. Virol.* **12,** 211–239.
2. Berns, K. I., and Bohenzky, R. A. (1987) Adeno-associated viruses: an update. *Adv. Virus Res.* **32,** 243–306.
3. Berns, K. I., and Linden, R. M. (1995) The cryptic life style of adeno-associated virus. *Bioessays* **17,** 237–245.

4. Smith, R. H., and Kotin, R. M. (2002) Adeno-associated virus, in *Mobile DNA II* (Craig, N. L., Craigie, R., Gellert, M., and Lambowitz, A., eds.), American Society for Microbiology, Washington, D. C., pp. 905–923.
5. Kronenberg, S., Kleinschmidt, J. A., and Bottcher, B. (2001) Electron cryo-microscopy and image reconstruction of adeno-associated virus type 2 empty capsids. *EMBO Rep.* **2,** 997–1002.
6. Walters, R. W., Agbandje-McKenna, M., Bowman, V. D., Moninger, T. O., Olson, N. H., Seiler, M., Chiorini, J. A., Baker, T. S., and Zabner, J. (2004) Structure of adeno-associated virus serotype 5. *J. Virol.* **78,** 3361–3371.
7. Xie, Q, Bu, W., Bhatia, S., Hare, J., Somasundaram, T., Azzi, A., and Chapman, M. S. (2002) The atomic structure of adeno-associated virus (AAV-2), a vector for human gene therapy. *Proc. Natl. Acad. Sci. U. S. A.* **99,** 10405–10410.
8. Auricchio, A., O'Connor, E., Hildinger, M., and Wilson, J. M. (2001) A single-step affinity column for purification of serotype-5 based adeno-associated viral vectors. *Mol. Ther.* **4,** 372–374.
9. Auricchio, A., Hildinger, M., O'Connor, E., Gao, G.-P., and Wilson, J. M. (2001) Isolation of highly infectious and pure adeno-associated virus type 2 vectors with a single-step gravity-flow column. *Hum. Gene Ther.* **12,** 71–76.
10. Blouin, V., Brument, N., Toublanc, E., Raimbaud, I., Moullier, P., and Salvetti, A. (2004) Improving rAAV production and purification: towards the definition of a scaleable process. *J. Gene Med.* **6,** S223–S228.
11. Brument, N., Morenweiser, R., Blouin, V., Toublanc, E., Raimbaud, I., Cherel, Y., Folliot, S., Gaden, F., Boulanger, P., Kroner-Lux, G., Moullier, P., Rolling, F., and Salvetti, A. (2002) A versatile and scalable two-step ion-exchange chromatography process for the purification of recombinant adeno-associated virus serotypes-2 and -5. *Mol. Ther.* **6,** 678–686.
12. Clark, K. R., Liu, X., McGrath, J. P., and Johnson, P. R. (1999) Highly purified adeno-associated virus vectors are biologically active and free of detectable helper and wild-type viruses. *Hum. Gene Ther.* **10,** 1031–1039.
13. Davidoff, A. M., Ng, C. Y. C., Sleep, S., Gray, J., Azam, S., Zhao, Y., McIntosh, J. H., Karimipoor, M., and Nathwani, A. C. (2004) Purification of recombinant adeno-associated virus type 8 by ion exchange chromatography generates clinical grade vector stock. *J. Virol. Methods* **121,** 209–215.
14. Gao, G., Qu, G., Burnham, M. S., Huang, J., Chirmule, N., Joshi, B., Yu, Q.-C., Marsh, J. A., Conceicao, C. M., and Wilson, J. M. (2000) Purification of recombinant adeno-associated virus vectors by column chromatography and its performance in vivo. *Hum. Gene Ther.* **11,** 2079–2091.
15. Ioffe, E., and Burova, E. (2005) Chromatographic purification of recombinant adenoviral and adeno-associated viral vectors: methods and implications. *Gene Ther.* **12,** S5–S17.
16. Kaludov, N., Handelman, B., and Chiorini, J. A. (2002) Scalable purification of adeno-associated virus type 2, 4, or 5 using ion-exchange chromatography. *Hum. Gene Ther.* **13,** 1235–1243.

17. O'Riordan, C. R., Lachapelle, Vincent, K. A., and Wadsworth, S. C. (2000) Scaleable chromatographic purification process for recombinant adeno-associated virus (rAAV). *J. Gene Med.* **2,** 444–454.
18. Smith, R. H., Ding, C., and Kotin, R. M. (2003) Serum-free production and column purification of adeno-associated virus type 5. *J. Virol. Methods* **114,** 115–124.
19. Zolotukhin, S., Byrne, B. J., Mason, E., Zolotukhin, I., Potter, M., Chesnut, K., Summerford, C., Samulski, R., J., and Muzyczka, N. (1999) Recombinant adeno-associated virus purification using novel methods improves infectious titer and yield. *Gene Ther.* **6,** 973–985.
20. Zolotukhin, S., Potter, M., Zolotukhin, I., Sakai, Y., Loiler, S., Fraites, T. J., Jr., Chiodo, V. A., Phillipsberg, T., Muzyczka, N., Hauswirth, W., W., Flotte, T. R., Byrne, B. J., and Snyder, R. O. (2002) Production and purification of serotypes 1, 2, and 5 recombinant adeno-associated viral vectors. *Methods* **28,** 158–167.
21. Atha, D. H., and Ingham, K. C. (1981) Mechanism of precipitation of proteins by polyethylene glycols. *J. Biol. Chem.* **256,** 12108–12117.
22. Xie, Q., Hare, J., Turnigan, J., and Chapman, M. S. (2004) Large-scale production, purification and crystallization of wild-type adeno-associated virus-2. *J. Virol. Methods* **122,** 17–27.

5

Spectroscopic Methods for the Physical Characterization and Formulation of Nonviral Gene Delivery Systems

Salvador F. Ausar, Sangeeta B. Joshi, and C. Russell Middaugh

Summary

Currently, with the exception of naked DNA formulations, most pharmaceutical preparations of plasmid DNA employ some type of polycationic delivery vector such as synthetic cationic polymers and lipids to enhance delivery. A number of biophysical techniques are readily available for the structural characterization of plasmid DNA within synthetic gene delivery complexes. Here we present applications of ultraviolet (UV) absorption, circular dichroism (CD), infrared (IR), and fluorescence spectroscopies as well as dynamic light scattering to the structural analysis of the oligonucleotide component of nonviral gene delivery vectors. We also illustrate this approach for the investigation of the formulation of lipoplex and polyplex-based gene delivery systems. To summarize such data, we show how the macromolecular complexes can be represented as vectors in a highly dimensional space in which the components of the vector consist of normalized values of experimental parameters measured as a function of different solution conditions such as pH and ionic strength.

Key Words: Dynamic light scattering; circular dichroism; UV absorption spectroscopy; FTIR; pDNA formulation; fluorescence; empirical phase diagram.

1. Introduction

Nonviral gene delivery complexes are being considered as alternatives to virally based systems because of their presumed safety and ease of manufacture. In such systems, DNA containing a therapeutic gene is encoded in a bacterial plasmid (pDNA) containing an appropriate promoter. A covalently closed circular DNA molecule in negatively supercoiled form is generally considered to be the best state for effective delivery. This topological conformation has

From: *Methods in Molecular Biology, vol. 434: Volume 2: Design and Characterization of Gene Transfer Vectors*
Edited by: J. M. Le Doux © Humana Press, Totowa, NJ

been reported to produce higher level of gene expression than the relaxed circular or open linear DNA states ***(1)***. During plasmid extraction, purification, and formulation, supercoiled molecules can become nicked on one or both strands to form the open circular or linear topoisomers, respectively. Less defined multimeric species (e.g., concatamers), aggregates, and small DNA fragments have been also reported in plasmid preparations ***(2–4)***. Such structural changes as well as covalent changes (e.g., oxidation) may affect formulation consistency and stability, and ultimately result in lower transfection efficiency.

The most common DNA secondary structure is referred to as the B-form. This right-handed form is very flexible and can be easily bent or supercoiled with minor changes in structure. The B-form contains 10.5 bp/turn and 3.4 Å axial rises between the planar bases, which are tilted at about a 6° angle. The A-form of DNA is also sometimes seen, especially in dehydrated DNA samples. A third state known as Z-DNA (11 bp/turn) consists of a left-handed helix. This form has a single large groove compared with the major and minor grooves of the A and B forms and is most often observed in oligonucleotides with alternating pyrimidine/purine bases. Other less common forms (i.e., C-, D-, and T-DNA) have been reported in certain specific conditions that are unlikely to be relevant to pDNA ***(5)***. Recognition of the potential presence of different DNA conformations within nonviral gene delivery systems may turn out to be of some importance, but this has yet to be definitively established. As a consequence of the interactions with carriers commonly used for plasmid condensation and delivery, pDNA molecules could be induced into novel conformational states that could affect their processing and transcription within cells ***(6–8)***.

Transfection of target cells with naked pDNA is generally an inefficient process requiring large amounts of material for a high level of gene transfer ***(9,10)***. Thus, most delivery systems are based on the use of cationic lipids or polymers with the exception of vaccines where amplification by the immune system can result in larger effects ***(11)***. Positively charged entities collapse pDNA into more globular and compact structures that are thought to be processed more efficiently by endocytosis, which appears to be the primary route of entry of such complexes ***(12,13)***. Furthermore, their positively charged structure facilitates their interaction with proteoglycans on the cell surface, a presumed early step in the transfection process ***(14)***.

From a pharmaceutical perspective, a number of issues still need to be comprehensively addressed concerning the characterization and formulation of the numerous and variable systems currently proposed to deliver genetic material into cells ***in vivo (15)***. Spectroscopic and calorimetric methods are valuable tools for the development and characterization of the pharmaceutical quality of nonviral gene delivery systems. They are useful for both analysis of the structure of the DNA component and for characterization of nucleic acid/carrier interactions and the resultant complex itself ***(15,16)***. In this chapter, we provide a description of spectroscopic methods that are useful for the characterization of the nucleic acid components of nonviral gene delivery systems as well as an empirical phase diagram approach that can be used for the rational formulation of lipoplexes and polyplexes. The use of calorimetric methods is described elsewhere ***(17,18)***.

2. Materials

2.1. Reagents

1. Plasmid DNA >95% supercoiled. Material from Valentis, Inc. (Burlingame, CA, USA) was used in all the studies described below.
2. Branched polyethyleneimines (PEI) (MW 2, 25, and 750 kDa) (Sigma-Aldrich, Milwaukee, WI, USA). Linear PEI (Polysciences, Inc., Warrington, PA, USA).
3. Dimethyldioctadecylammonium bromide (DDAB), Cholesterol (CHOL), 1,2-dioleoyltrimethylammoniumpropane (DOTAP), 1,2- distearyltrimethylammoniumpropane (DSTAP), and 1,2-dioleoyl-*sn*-glycero-3-phosphatidylethanolamine (DOPE) (Avanti Polar Lipids, Inc., Alabaster, AL, USA).
4. 2-(*N*-morpholino) ethanesulfonic acid (MES) and Tris (hydroxymethyl) aminomethane were from Sigma (Saint Louis, MO, USA) and piperazine-*N,N'*-bis(3-propanesulfonic acid) (PIPPS) (Calbiochem, San Diego, CA, USA).
5. YOYO-1 and ethidium bromide (ETBR) (Molecular Probes, Eugene, OR, USA).
6. Nucleopore (100 nm) polycarbonate filter membranes (Whatman, Clifton, NJ).
7. 10 mM MES-PIPPS buffer was prepared by adding the appropriate volume of stock solutions of 0.1 M MES to 0.1 M PIPPS to reach the desired pH. The concentrated buffer was finally diluted 10 times with ultrapure water. The final pH was adjusted with HCl or NaOH and the ionic strength with 1 M NaCl.

2.2. Instrumentation

1. UV-visible diode array spectrophotometer: Agilent 8453 (Agilent Technologies, Palo Alto, CA, USA) equipped with a Peltier temperature controller device.
2. Spectropolarimeter: Jasco 810 (Jasco Inc., Easton, MD, USA) equipped with a Peltier temperature controller unit and six-position cuvette holder.
3. Fourier Transform Infrared Spectroscopy (FTIR) spectrometer: Nicolet 560 ESP (Nicolet Instrument, Madison, WI, USA) equipped with a mercury-cadmium-telluride (MCT) detector and a zinc selenide attenuated total reflectance (ATR) cell (Thermal A.R.K., SpectraTech, Shelton, CT, USA).
4. Spectrofluorometer: PTI QuantaMaster (PTI, Monmouth, NJ, USA) equipped with a Peltier temperature controller device.
5. Light-scattering system (Brookhaven Instrument Corp., Holtzille, NY, USA) equipped with a 50-mW helium–neon laser operating at 532 nm and an EMI9863 PMT mounted on a BI-200M goniometer.

3. Methods

3.1. Preparation of Liposomes

1. Dissolve the lipids in chloroform at approximately 10 mg lipid/ml chloroform. Transfer the lipid solution into glass vials and evaporate the chloroform under a stream of nitrogen gas in a fume hood using a pipette tip attached to a tygon tube. To remove traces of solvent, place the vials in a desiccator under vacuum for at least 3 h.
2. Disperse the lipid film by vortexing in 10 mM Tris/HCl pH 7.0 at a temperature above the phase transition temperature of the lipids. Hydration times vary from 30 to 60 min.
3. Extrude the liposomes 10–15 times through 100 nm polycarbonate membranes.
4. Store the liposome preparation at 4°C and use within 3 days (*see* **Note 1**).

3.2. Preparation of Complexes

1. Before complex preparation, dilute both DNA and cationic carrier in 10 mM MES-PIPPS buffer at the desired pH.
2. Prepare the complexes by adding the component of lesser quantity to that of the greater. By using this method, passage through charge neutrality is avoided (*see* **Note 2**).
3. Adjust the desired nitrogen (N) to DNA phosphate (P) molar ratios by varying the amount of cationic carrier and holding the amount of DNA constant at approximately 100 μg/ml. The concentration of DNA is determined by UV absorbance at 260 nm (*see* **Subheading 3.3.**).
4. Before performing the experiments, allow the complexes to equilibrate for 20–30 min.

3.3. Ultraviolet Absorption Spectroscopy

A variety of ultraviolet (UV)-visible spectrophotometers are commercially available with different capabilities and resolution. For the analysis of pDNA, we find that a relatively inexpensive and versatile diode-array instrument is optimal. In such instruments, the polychromatic spectrum of the incident light source is passed through a quartz cuvette containing the sample. The transmitted light is dispersed by a polychromator and detected by an array of photodiodes set at regular intervals capable of detecting a narrow bandwidth of light (typical diode spacing ≥0.5 nm). The advantage of this type of spectrophotometer is the ability to record rapidly the full spectrum of the sample of interest. The nominal resolution, restricted by the spacing of photodiodes, can be increased by using interpolation algorithms permitting an effective resolution of 0.01 nm *(19)*.

UV spectroscopy is widely used to assess the concentration of nucleic acids by measurement of the absorbance intensity at 260 nm, which roughly corresponds to the maximum in the absorbance spectrum (*see* **Fig. 1**). Knowing

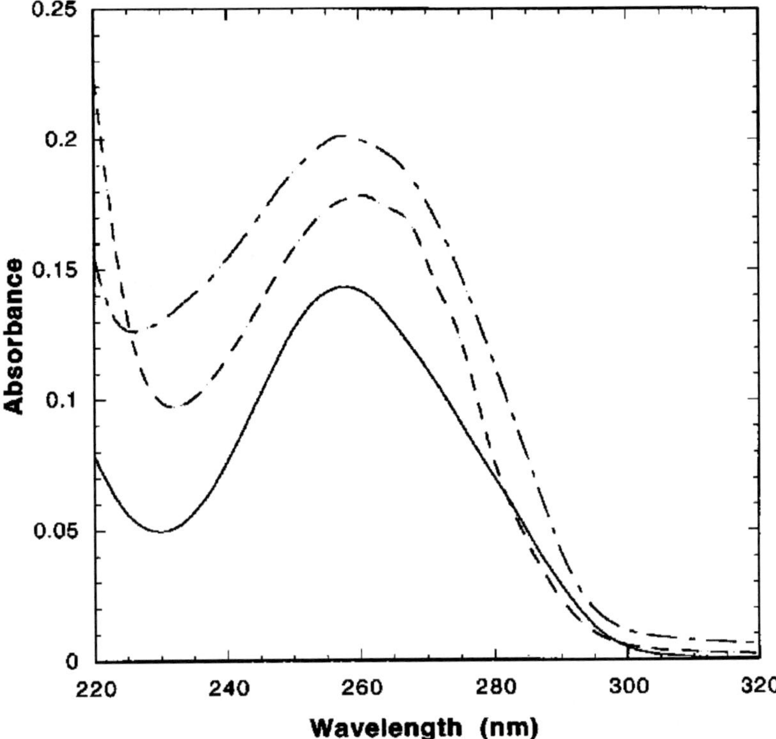

Fig. 1. UV absorption spectra of the same concentrations (weight/volume) of a mixture of mononucleotides (— • —) and single-stranded (– – –) and native double-stranded M13mp18 DNA (—). *J Pharm Sci* 87,130–46. Copyright 1998, reproduction with permission of Wiley-Liss, Inc. and the American Pharmaceutical Association.

the extinction coefficient of the polynucleotide of interest, the concentration is easily calculated by the Beer–Lambert law:

$$A = \varepsilon bC \tag{1}$$

where A is the absorbance at the specific wavelength, ε is the molar absorbtivity at the wavelength, b is the path length of the sample in centimeters, and C is the sample concentration. To determine the concentration of DNA by UV absorption spectroscopy, proceed as follows:

1. Turn on the instrument and set the temperature at 20°C. The warming period for the lamp is typically about 15 min.
2. Blank the spectrophotometer with appropriate buffer. Use quartz cuvettes and make sure they are clean.

3. Dilute a known volume of nucleic acid or complexes in appropriate buffer (i.e., PBS or Tris pH 7.4) at an absorbance value between 0.1 and 1.0. It is possible to work outside this range. With suitable care more concentrated solutions can be measured by reducing the pathlength of the cell. All experiments must be performed at least in triplicate.
4. Obtain the spectrum from 200 to 400 nm and correct for light scattering (*see* **Note 3** and **Fig. 2**). For optimal collection of high-resolution spectrum, use an integration time of 20 s.
5. Calculate the concentration of the original solution using equation ***(1)*** and multiply by the dilution factor. If ε is not known, the concentration can be calculated by using the following approximation: 1 A_{260} U (dsDNA) = 50 μg/ml, 1 A_{260} U (ssDNA) = 37 μg/ml, 1 A_{260} U (ssRNA) = 40 μg/ml (*see* **Note 4**).
6. To evaluate the purity of the preparation, calculate the ratio $A_{260} : A_{280}$. For a pure DNA preparation $A_{260} : A_{280} = 1.8$ and for pure RNA $A_{260} : A_{280} = 2.0$.

An alternative approach for determining the concentration and purity of pDNA involves the calculation of second derivative spectra ***(20)***. For this purpose, the second derivative trace is numerically calculated and then interpolated using a spline function with a final data spacing of 0.01 nm. Since the second derivative spectrum obeys the Beer–Lambert law, the concentration can be calculated from a calibration curve using the intensity of a selected negative peak (i.e., 252, 260, or 271 nm). Although the zero-order UV spectra of nucleic acids do not impart fine structure information regarding the conformational

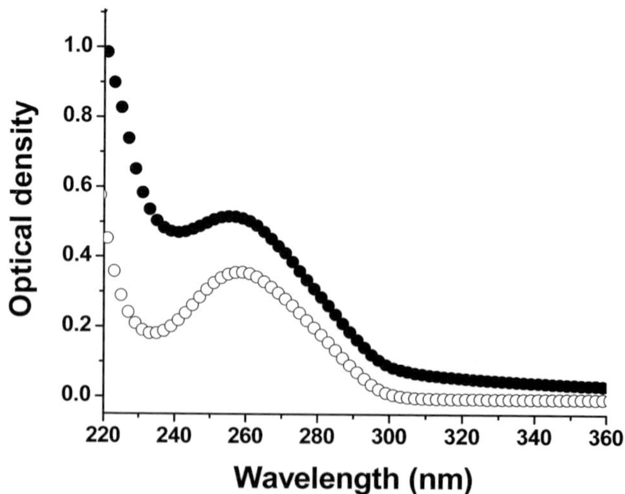

Fig. 2. Absorption spectra of a pDNA sample with a significant light scattering component (filled symbols) and after scattering correction (empty symbols).

features of pDNA, derivative analysis has the potential of providing more detail in this regard. More extensive explanations are described elsewhere ***(21)***.

Other common applications of UV-visible spectroscopy include the analysis of thermal melting of polynucleotides. The peak intensity at 260 nm is particularly sensitive to base stacking interactions. The absorbance of double-stranded DNA is approximately 40% lower in intensity than its single-stranded counterpart (*see* **Fig. 1**). The high intrinsic melting temperature (usually >90°C) observed for pDNA, however, often precludes the use of UV-visible spectroscopy for these type of analyses in the applications of interest here.

3.4. Circular Dichroism Spectroscopy

The circular dichroism (CD) spectrum of a macromolecule arises from the difference in absorption of left- and right-handed circularly polarized light. In nucleic acids, the bases possess plane symmetry, which make them only very weakly optically active. The base-stacking interactions and the helicity of DNA, however, dramatically enhance this optical activity. As the peak position, polarity, and intensity of these CD signals are functions of base-stacking interactions and helicity, this technique is an excellent tool with which to study the secondary structure of DNA ***(22)***. The spectral region of interest is from 190 to 300 nm over which the purine and pyrimidic bases absorb. Thus, the major secondary structures types of DNA can be recognized by their unique spectra in this region. Representative near-UV CD spectra of the various forms of DNA are illustrated in **Fig. 3**. B-form DNA (10.4 bp/turn) exhibits an intense positive band at 275 nm, a negative peak around 240 nm and a crossover point around 260 nm (*see* **Fig. 3**). A second more tightly wound B-form having 10.2 bp/turn has similar spectral characteristics with the exception of the disappearance of the 275-nm peak. B-form spectra seen in pDNA/Cationic polymer complexes were originally assigned to C-form DNA, but vibrational spectroscopy and molecular dynamics studies corrected this error ***(23,24)***. The A-form (11 bp/turn) can be most easily induced by dehydration of DNA with ethanol. This form is characterized by a strong positive signal at 260 nm and a weak 245 nm band with strong negative ellipticity at around 210 nm (*see* **Fig. 3**). The C-form structure (8.8 bp/turn) that can be induced by lithium and ethanol (90% and higher) and manifests a CD spectrum with a reduced 245 nm and an intermediate 275-nm signal between the values for the A- and B-forms (not shown). The left-handed Z-form (12 bp/turn) is characterized by a negative peak at 290 nm, a positive ellipticity at around 260 nm with a crossover at 280 nm and a negative band at about 200 nm. This last band is the best diagnostic feature of the left-handed DNA helix. The degree of supercoiling (tertiary structure) also affects the CD spectrum of DNA. It has been reported that the

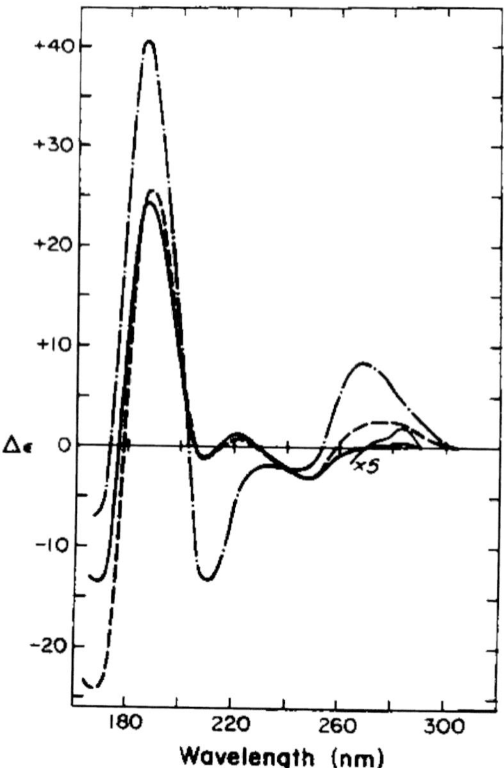

Fig. 3. CD spectra of bacterial DNA of various secondary structures: 10.4B form in 10 mM sodium phosphate buffer pH 7 (– – –): 10.2B-form in 6 M ammonium fluoride, 10 mM sodium phosphate buffer (–––), and as the A-form in 80% trifluoroethanol (–•–•–). *Biopolymers*, *18*, 1009–19. Copyright 1979, reproduction with permission of John Wiley & Sons, Inc.

ellipticity near 275 nm is around 20% greater in negatively supercoiled plasmids than in the relaxed forms ***(25,26)***. This change is attributed to modifications in the rotation angle ($\Delta\theta$) associated with negative supercoiling ***(25,27)***. The following is a general protocol used for the study of the secondary structure of pDNA within complexes using CD spectroscopy:

1. Turn on N_2 gas flow and purge the lamp compartment for about 5 min. Turn on the spectropolarimeter and set up the desired temperature. The required warm up period for the light source in a Jasco J-810 spectropolarimeter (air cooled 150 W Xenon arc lamp) is about 30 min.
2. For most studies, the use of rectangular quartz cuvettes with pathlength of 0.1 cm are satisfactory although cylindrical cells work as well. The pathlength and sample concentration should be varied to obtain an ideal signal to noise ratio (SNR).

Typical concentrations of pDNA (within or without complexes) suitable for most studies using 0.1 cm cell are on the order of 50 μg/ml. To obtain an optimal SNR, the OD of the sample needs to be kept below 1.0. The high tension voltage (HT) of the spectropolarimeter, which is related to the OD of the sample, can be used to optimize the sample concentration. Usually, the optimum SNR is satisfied at HT values between 300 and 350 V. The SNR can also be improved by increasing the number of scans (n), because the SNR is improved by a factor of $n^{1/2}$.

3. Set the instrumental parameters to record the spectrum. As an initial guideline, use a bandwidth of 1 nm, 1-s response time, measurement range of 200–350 nm, data pitch of 0.5 nm, a continuous scanning mode, and a scanning speed 50 nm/min and at least 5 accumulations (*see* **Note 5**).
4. Collect the sample and buffer spectra. The sample spectrum must be corrected by buffer subtraction. The signal of the second channel must be monitored to detect and prevent saturation of the photomultiplier. Detectors typically saturate at HT values >700 V. Additionally, saturation is indicated by an increase in the noise of the signal. Investigators must be especially aware of two common CD artifacts in spectral interpretation: Absorption flattening and differential scattering (*see* **Note 6**).
5. Additional smoothing using (for example) a seven-point Savitsky–Golay algorithm may be needed for an accurate determination of peak positions. This and other smoothing algorithms are included with the instrument software package (Jasco Spectra Manager). This could also be done using commercial graphing and data analysis software (e.g., Origin, Sigmaplot).
6. Convert the observed ellipticity in mdeg to molar ellipticity. Molar ellipticity [θ] is defined as:

$$[\theta] = 1000\theta/Cl \qquad (2)$$

where C is the molar concentration, θ is the observed ellipticity, and l the pathlength in cm. Molar ellipticities can be reported as deg cm^2/dmol or deg/M/m. The exact molar concentration of the sample is needed for this step. The approximate molecular weight of 1 kb dsDNA = 6.5×10^5 g/mol.

7. Peak positions can be calculated using a peak-finding algorithm such as that provided with Origin 7.0 software.

The most frequent way of studying the effect of a given vehicle on the structural properties of pDNA is by performing titration of the gene delivery agent into a known amount of pDNA as described in **Subheading 3.2**. Here we present as an example the titration of a cationic polymer (PEI). The changes in the CD spectra of pDNA/PEI complexes containing four different PEI polymers are shown in **Fig. 4**. Upon complexation with the polymer, both the 247 nm positive peak and the negative minimum near 246 nm are altered. In general, as

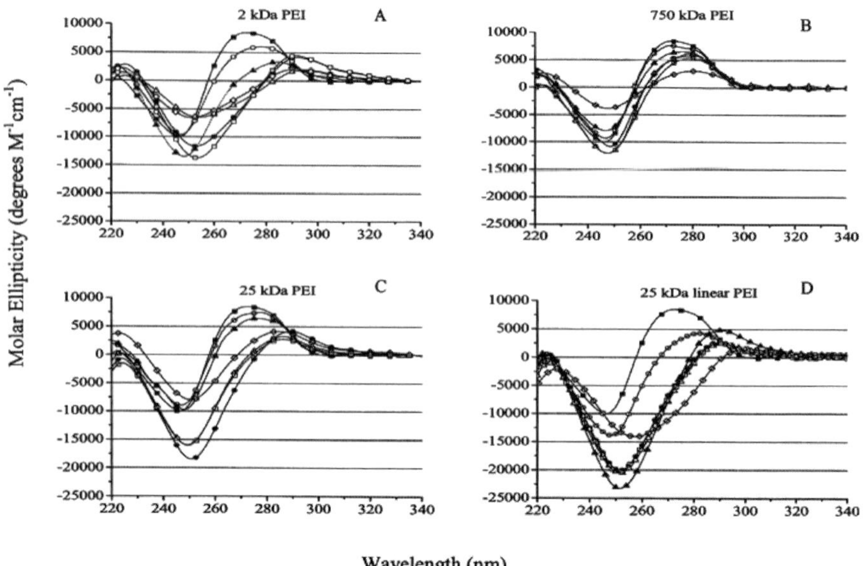

Fig. 4. Effect of PEI MW and Nitrogen/Phosphate ratio (N/P) on the CD spectra of pDNA (pMB 290) and PEI/DNA complexes. (**A**) PEI (2 kDa)/DNA complexes; (**B**) PEI (750 kDa)/DNA; (**C**) PEI (25 kDa)/DNA complexes; and (**D**) PEI (25 kDa linear)/DNA. Legend: black square = plasmid DNA; open circles = N/P 0.5; black triangle = N/P 1; open diamond = N/P 2; black circle = N/P 4; open triangle = N/P 6; and open square = N/P 10. *J Pharm Sci,* 92, 1710–22. Copyright 2003, reproduction with permission Wiley-Liss, Inc and American Pharmacist Association.

the fraction of PEI in the complexes is increased, the spectra show a decrease in the ellipticity of the positive peak and a red shift (*see* **Fig. 4**). The negative band also shows a red shift with increasing N/P ratio. In general, the branched 25 kDa PEI induces more gradual change in the intensity and red shift of the positive peak at 276 compared with its linear counterpart. The formation of complexes between pDNA and cationic lipids (i.e., DDAB, DOTAP, etc.) also induces significant changes in the CD spectrum of native DNA. For instance, an increase in the charge ratio of DDAB/DNA polyplexes results in a red shift of the positive (~275 nm) and negative band (~245 nm) and a significant reduction in the intensity of the positive band ***(28)***. Similar changes are seen with other polycations such as polylysine, peptoids, and aminodendrimers ***(29–31)***. The spectroscopic changes observed for most types of complexes can be explained by limited local structural perturbations of the bases induced by direct interaction between carrier molecules and the DNA ***(28,32)***.

3.5. Fourier Transform Infrared Spectroscopy

Infrared (IR) spectroscopy involves the absorption of light because of vibrational and rotational transitions within molecules. Only vibrations that result in an overall change in the dipole moment of the molecules are active in the IR region. IR analysis of nucleic acids is now almost entirely performed with Fourier transform instruments. These instruments do not use a monochromator. Rather, all IR frequencies are recorded simultaneously with a Michelson or often dynamic interferometer and Fourier transformed simultaneously by a computer. Thus, the IR absorption spectrum of a molecule can be obtained within seconds. This technique has the ability to provide information regarding the individual bases, phosphate backbone, base pairing, sugars, and DNA structural polymorphisms as well as moieties within cationic vehicles.

The IR region of greatest interest for the structural analysis of DNA is between 1800 and 700 cm^{-1}. The 1700–1300 cm^{-1} region contains absorption signals for the bases and has the potential to provide valuable information about hydrogen bonding between the bases and chemical modifications of DNA that may occur during plasmid manipulation, purification, and storage. The different topological conformations adopted by DNA molecules can also be studied by FTIR spectroscopy because the IR spectra of these various forms contain several distinctive bands that permit discrimination between each of the various helical geometries. Some different vibration modes sensitive to DNA conformation are summarized in **Table 1**. The FTIR bands traditionally taken as markers of B form DNA (the most common form adopted by plasmids) are the guanine/thymidine carbonyl stretching band (~1715 cm^{-1}), asymmetric phosphate stretching (~1225 cm^{-1}) and the coupled sugar–phosphodiester signal (~970 cm^{-1}). These frequencies can be used to detect non-B forms in a given sample and to probe potential conformational changes upon interaction with cationic lipids and polymers. The following is a general protocol for analysis of the structure of pDNA nonviral gene delivery systems using ATR-FTIR:

1. To minimize water vapor and carbon dioxide absorption, purge the instrument continuously with dry air or nitrogen. Before starting experiments, purge the sample compartment for at least 15 min.
2. Set up the scanning parameters. As a starting guideline set the instrument in absorbance mode, a spectral range 3000–700 cm^{-1}, resolution at 4 cm^{-1} and use of co-addition of 256 scans (*see* **Note 7**). Cool down the MCT detector by filling the detector cooling system with liquid nitrogen.
3. Before sampling, collect background spectra. Evaluate the background scan for significant contamination by water vapor (a very sharp group of peaks in the region 1400–1800 cm^{-1}) and CO_2 (~2340 cm^{-1}). If significant contamination is present, purge the instrument for longer period of time and check the quality of the dry air source. Some subtraction of residual peaks is possible, however.

Table 1
Conformational Sensitive Bands in DNA (cm^{-1})[a]

A-DNA	B-DNA	C-DNA	Z-DNA	Assignment
1705	1715	1710	1695	Basc carbonyl indicative of base pairing
1418	1425	1425	1408	Deoxyribose
1375	1375	1375	1355	dGdA
1335	1344	1344		dA
	1328	1328		dT
			1320	dC
1275	1281			dT
		1265		
1240	1225	1230-1217	1215	Antisymmetric phosphate stretching
		1069	1065	Deoxyribose
			1013	
970(triplet)	970(singlet)	968	929	Doxyribose
	840	833		Deoxyribose
806				Deoxyribose

[a]Adapted from **ref.** ***34***.

4. Cover the ATR trough with at least 1 ml of sample. Typical pDNA concentrations for ATR measurements are in the order of 1.5 mg/ml (*see* **Note 8**). It is important to emphasize that to obtain data of sufficient resolution in peak position for analytical purposes, multiple measurements of independent samples ***(3–6)*** must be made.
5. Data analysis can usually be performed with commercial software (Omnic 4.0, Nicolet Instrument). To produce the final corrected spectrum, the buffer scan is subtracted from the sample spectrum using the water association band near 2200 cm^{-1}. The goal of this procedure is to achieve a flat region around 2200 cm^{-1} by modifying the subtraction factor. It is not generally possible to subtract the spectrum of the cationic vehicle. Therefore, emphasis is placed on windows in the spectrum where either signal from the DNA or polycation can be analyzed at least semi-independently of the other component ***(33)***.
6. Baseline correction between 1800 and 900 cm^{-1}should be performed to correct a sloping, curving, shifted or otherwise undesirable baseline of the spectrum. If necessary, smooth the spectrum using an appropriate method (e.g., 5–7 points Savitsky–Golay).
7. The peak position is finally obtained using a peak finding algorithm provided with the software. Emphasis is placed on peak position rather that intensities because of the better reproducibility and higher information content of such data.

As an example of the utility of this technique to study DNA–carrier interactions, changes in vibrational bands of DNA on addition of DOTAP and DDAB liposomes are illustrated in **Fig. 6** *(34)*. The small changes in the base carbonyl stretching band (1715 cm^{-1}) and the asymmetric phosphate band (1225 cm^{-1}) are consistent with very little if any alteration of the normal B-form conformation of pDNA. Similarly, pDNA remains in the canonical B form after complexation with PEI *(32)*. Electrostatic interactions between the lipid cationic head groups and pDNA are indicated by the increased frequency of the asymmetric phosphate stretching vibration (*see* **Fig. 5C and F**). The guanine/cytosine peak at 1492, on the other hand, manifests an overall decrease in the vibrational frequency with an increment in the lipid/DNA ratio, reaching saturation at a weight ratio of 0.6 for both cationic lipids under study (*see* **Fig. 5B and E**). The carbonyl stretching band at 1715 cm^{-1} manifests a shift to higher frequencies as the lipid/DNA ratio is increased, with a discontinuity in the trend around 0.5 weight ratio (*see* **Fig. 5C and F**). This can be interpreted as an alteration in the hydrogen bonding of the base carbonyl groups either by alterations in lipid–base interactions or changes in the hydration state of the DNA molecules.

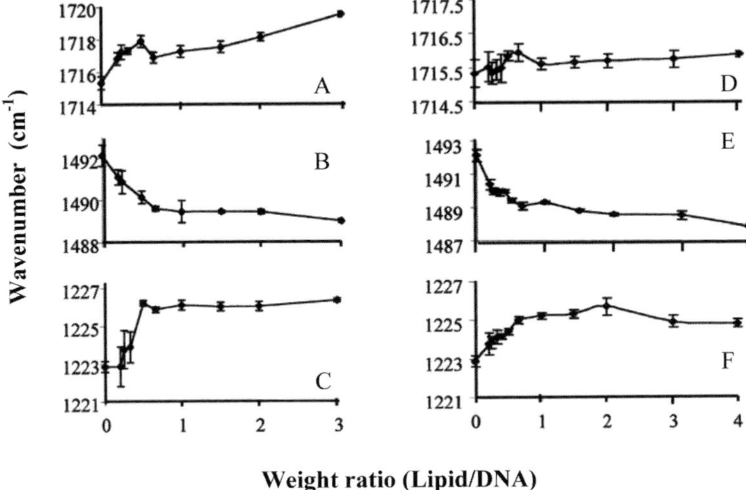

Fig. 5. The effect of DOTAP–DNA (A-C) weight ratio and DDAB–DNA (D-F) weight ratio on DNA and lipid vibrational modes. The peak positions were plotted versus the weight ratio of lipid to DNA with 1 mg/ml of DNA present in each sample. **A** and **D**, guanine/thymidine carbonyl stretching band; **B** and **E**, in-plane stretching band of guanine/cytosine; **C** and **F**, DNA asymmetric phosphate stretching band. *J Biol Chem* 276: 8037–43.Copyright 2001, reproduction with permission of The American Society for Biochemistry and Molecular Biology.

3.6. Analysis of Particle Size by Dynamic Light Scattering

Dynamic light scattering (DLS) is commonly used to analyze the mean diameter and particle size distribution of DNA, nonviral gene delivery vehicles such as cationic lipids and polymers as well as complexes of DNA with these vectors. DLS uses fluctuations in the intensity of scattered light to measure the rate of diffusion (Brownian motion) of particles. The time dependence of the intensity of light scattered from a very small region of solution, over a time range from tenths of a microsecond to milliseconds, is measured. These fluctuations in the intensity of the scattered light are analyzed by autocorrelation methods to directly give the diffusion coefficients of the particles. The data are further processed to calculate the size of the particles (*see* **Note 9**).

Various DLS instruments are now commercially available. Typically, these instruments are equipped with a laser light source (15–5000 mW), a photomultiplier tube (often movable to vary the scattering detection angle with 90° as the typical default angle) and a digital autocorrelator that analyzes the intensity fluctuations and generates an autocorrelation function. These instruments usually accommodate various sample cells and the loading concentrations required are dependent primarily on the size of the scattering particles and the output of the laser.

A major obstacle associated with obtaining interpretable and reproducible results from light scattering instruments is sample purity (*see* **Note 10**). In the case of plasmid DNA, contaminants may arise during isolation and purification steps. If the DNA is purified from bacterial cell culture, contamination may arise from RNA and bacterial polysaccharides, among other substances. The isolation of DNA by gel electrophoresis may sometimes lead to the presence of residual agarose as a major impurity in DNA samples. Another problem associated with this technique involves the potential intrinsic polydispersity of macromolecular samples. If the particles are truly homogenous in size and the exclusion of dust and other particulate contamination is optimal, it is fairly straightforward to characterize particle size. In the case of macromolecular complexes, such as nonviral gene delivery vectors, however, a uniform population of particles is highly unlikely. To address this problem, numerous algorithms have been developed to further analyze DLS data. The method of cumulants (Gaussian analysis) permits the determination of both a mean diameter and a polydispersity index *(35)*. For multimodal distributions, a number of methods are available to deconvolute the multiexponential data to yield estimates of sizes of each component *(36)*. Alternatively, a DLS detector could be used in a chromatography mode, making possible the analysis of particles separated by either size exclusion or asymmetrical flow field-flow fractionation (AFFF). Concentration is an important parameter when making light scattering measurements of DNA–polycation complexes. In general, a useful DNA concentration range for initial evaluation is between 10 and 100 μg/ml. The DNA concentration may need to be higher if DNA is to be evaluated in the absence of complexing agents. Ideally, samples should be prepared in a concentration range in which the intensity of scattered light is linearly dependent on the concentration assuming that the physical properties of the complex (size, shape, etc.) do not change over this concentration range. It must be remembered, however, that the intensity of scattered light is proportional to the square of the refractive index increment that can change dramatically depending on the extent of collapse of DNA/polycation particles. A representative mean hydrodynamic diameter plot of PEI/DNA complexes in 10 mM Tris buffer, pH 7.4 is shown in **Fig. 6**. The DNA concentration was held constant at 100 μg/ml while the ratios of PEI nitrogen (N) to DNA phosphate (P) of the PEI/DNA complexes were varied. The mean hydrodynamic diameters shown in **Fig. 6** were determined by cumulant analysis. The average size of the complexes was approximately 100 nm at the highest and lowest N/P ratios tested and independent of PEI molecular weight (*see* **Fig. 6**). At intermediate N/P ratios, an approximately twofold increase in size was observed with several of the polymers.

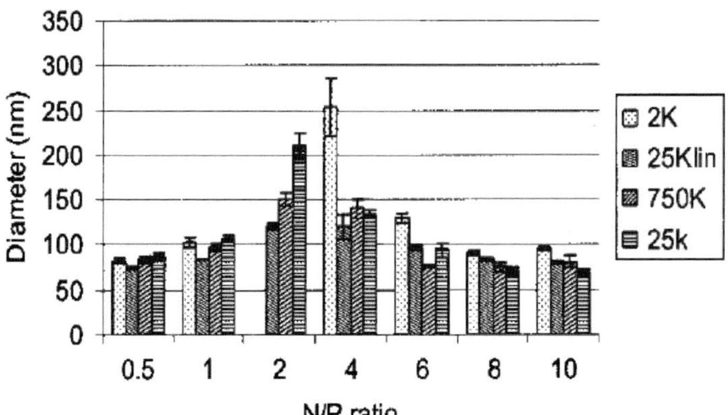

Fig. 6. Hydrodynamic size of PEI/DNA complexes as a function of increasing N/P molar ratio. Branched PEIs (MW 750, 25 and 2 kDa) and linear PEI (MW 25 kDa) were used. Complexes were formed in 10 mM Tris buffer, pH 7.4. The data represents mean and standard error of at least three separate measurements. *J. Pharm. Sci.* 92, 1710-1722 Copyright 2003, reproduction with permission of Wiley-Liss, Inc. and American Pharmaceutical Association.

The following protocol can be used to determine the size of DNA–polycation complexes using DLS:

1. Use chemical methods and sonication to clean glass tubes; rinse with ultra pure water and dry in an oven.
2. A 1-ml sample volume is recommended. Smaller sample volumes can be used as long as a clear light path is present.
3. Ensure that the refractive index matching fluid (e.g., decalin) covers the sample cell if such a bath is used.
4. Adjust the detection aperture to the proper size; this will have to be done experimentally by monitoring the signal intensity. The average count rate, usually expressed in thousands of counts per second (KCPS), is a good measure of the signal intensity (*see* **step 6**).
5. Turn on the PMT and laser and wait for at least 15 min before data collection.
6. Adjust available intensity filters and the aperture to get a count of about 100 Kcps. Anything between 50 and 200 Kcps is generally viable.
7. Enter run parameters for software such as number of runs, duration of run, and temperature. Enter the appropriate viscosity and refractive index medium values for the solvent that is being used.
8. Check the system using standard latex beads. Standards usually need to be diluted approximately 1000-fold in water to be at a suitable concentration for measurement.
9. Measure the mean diameter and polydispersity of the complexes at a concentration of 20–100 μg/ml. If the intensity of the signal is too low for accurate data, increase the time of run or change the intensity filter or the aperture. Check reproducibility of the data.

3.7. Fluorescence Spectroscopy

Fluorescence-based methods are widely used in DNA analysis. Because the intrinsic fluorescence from DNA bases is very weak (quantum yields of 10^{-4} to 10^{-5}), almost all such methods applied to the study of nucleic acids rely on the use of extrinsic fluorescent probes. Because of its high sensitivity, fluorescence spectroscopy is commonly employed as a preferred method for detection of very minute amounts of sample, often allowing DNA concentrations in the nanomolar range to be quantified.

In fluorescence spectroscopy, absorption of a photon triggers the emission of a lower-energy photon. The fluorescence quantum yield is a measure of the efficiency of the fluorescence process and is defined as the ratio of the number of photons emitted to the number absorbed. The maximum fluorescence quantum yield is 1.0 (100%); i.e., every photon absorbed results in a photon emitted. The emission spectral ranges and quantum yields of fluorescent probes that are of practical use are generally in the range of 400–700 nm and 0.05–1.0, respectively. Fluorescence instruments (spectrofluorometers) use UV to visible light to excite the sample and usually observe the emitted fluorescence at right angles. A monochromator before the sample selects the wavelength of exciting light, and a second monochromator after the sample scans the various wavelengths of emitted light before quantitation by a photomultiplier tube. Variable slits on both excitation and emission sides permit the control of the intensity and bandwidth of the light.

Extrinsic fluorescent probes have found numerous applications in nucleic acid research. Fluorophores can be added covalently or non-covalently to permit DNA detection on agarose gels, DNA sequencing, fluorescence in situ hybridization, and for reading nucleic acids arrays to analyze gene expression *(37)*. One of the most common approaches used to characterize the interaction of nonviral gene delivery systems (cationic lipids and polymers) with DNA involves competitive displacement assays using extrinsic fluorescent probes such as ETBR or other intercalating dyes such as YOYO-1, TOTO-1, acridine orange, etc. Such assays that monitor the displacement of the dyes from DNA by various cationic delivery systems are routinely used to verify the condensed state of DNA in these complexes *(38)*. Addition of intercalating dyes to solutions of nucleic acids results in the formation of highly fluorescent DNA–dye complexes and causes large increases in the fluorescence signal. Upon addition of the polycations, dye molecules are displaced from the DNA, causing a corresponding reduction in the fluorescence signal. **Fig. 7A** shows the displacement of ETBR, when DNA pre-equilibrated with ETBR at a ratio of 1:4 dye/DNA base pairs is complexed to DOTAP or DDAB. Approximately 85–90% of the dye is displaced as shown by the fluorescence decrease. Incorporation of equimolar amounts of the helper lipids DOPE or CHOL into DOTAP complexes, however, results in a significant reduction in the amount of ETBR displaced. The relative ability of the cationic polymer PEI to displace ETBR is shown in **Fig. 7B**. Complex formation of DNA with increasing amounts of PEI results in an almost complete displacement of the dye from the DNA when added above an N/P ratio of 3:1. The following protocol is typical for such competitive dye displacement methods:

1. Before DNA/polycation complex formation, add ETBR to the DNA solution at a ratio of 1 dye molecule to 4 DNA base pairs in a 1-cm path length quartz cuvette and incubate for 1 h. Typical DNA concentration range from 20 to 100 μg/ml.
2. Measure the fluorescence of DNA alone (F_{DNA}) at an excitation wavelength of 518 nm to obtain both the fluorescence intensity and the wavelength of the emission maximum. Set the excitation and emission slits for a bandpass of 3 nm and obtain spectra at an appropriate temperature over a range of 545–695 nm by scanning every 1 nm using a 1-s integration time.

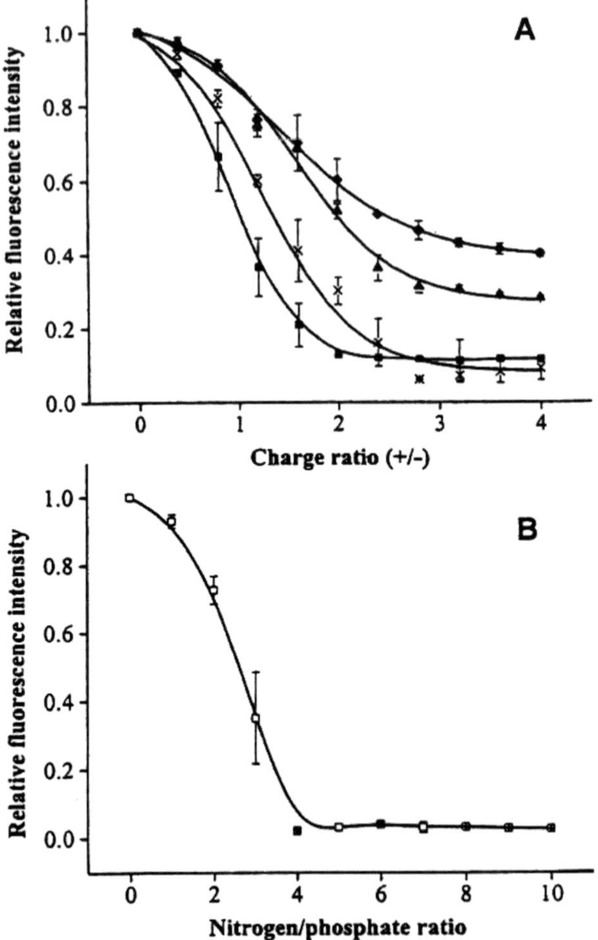

Fig. 7. Displacement of ETBR by cationic lipids or PEI. DNA was pre-equilibrated with ETBR at a dye/DNA base pair ratio of 1:4 in 10 mM Tris pH 7.4. (**A**) displacement by lipids. DOTAP (squares), DOTAP/DOPE (diamonds), DOTAP/CHOL (triangles), and DDAB (crosses). (**B**) displacement by PEI. Data represent the average and SEM of at least three replicates. Lines represent sigmoidal fits to the data and are meant only as visual guide. *J. Pharm. Sci.* 92, 1272–85 Copyright 2003, reproduction with permission of Wiley-Liss, Inc. and American Pharmaceutical Association.

3. Add the required amount of cationic partner to the labeled DNA with constant stirring for 20 s. Allow the complexes to equilibrate for 20 min before performing fluorescence measurements.
4. Measure the background fluorescence of ETBR alone (also *see* **Notes 12** and **13**).

5. The percentage of dye displaced upon polycation binding is calculated using the following equation:

$$Relative\ displacement = (F_{obs} - F_0)/(F_{DNA} - F_0) \quad (3)$$

where $F_{obs,}$ F_0, and F_{DNA} are the fluorescence intensities of a given sample, dye in buffer alone, or the dye complexed to DNA alone, respectively.

3.8. Empirical Phase Diagram Approach to the Formulation of Nonviral Gene Delivery Complexes

A structural characterization of nonviral gene delivery complexes is generally performed by a combination of the techniques described in this chapter. Additionally, the thermal stability, binding affinity, and stoichiometry of binding can be analyzed by a combination of differential scanning and isothermal titration calorimetry ***(18,39)***. Because of their complexity and heterogeneity, it is apparent that none of these methods alone can supply definitive structural information about the complexes. Their use in combination, however, has the potential to provide a comprehensive view of a wide variety of structural perturbations of pDNA upon complexation. A number of environmental stress variables (pH, temperature, ionic strength, redox potential, etc.) are also expected to have an impact on the structure of the complexes and need to be considered to provide a comprehensive picture of the behavior of gene delivery complexes. An extensive biophysical characterization based on this idea generates very large sets of data that complicate their interpretation. We have recently developed a mathematical approach that employs a variety of such experimental measurements to develop an empirical phase diagram to portray the behaviors of macromolecular complexes ***(19,40–44)***. In this approach, different physical states of macromolecules and their complexes are represented by vectors in a multidimensional vector space. Each component of the resultant vectors consists of the normalized values of multiple experimental parameters measured as a function of different environmental conditions (pH, temperature, ionic strength, etc.). To simplify the results, the multidimensional vectors obtained are truncated to the three most influential components and are represented by a red, green, and blue (RGB) color scheme. A well-defined structural state is then represented by a uniformly colored region in the phase diagram, while changes in states are reflected by abrupt color changes. More detailed information regarding this mathematical procedure is provided elsewhere ***(19)***. The rationale for this approach is to synthesize large set of data into a single graphic image that provides a more intuitive and comprehensive picture of the complexes behavior as well as changes in the physical stability of nonviral gene delivery complexes under different solution conditions. A representative ionic strength–pH empirical phase diagram for DNA/DOPE complexes is illustrated in **Fig. 8**. DLS, CD, and the binding of YOYO-1 were selected for this study to monitor different structural aspects related to size, secondary structure of the DNA, and the extent of condensation, respectively ***(44)***. The environmental variables pH and ionic strength were chosen because of the electrostatic nature of the interaction between the pDNA and the cationic carriers. An empirical phase diagram for DNA/DOPE at high charge ratio produces a fairly homogenous phase from pH 5 to pH 8 at all ionic strengths. At low pH, however, the phase diagram manifests a unique phase between 10 and 100 mM NaCl. Although the phase diagram suggests the presence of distinct structural forms of the complexes, it does not directly identify the nature of these different structural states. For this purpose, it is necessary to evaluate the individual data generated by each technique responsible for the change in empirical phases observed. Note that these are not true "thermodynamic phases" but are simply empirical states defined by the experimental methods employed. Thus, the empirical phase diagram approach provides a global picture of the complexes as they are perturbed by different solution conditions. Although the structural resolution of the various techniques used here is low, when used in combination, they are powerful tools with which to characterize the structure of the complexes over a wide range of experimental conditions under which the use of higher resolution techniques such as NMR or X-ray crystallography is difficult or impossible. The identification of "apparent" phase boundaries is one

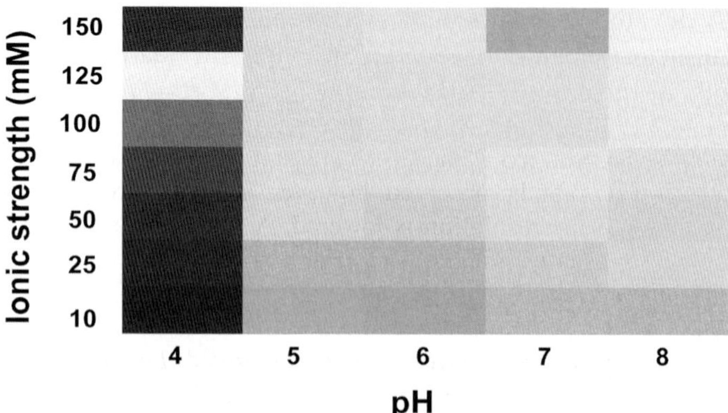

Fig. 8. Ionic strength versus pH empirical phase diagram of DNA/DOTAP complexes at ± charge ratio of 4. Phase diagram was generated from DLS, CD, and fluorescence studies. See reference ***(44)*** for color versions of these phase diagrams.

of the most important applications of the approach described here. This locates conditions under which the complexes are marginally stable or undergo critical structural changes. These conditions can be employed to accelerate degradation pathways and used to develop screening assays for potential stabilizers for the rational formulation of nonviral gene delivery systems.

4. Notes

1. Storage of liposome preparations for longer periods of time or at freezing temperatures may cause aggregation and fusion of the liposomes.
2. Passage through charge neutrality needs to be avoided to prevent precipitation of the complexes.
3. To obtain an accurate measurement of the DNA concentration, any light scattering contribution introduced by the particulate nature of complexes in suspension must be subtracted. The intensity of the scattered light for small particles ($d < 1/20\ \lambda$) is approximately proportional to the inverse of the four power of the wavelength of the light. The scattering component can thus be at least partially eliminated by plotting log A vs. log λ in non-absorbing regions of the spectrum (320–400 nm) and then extrapolating this signal across the absorbing region. Contributions of light scattering can then be subtracted from the observed spectrum (*see* **Fig. 2**). Software packages included in most modern spectrophotometers permit scattering corrections by this and other methods. Note that scattering problems can also be partially eliminated by the use of derivative methods because broad scattering components often do not affect derivative spectra.
4. The absorbance of nucleic acids is sensitive to changes in pH and ionic strength of the media (e.g., the extinction coefficient of DNA is decreased in the presence of salts compared with pure water).
5. Optimal machine settings depend on the instrument and sample characteristics and must be established for each case. By increasing the bandwidth, the STN ratio is increased because more light is incident on the sample and hence the photomultiplier, but the spectral resolution will decrease. The optimal time period over which data are collected at each wavelength depends on the type of instrument and usually is set between 1 and 2 s. Increasing this time will increase the SNR. Spectral distortion may be observed, however, if the spectrum is collected at too high a speed or if the resolution is lowered. A resolution or data pitch between 0.5 and 1 nm is usually suitable for spectral analysis of nucleic acids. If higher resolution in the peak position is needed, the resolution could be increased up to 0.1–0.2nm. By increasing the number of scans and decreasing the scanning speed, the quality of the spectrum will be improved. Instrumental drift, sample stability, and the time required for each experiment should be carefully considered, however.
6. Absorption flattening is a well-recognized phenomenon that occurs when chromophores are closely packed resulting in their non-homogeneous distribution. Thus, the probability of any absorbing species encountering a photon is reduced,

and consequently, the absorption is "flattened" (i.e., reduced in intensity). The extent of flattening depends on the size of the particles in suspension and the concentration of chromophores within the particles. Reduction of CD signals because of absorption flattening is maximal at the wavelengths corresponding to the absorption peaks and may result in significant distortions in the shape of the CD spectrum. It can be most simply recognized by varying concentration and pathlength. Differential scattering, on the other hand, results from the difference in the extent of scattering of the left and right circularly polarized light by the optically active molecules. This is detected by the instrument as apparent differential absorption of circularly polarized light consequentially producing significant distortion in the spectra. Details of this phenomena are described elsewhere ***(44,45)***.

7. Spectral range, resolution, and number of co-added scans should be adjusted according to the experimental requirements and sample characteristics. In general, the main vibrational features of nucleic acids and cationic vehicles are analyzed between 3000 and 700 cm^{-1}. For the analysis of naked pDNA preparations, data between 1800 and 700 cm^{-1} are adequate. As mentioned for CD spectroscopy, the SNR is proportional to the square root of the number of co-added scans. Thus, the quality of the spectra can be increased by co-adding many scans. A tradeoff between SNR and resolution need to be considered when setting up resolution parameters and the number of scans employed. Increasing the spectral resolution will require increasing the number of co-added scans to maintain the same SNR.
8. If insufficient sample is available for the analysis, spectra can be easily acquired by transmission experiments performed on solid DNA films. In this method, samples (typically 20–50 μl, 1.5 mg/ml) are dried on a suitable polymer membrane such as polyethylene and directly examined ***(46)***.
9. The size of the particles is determined by the Strokes–Einstein relationship between diffusion coefficients (D) and particle size, assuming that the particles in the solution are spherical:

$$D = \frac{k_B T}{3\pi\eta(t)d} \tag{4}$$

where k_B is Boltzmann's constant, T is the temperature in Kelvin, $\eta(t)$ is the viscosity of the liquid in centipoise, and d is the particle diameter. This method is only suitable for analyzing the size of spherical particles, and inaccurate results may be obtained for extended structures. Other methods such as electron microscopy, static light scattering, sedimentation data, etc., should be employed to determine the particle morphology.

10. The presence of dust or other particulates in samples is the most common problem in obtaining accurate measurements. All solutions and cells therefore need to be essentially free from even trace amounts of dust or other contaminants. To this end, samples must be repetitively filtered and/or centrifuged before

measurement. Clarity can easily be confirmed by the use of a small hand held laser (e.g., laser pointer).

11. One of the key strengths and applications of DLS is the detection of very small amounts of higher mass species (aggregates <0.01% by weight). This technique can therefore be quite useful in comparing the stability of different formulations of nucleic acids.
12. Buffer baselines in fluorescence measurements should always be subtracted from each spectrum before data analysis, especially to eliminate the ubiquitous water Raman peak.
13. Use appropriate cut-off filters to help reduce the turbidity/light scattering signal because turbidity in the samples may interfere with the fluorescence signal. Thus, changes in scattering can be confused with changes in conformation.

References

1. Hirose, S., Tsuda, M., and Suzuki, Y. (1985) Enhanced transcription of fibroin gene*in vitro* on covalently closed circular templates. *J Biol Chem 260*, 10557–62.
2. Levene, S. D., Donahue, C., Boles, T. C., and Cozzarelli, N. R. (1995) Analysis of the structure of dimeric DNA catenanes by electron microscopy. *Biophys J 69*, 1036–45.
3. Adami, R. C., Collard, W. T., Gupta, S. A., Kwok, K. Y., Bonadio, J., and Rice, K. G. (1998) Stability of peptide-condensed plasmid DNA formulations. *J Pharm Sci 87*, 678–83.
4. Tumanova, I., Boyer, J., Ausar, S. F., Burzynski, J., Rosencrance, D., White, J., Scheidel, J., Parkinson, R., Maguire, H., Middaugh, C. R., Weiner, D., and Green, A. P. (2005) Analytical and biological characterization of supercoiled plasmids purified by various chromatographic techniques. *DNA Cell Biol 24*, 819–31.
5. Ghosh, A., and Bansal, M. (2003) A glossary of DNA structures from A to Z. *Acta Crystallogr D Biol Crystallogr 59*, 620–6.
6. Jose, D., and Porschke, D. (2005) The dynamics of the B-A transition of natural DNA double helices. *J Am Chem Soc 127*, 16120–8.
7. Lu, X. J., Shakked, Z., and Olson, W. K. (2000) A-form conformational motifs in ligand-bound DNA structures. *J Mol Biol 300*, 819–40.
8. Jose, D., and Porschke, D. (2004) Dynamics of the B-A transition of DNA double helices. *Nucleic Acids Res 32*, 2251–8.
9. Roth, C. M., and Sundaram, S. (2004) Engineering synthetic vectors for improved DNA delivery: insights from intracellular pathways. *Annu Rev Biomed Eng 6*, 397–426.
10. Wolff, J. A., Malone, R. W., Williams, P., Chong, W., Acsadi, G., Jani, A., and Felgner, P. L. (1990) Direct gene transfer into mouse muscle*in vivo*. *Science 247*, 1465–8.
11. De Laporte, L., Cruz Rea, J., and Shea, L. D. (2006) Design of modular non-viral gene therapy vectors. *Biomaterials 27*, 947–54.

12. Wiethoff, C. M., and Middaugh, C. R. (2003) Barriers to nonviral gene delivery. *J Pharm Sci 92*, 203–17.
13. Feng, M., Lee, D., and Li, P. (2006) Intracellular uptake and release of poly(ethyleneimine)-co-poly(methyl methacrylate) nanoparticle/pDNA complexes for gene delivery. *Int J Pharm 311*, 209–14.
14. Wiethoff, C. M., Smith, J. G., Koe, G. S., and Middaugh, C. R. (2001) The potential role of proteoglycans in cationic lipid-mediated gene delivery. Studies of the interaction of cationic lipid-DNA complexes with model glycosaminoglycans. *J Biol Chem 276*, 32806–13.
15. Middaugh, C. R., Evans, R. K., Montgomery, D. L., and Casimiro, D. R. (1998) Analysis of plasmid DNA from a pharmaceutical perspective. *J Pharm Sci 87*, 130–46.
16. Welz, C., and Fahr, A. (2001) Spectroscopic methods for characterization of nonviral gene delivery systems from a pharmaceutical point of view. *Appl Spectros Rev 36*, 333.
17. Lobo, B. A., Davis, A., Koe, G., Smith, J. G., and Middaugh, C. R. (2001) Isothermal titration calorimetric analysis of the interaction between cationic lipids and plasmid DNA. *Arch Biochem Biophys 386*, 95–105.
18. Lobo, B. A., Rogers, S. A., Choosakoonkriang, S., Smith, J. G., Koe, G., and Middaugh, C. R. (2002) Differential scanning calorimetric studies of the thermal stability of plasmid DNA complexed with cationic lipids and polymers. *J Pharm Sci 91*, 454–66.
19. Kueltzo, L. A., Ersoy, B., Ralston, J. P., and Middaugh, C. R. (2003) Derivative absorbance spectroscopy and protein phase diagrams as tools for comprehensive protein characterization: a bGCSF case study. *J Pharm Sci 92*, 1805–20.
20. Wilfinger, W. W., Mackey, K., and Chomczynski, P. (1997) Effect of pH and ionic strength on the spectrophotometric assessment of nucleic acid purity. *Biotechniques 22*, 474–6, 478–81.
21. Mach, H., Middaugh, C. R., and Lewis, R. V. (1992) Detection of proteins and phenol in DNA samples with second-derivative absorption spectroscopy. *Anal Biochem 200*, 20–6.
22. Sprecher, C. A., Baase, W. A., and Johnson, W. C., Jr. (1979) Conformation and circular dichroism of DNA. *Biopolymers 18*, 1009–19.
23. Johnson, W. C. (1996) *Determination of the Conformation of Nucleic Acids by Electronic CD. In Circular Dichroism and the Conformational Analysis of Biomolecules*. Plenum Press, New York.
24. Zimmerman, S. B., and Pheiffer, B. H. (1980) Does DNA adopt the C form in concentrated salt solutions or in organic solvent water mixtures? An x-ray diffraction study of DNA fibers immersed in various media. *J Mol Biol 142*, 315–30.
25. Serban, D., Benevides, J. M., and Thomas, G. J., Jr. (2002) DNA secondary structure and Raman markers of supercoiling in Escherichia coli plasmid pUC19. *Biochemistry 41*, 847–53.

26. Thumm, W., Seidl, A., and Hinz, H. J. (1988) Energy-structure correlations of plasmid DNA in different topological forms. *Nucleic Acids Res 16*, 11737–57.
27. Vologodskii, A. V., and Cozzarelli, N. R. (1994) Conformational and thermodynamic properties of supercoiled DNA. *Annu Rev Biophys Biomol Struct 23*, 609–43.
28. Braun, C. S., Jas, G. S., Choosakoonkriang, S., Koe, G. S., Smith, J. G., and Middaugh, C. R. (2003) The structure of DNA within cationic lipid/DNA complexes. *Biophys J 84*, 1114–23.
29. Lobo, B. A., Vetro, J. A., Suich, D. M., Zuckermann, R. N., and Middaugh, C. R. (2003) Structure/function analysis of peptoid/lipitoid:DNA complexes. *J Pharm Sci 92*, 1905–18.
30. Braun, C. S., Vetro, J. A., Tomalia, D. A., Koe, G. S., Koe, J. G., and Middaugh, C. R. (2005) Structure/function relationships of polyamidoamine/DNA dendrimers as gene delivery vehicles. *J Pharm Sci 94*, 423–36.
31. Ruponen, M., Braun, C. S., and Middaugh, C. R. (2006) Biophysical characterization of polymeric and liposomal gene delivery systems using empirical phase diagrams. *J Pharm Sci. 95* (10), 2101–14.
32. Choosakoonkriang, S., Lobo, B. A., Koe, G. S., Koe, J. G., and Middaugh, C. R. (2003) Biophysical characterization of PEI/DNA complexes. *J Pharm Sci 92*, 1710–22.
33. Choosakoonkriang, S., Wiethoff, C. M., Anchordoquy, T. J., Koe, G. S., Smith, J. G., and Middaugh, C. R. (2001) Infrared spectroscopic characterization of the interaction of cationic lipids with plasmid DNA. *J Biol Chem 276*, 8037–43.
34. Choosakoonkriang, S., Wiethoff, C.M., Kueltzo, L.A., and Middaugh CR. (2001) Characterization of synthetic gene delivery vectors by infrared spectroscopy, in *Methods in Molecular Medicine: Nonviral Vectors for Gene Therapy* (Findeis, M. A., Ed.) pp. 285–317, Humana Press, Totowa, NJ.
35. Koppel, D. E. (1972) Analysis of macromolecular polydispersity in intensity correlation spectroscopy: The method of cumulants. *J. Chem. Phys. 57*, 4814–20.
36. Stock, R. S., and Ray, W. H. (1985) Interpretation of photon correlation spectroscopy data: A comparison of analysis methods. *J. Polym. Sci.-B POlym Phys. 23*, 1393–447.
37. Lakowicz, J. R., Shen, B., Gryczynski, Z., D'Auria, S., and Gryczynski, I. (2001) Intrinsic fluorescence from DNA can be enhanced by metallic particles. *Biochem Biophys Res Commun 286*, 875–9.
38. Wiethoff, C. M., Gill, M. L., Koe, G. S., Koe, J. G., and Middaugh, C. R. (2003) A fluorescence study of the structure and accessibility of plasmid DNA condensed with cationic gene delivery vehicles. *J Pharm Sci 92*, 1272–85.
39. Lobo, B. A., Koe, G. S., Koe, J. G., and Middaugh, C. R. (2003) Thermodynamic analysis of binding and protonation in DOTAP/DOPE (1:1): DNA complexes using isothermal titration calorimetry. *Biophys Chem 104*, 67–78.
40. Fan, H., Ralston, J., Dibiase, M., Faulkner, E., and Russell Middaugh, C. (2005) Solution behavior of IFN-beta-1a: an empirical phase diagram based approach. *J Pharm Sci 94*, 1893–911.

41. Ausar, S. F., Rexroad, J., Frolov, V. G., Look, J. L., Konar, N., and Middaugh, C. R. (2005) Analysis of the thermal and pH stability of human respiratory syncytial virus. *Mol Pharm 2*, 491–9.
42. Rexroad, J., Evans, R. K., and Middaugh, C. R. (2006) Effect of pH and ionic strength on the physical stability of adenovirus type 5. *J Pharm Sci 95*, 237–47.
43. Ausar, S. F., Foubert, T. R., Hudson, M. H., Vedvick, T. S., and Middaugh, C. R. (2006) Conformational stability and disassembly of Norwalk virus-like particles. Effect of pH and temperature. *J Biol Chem 281*, 19478–88.
44. Bustamante, C., Tinoco, I., Jr., and Maestre, M. F. (1983) Circular differential scattering can be an important part of the circular dichroism of macromolecules. *Proc Natl Acad Sci U S A 80*, 3568–72.
45. Bustamante, C. (1984) Contribution of differential scattering of circularly polarized light to the optical rotatory dispersion of a sample. *J Opt Soc Am A 1*, 1114–9.
46. Choosakoonkriang, S., Wiethoff, C. M., Koe, G. S., Koe, J. G., Anchordoquy, T. J., and Middaugh, C. R. (2003) An infrared spectroscopic study of the effect of hydration on cationic lipid/DNA complexes. *J Pharm Sci 92*, 115–30.

6

Real-Time Multiple Particle Tracking of Gene Nanocarriers in Complex Biological Environments

Samuel K. Lai and Justin Hanes

Summary

Complex biological fluids, such as the vast and molecularly crowded cell cytoplasm and the highly viscoelastic mucus that protects many entry ways to the body, pose significant barriers to efficient gene delivery. Understanding the dynamics of gene carriers in such environments allows insight that leads to rational improvements in gene vector design. Fluorescence techniques that provide only ensemble-averaged transport characteristics do not provide detailed information related to the nature of various barriers to efficient gene vector transport to target cell nuclei. Multiple particle tracking (MPT) allows the tracking of the real-time motion of up to hundreds of individual particles simultaneously with high temporal and spatial resolution. We have adapted MPT to study gene carrier transport in live cells and in fresh, undiluted human mucus. By analyzing the displacements of gene vectors as a function of time scale, this technique provides, on a per particle basis, highly quantitative measurements of the transport rates and transport mechanisms, as well as biophysical information of the complex biological environments. Combining MPT with confocal microscopy (confocal particle tracking) allows dynamic and quantitative co-localization determination of gene carriers with various cellular structures, such as endosomes, lysosomes, the endoplasmic reticulum, and Golgi. We have applied MPT to enhance understanding of critical extracellular and intracellular bottlenecks to gene transfer.

Key Words: mucus; intracellular transport; polyethylenimine; polymers; polyplexes; fluorescence; drug delivery systems; active transport; diffusion.

From: *Methods in Molecular Biology, vol. 434: Volume 2: Design and Characterization of Gene Transfer Vectors*
Edited by: J. M. Le Doux © Humana Press, Totowa, NJ

1. Introduction

1.1. Extracellular and Intracellular Barriers to Gene Transfer

Biological environments present a multitude of barriers that can hinder the transfer of genes to the nucleus of targeted cells *(1)*. The gene that could correct the CFTR defect responsible for Cystic Fibrosis (CF) was discovered in 1989, yet no patient has been cured *(2)*. The Achilles heel to gene therapy for CF, as with most other diseases for which gene therapy holds promise, is the effective delivery of corrective genes through hordes of biological obstacles *(3,4)*. For example, for gene therapy to be effective for CF in the lungs, gene vectors must first traverse lung mucus linings that typically serve to protect the underlying epithelia from the entry of foreign particles, such as pathogens and toxins *(5,6)*. Mucus in the lungs, gastrointestinal tract, and cervicovaginal tract poses both steric obstruction, because of dense mesh structures, and adhesive interactions between mucin fibers (with both anionic and hydrophobic domains) and polymers and lipids often used for condensation of cargo plasmids *(7)*. Upon cell entry, gene vector transport to the cell nucleus may also be hindered by the highly crowded cytoplasm, including dense cytoskeletal networks of actin and microtubules *(8)*. The cell cytoplasm is particularly crowded near the cell nucleus, which may significantly hamper gene delivery even with particles that rapidly accumulate near the cell nucleus *(8)–(10)*. For example, the transport of naked DNA larger than 2000 bp in the cytoplasm is strongly retarded, rendering even modest-sized DNA molecules essentially immobile *(11)*.

In this chapter, multiple particle tracking (MPT) of gene carriers in extracellular and intracellular environments is described. The procedures cover the formation of fluorescent polymeric gene nanocarriers based on polyethylenimine (PEI), preparation of microscopy samples, collection of videos with high spatiotemporal resolution, and analysis of the motion of gene nanocarriers that yield information about their transport rates and mechanisms.

1.2. MPT: Theory and Applications

MPT involves resolving the dynamic motions of tens to hundreds of individual particles simultaneously in a given environment, such as a cell, using fluorescence video microscopy. The dynamic behavior of each gene vector yields quantitative information (e.g., effective diffusivities, maximum velocities, directionality) and qualitative information (e.g., mechanism of transport) that characterizes the transport of individual gene carriers. Tracking of individual particles also allows critical quantification of particle–environment interactions and biological processes that may dictate bulk transport properties.

In practice, because of current limitations in microscopy, it remains difficult to obtain 3D positional data with high temporal resolution. Therefore, the motions of gene carriers are typically tracked in 2D. 2D tracking provides accurate information on particle behavior if the biological environment (e.g., cell cytoplasm) is locally isotropic, but not necessarily homogeneous ***(1)***. Briefly, displacements of particles along the x, y, and z axes (Δx, Δy, Δz) are uncorrelated in a locally isotropic medium; hence, timescale-dependent ensemble 2D mean squared displacements [$<\Delta r^2(\tau)>$ or MSD],

$$< \Delta r^2(\tau) >_{2D} = < \Delta x^2 > + < \Delta y^2 > \quad (1)$$

is exactly equal to two-thirds of the timescale-dependent ensemble 3D MSD,

$$< \Delta r^2(\tau) >_{3D} = < \Delta x^2 > + < \Delta y^2 > + < \Delta z^2 > \quad (2)$$

because the ensemble displacements $<\Delta x^2> = <\Delta y^2> = <\Delta z^2>$.

Single particle tracking (SPT) was previously used to quantify the lateral movements of proteins and receptors on the cell surface ***(12,13)***, and the intracellular dynamics of adeno-associated viruses (AAV) ***(14)***. Over the past decade, advances in computational processing and image analysis software have afforded the development of MPT as a high throughput analog of SPT for rapid analysis of the movements of hundreds of particles simultaneously. MPT has been applied to the characterization of viscoelastic properties of various complex biological fluids ***(15)*** and the dynamic transport of gene complexes and model gene carriers in live cells and in fresh human mucus ***(5,6,8)***.

1.3. Modes of Transport in Biological Environments

The Brownian motion of particles in a purely viscous (i.e., not elastic) environment tracked in 2D can be described by the equation

$$< \Delta r^2(\tau) >_{2D} = 4D_o\tau \quad (3)$$

where the time scale-independent diffusivity D_o is governed by the Stokes–Einstein relation $D_o = k_B T/6\pi\eta a$, where k_B is the Boltzmann's constant, T is the temperature, η is the fluid viscosity, and a is the particle radius. The key feature of unobstructed diffusion is that D_o is independent of time scale, τ. Thus, D_o can be obtained from $D_o = <\Delta r^2(\tau)>_{2D}/4\tau$ and purely diffusive Brownian motion can be identified by a slope of 1 on a MSD vs. τ plot.

In the presence of dense fiber networks, where the effective pore size approaches the size of gene carriers, particle motion can be impeded by steric obstruction. Adhesive interactions between gene vectors and fibers or other obstacles can further reduce diffusion. The hindered transport arising from these steric and adhesive forces can be characterized by

$$< \Delta r^2(\tau) >_{2D} = 4D_o^{\alpha}\tau \tag{4}$$

where the value of α quantifies the extent of impediment (0 < α < 1; smaller α values represent more strongly hindered transport) ***(16,17)***. In the case of strong interactions between gene vectors and fibers, the impediment may lead to displacements below the microscope's detection limit. We often refer to these vectors as immobile.

Gene vectors may also utilize motor proteins to achieve energy-dependent active transport. Active transport is characterized by large, directed

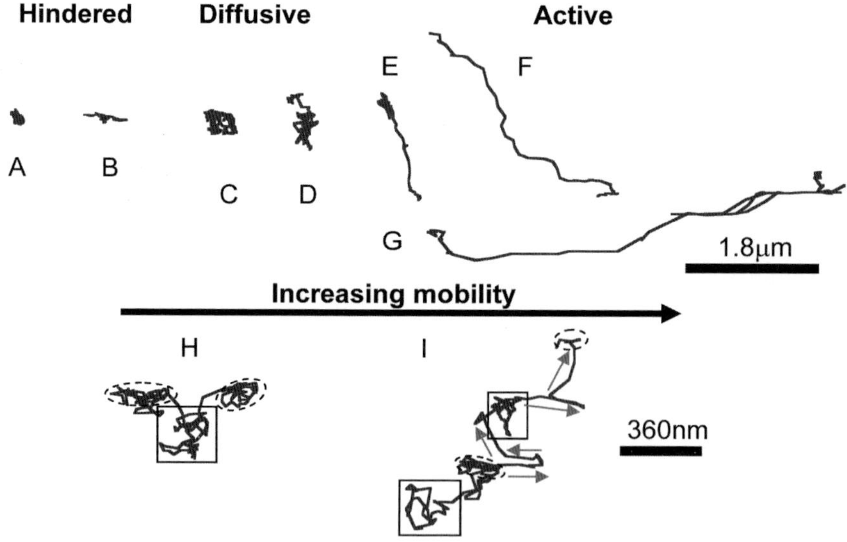

Fig. 1. Sample trajectories of nonviral polyethylenimine gene vectors undergoing different mechanisms of transport in live human cervical epithelial carcinoma HeLa cells. Gene vectors may undergo various mechanisms of transport: (**A** and **B**) diffusion that is substantially hindered by cellular components, (**C** and **D**) relatively unhindered diffusion, or (**E–G**) active transport characterized by directed displacements over great distances. (**H** and **I**) Gene vectors may display episodes of all three transport phenomena; hindered motions are outlined by dashed lines, diffusive motions by solid lines, and directed active motions by grey arrows. Scale bar is 1.8 μm for trajectories (**A–I**) and 360 nm for trajectories (**H–J**).

displacements and saltatory motions. The mean velocity, v, of facilitated transport is determined from the measured MSD of a particle through the relation

$$< \Delta r^2(\tau) >_{2D} = 4D_o\tau + v^2\tau^2 \quad (5)$$

In the absence of facilitated transport mechanisms, (5) can be used to describe displacements influenced by convective bulk fluid flow.

In biological environments, it is rare that displacements of a group of particles over an extended period of time can be described by just one of the aforementioned mechanisms of transport. Even individual particles may experience episodes of hindered transport, free diffusion, and facilitated active transport (Figure 1). Using MPT and the concept of relative change in diffusivity (RC) ***(9,18)***, it is possible to quantitatively assign appropriate transport modes to hundreds of individual particles.

For more information on particle tracking for drug and gene delivery applications, the reader is referred to our recent review ***(1)***.

2. Materials

2.1. Cell Culture

1. Dulbecco's Modified Eagle's Medium (DMEM) supplemented with 10% fetal bovine serum (FBS). Store at 4 °C.
2. Solution of trypsin (0.25%) and 1 mM ethylenediamine tetraacetic acid (EDTA). Store at 4 °C.
3. Phosphate-buffered saline (PBS): dissolve 0.23 g anhydrous NaH_2PO_4, 1.15 g anhydrous Na_2HPO_4, and 8.75 g NaCl in 900 ml ddH_2O, correct pH to 7.4 using 1 M NaOH or 1 M HCl, and adjust volume to 1 l with ddH_2O. Autoclave; store at room temperature.
4. 0.4% w/v Trypan Blue solution (Gibco/BRL, Bethesda, MD, USA). Store at room temperature in dark.

2.2. Extracellular Fluids

The collection of various bodily fluids, including CF lung sputum, nasal or cervical vaginal mucus, should be performed by experts and should conform to protocols approved by the Institutional Review Boards at respective universities. To maximize biological relevance, dilution of samples at any point during the collection process should be minimized or avoided because dilution of mucus may lead to drastic changes in the micro- and macro-rheological properties. Fresh samples should be used for microscopy within 2 h of collection. Alternatively, mucus may be frozen for long-term storage at –20 °C before use; however, the transport properties of control particles (100 nm PEG-modified polystyrene) should be determined on fresh samples and on each thawed sample to ensure that the mucus structure has not been altered by freezing.

2.3. Fluorescently-Labeled Gene Complexes

1. 25 kDa branched PEI (Polysciences, Warrington, PA, USA). Prepare a 100 mg/ml solution in ddH_2O, pH 7.0. Store at room temperature.
2. DNA from salmon testes (~2 kDa; Sigma, St. Louis, MO, USA) or, alternatively, any plasmid DNA that does not express fluorescent proteins. Dissolve in ddH_2O to 1 mg/ml. Store at 4 °C.
3. 1.0 M NaCl, pH 7.0. Store at room temperature.
4. 0.1 M sodium tetraborate, pH 9.3. Store at room temperature.
5. Oregon Green 514 carboxylic acid, succinimidyl ester (Molecular Probes, Eugene, OR, USA). Store at –20 °C in dark, dessicated.
6. Dimethylsilfoxane (DMSO). Store at room temperature.
7. 0.1 M sodium bicarbonate buffer. Store at room temperature.
8. G-75 Sephadex (Pharmacia, Piscataway, NJ, USA). Store at room temperature.
9. Trinitrobenzenesulfonic acid (TNBS; Sigma). Store at 4 °C in dark.
10. Microcon Centrifugal Filter (100,000 Da cutoff; Millipore, Bedford, MA, USA).

2.4. Confocal Microscopy

1. Opti-MEM® Reduced Serum Media (Phenol red-free; Invitrogen, Carlsbad, CA, USA) Store at 4 °C.
2. Hoechst 34580 Nucleic Acid label (Invitrogen). Store at 4 °C in dark.
3. Glass bottom 35-mm culture dishes (Mattek Corp, Ashland, MA, USA)
4. Zeiss LSM 510 Meta Confocal Microscope with 100 × objective (Numerical Aperture 1.4) and excitation and emission filters for FITC (488 nm *ex*, 530 nm *em*) and UV (392 nm *ex*, 440 nm *em*). Other inverted fluorescence microscopes, such as the Zeiss Axiovert 100A, may be substituted for some studies. For live cell microscopy, a motorized stage (for collection of z-stack images) and a live cell chamber, such as Bioptech FCS-2 Environmental Stages, are recommended. In this case, care must be taken to ensure that evaporation does not affect sample properties. Alternatively, a stage warmer, such as Nevtek, can be used to maintain temperature at 37 °C. Microscopes are assumed to be fully functional, well aligned, and dimensions of pixels of the camera carefully measured. The ability to capture movies with high temporal and spatial resolution is solely dependent on the individual microscope setup and capability; readers are recommended to contact the microscope manufacturers for optimizing the microscope for MPT.

2.5. MPT

1. Glycerol. Store at room temperature.
2. Fluorescent latex beads (200 nm, COOH-modified, yellow-green fluorescent; Invitrogen). Store at 4 °C in dark.

3. Labtek four-well Borosilicate chambered coverglass (Nalge Nunc International, Naperville, IL, USA).

2.6. Particle Tracking and Data Analysis Software

1. MetaMorph software 6.1 (Universal Imaging Co., Downingtown, PA, USA) with the Track Objects Application included. Other freely available software, such as ImageJ with the Particle Tracking plug-in, may be used instead.
2. Any programming software capable of manipulation of large data sets. Either Mathworks Matlab 7.0 or Microsoft Visual Studio C++ 6.0 is recommended.

3. Methods

3.1. Fluorescent Labeling of polyethylenimine

The experimental procedure given below describes the general steps needed for labeling common cationic polymers. It is possible to label the plasmid as well using various nucleic acid labels. Covalently attached fluorescent dyes are recommended for nucleic acids as intercalating dyes may label endogenous DNA instead. A few covalent DNA labels are listed in **Table 1**.

1. Preparing reagents:
   a. Reagent A: Dissolve 1 mg OG-488 in 100 μl DMSO. Use within 1 h.
   b. Reagent B: Dilute 10 μl stock 100mg/ml PEI in 490 μl 0.1 M Na_2CO_3.
2. Add Reagent A to Reagent B, vortex at lowest setting at room temperature in dark for 3 h. The end product is orange in color.
3. Prepare size exclusion chromatography column. Place 400 mg Sephadex in 125 ml ddH_2O, hydrate for 3 h at 90 °C with occasional swirl. Cool for 15 min, then carefully transfer to column. Prep column with 3 volumes of water.
4. Add sample to column along with 10 ml ddH_2O. Sample passes through the gel through gravity flow. Collect elutent in 8–10 volume fractions (~1 ml each).
5. Identify collected volume fractions containing PEI through TNBS assay. First, establish a PEI standard curve by preparing a series of diluted PEI solutions from stock 100 mg/ml PEI solutions using 0.1 M sodium tetraborate to the range of 2.5–25 μg/ml. 25 μl of 0.3 M TNBS solution should be added to 1 ml of PEI solution, and absorbance on a spectrophotometer should be read at 511 nm, resulting in a linear fit within this range. Subsequently, the same TNBS assay is performed on eluent fractions diluted 1:300 in 0.1 M sodium tetraborate. PEI concentrations can be calculated by comparison to the standard curve. To confirm the calculations, the total amount of PEI from all eluent fractions should roughly equal 1 mg.

Table 1
Select Samples of Covalent Nucleic Acid Fluorescent Labels

Product	Fluorescence (ex, em)	Remarks
Alexa Fluor™ succinimidyl ester	AF 546: 556 nm, 573 nm AF 647: 650 nm, 668 nm	Requires DNA to be amine modified by reverse transcription in the presence of aminoallyl dUTP before conjugation with amine reactive dyes. Exceptional resistance to photobleaching
Ethidium monoazide bromide	510 nm, 605 nm	Photocrosslinking agent covalently binds to nucleic acid upon UV excitation
Label IT Tracker™	Cy™ 3: 553 nm, 575 nm Cy™ 5: 651 nm, 674 nm	Non-enzymatic chemical labeling of predominantly the N7 position of guanine
pGeneGrip™	Rhodamine: 540 nm, 590 nm Fluorescein: 490 nm, 525 nm	Covalent linkage of fluorophores through peptide nucleic acid clamps; expression of plasmid is not affected
ULYSIS Labeling Kit	OG488: 495 nm, 520 nm Pacific Blue: 410 nm, 455 nm	Similar labeling mechanism as Label IT Tracker™ and photobleaching resistance as Alexa Fluor™ succinimidyl ester

6. The degree of conjugation (DOC) can be quantified using the following equation:

$$DOC = {}^{(A_{511})\,(DilutionFactor)}\!\big/_{(70000)\,[PEI]} \quad (6)$$

where [PEI] is expressed in moles per liter. Samples should be diluted 50× for absorbance measurements. Typically, DOC is in the range of 10–20 dye molecules per molecule PEI.

3.2. Tracking Latex Beads in Glycerol (To Determine Tracking Resolution)

1. Prepare a small aliquot of 1:100 diluted 200 nm latex beads from stock solution in ddH_2O (final w/v 0.02%).
2. Add 25 μl of diluted particle solution to 1 ml glycerol in an eppendorf tube. The exact volume of glycerol is not critical. Wrap the sample in aluminum foil and vortex overnight.
3. Carefully transfer 600 μl of glycerol with particles to one of the chambers in a 4-well glass chamber slide, taking care to minimize formation of air bubbles. Equilibrate at room temperature in dark for 2 h.
4. Visualize the sample under fluorescence microscope (add oil if using an objective that requires lens oil). The plane of focus should be at least 10 μm above the bottom glass surface to ensure microscopy of particles in glycerol. Carefully look for convective motions along any one direction, if any. Using a 512 × 512 silicon-intensified target camera (0.23 μm/pixel), approximately 10–20 particles should be in view of focus. Adjust fluorescence intensity and detector gain to maximize signal-to-noise and minimize intensity saturation.
5. Carefully record 20–40-s movies at maximum temporal resolution. Very few, if any, particles become out of focus over the duration of microscopy (*see* **Note 1**). Pixel binning should be avoided to maximize the tracking resolution. A minimum of 20–30 movies should be collected for analysis. Movies should be saved as 16-bit TIFF format to prevent loss of information.

3.3. Preparation of Cells for Microscopy

1. Human cervical epithelia carcinoma cells (HeLa) should be passed and plated onto 35-mm glass bottom culture dishes at least 24 h before microscopy. For imaging of spread-out adherent cells, a recommended seeding density is 90,000 cells in 2 ml of DMEM supplemented with 10% FBS (*see* **Note 2**).
2. Confluency of HeLa cells should not exceed 50% before addition of gene vectors.

3.4. Preparation of Fluorescent Gene Carriers

Nonviral PEI/DNA gene complexes are prepared at nitrogen (NH_3^+) to phosphate (PO_4^-) (N/P) ratio of 10 in some preferred formulations; however, other ratios can be used. In general, excess N/P ratio beyond 6 affords rapid formation of uniformly small gene complexes in the range of 100–200 nm. The size distribution of gene vectors should be measured using, for example, dynamic light scattering.

1. Prepare working PEI solution (65 μg/ml, 150 mM NaCl): 6.5 μl of 1 mg/ml fluorescently labeled 25 k PEI, 7.5 μl 1 M NaCl, 36 μl ddH_2O.
2. Prepare working DNA solution (50 μg/ml, 150 mM NaCl): 2.5 μl of 1 mg/ml DNA, 7.5 μl 1 M NaCl, 40 μl ddH_2O.

3. Quickly add PEI drop wise to DNA, vortex for 6–8 s at medium setting. Incubate at room temperature for 15 min.
4. Remove unconjugated PEI by ultra-filtration (*see* **Note 3**). The solution of complexes may be filtered to remove free-PEI using any filtration device with 100,000 Da MWCO. Two additional wash and filter steps with 150 mM NaCl ensures complete removal of free-PEI. Up to 30–40% of the complexes may be lost during this process (*see* **Note 4**).

3.5. MPT of Nonviral Gene Complexes in Fresh Human Mucus

1. Collect fresh mucus samples (refer to **Subheading 2.2.** for details). Store mucus at 4 °C in dark before use (for fresh mucus) or at –20 °C (for long-term storage). For fresh mucus, do not store sample for more than 2 h before use.
2. On the basis of the total volume of the mucus sample, add up to 3% v/v of particle solution into the core of the sample. Do not attempt to facilitate mixing of particles by stirring with pipet tips, as rigorous stirring may disrupt and shear mucus fiber networks (*see* **Note 5**).
3. Carefully transfer 400–500 μl of the sample to one of the chambers in a 4-well glass chamber slide and minimize formation of air bubbles (*see* **Note 6**). Equilibrate at room temperature in dark for 2 h.
4. Perform microscopy as outlined in **Subheading 3.2, steps 4–5**. At high gene vector concentrations, the gene vector may exert sufficient attractive forces with mucus fibers to collapse the mucus network into bundles, which can be observed directly under the microscope. In such cases, the concentration of gene vectors should be diluted.

3.6. MPT of Nonviral Gene Complexes in Live Cells

1. Add 20 μl of freshly prepared DNA/polymer nanocomplexes to cell media. Gently swirl the glass bottom chamber to facilitate mixing of complexes in media. Place back in cell culture incubator for the duration of incubation desired. For nonviral gene vectors, a minimum of 15–30 min is typically required.
2. Add 2 μl of Hoechst 34580 30 min before microscopy (*see* **Notes 7 and 8**).
3. Warm the observation area of the microscope with a stage warmer (preferably a live cell chamber) to 37 °C.
4. Gently wash cells (2 ml) thrice with warm (37 °C) PBS to remove non-internalized gene vectors (*see* **Note 9**).
5. Replace media with phenol red-free OptiMEM (2 ml).
6. Place sample under microscope for observation. Ensure that the temperature (37 °C) is maintained if using a blower. Identify the cell layer.
7. Perform a z-stack with dual channel imaging of UV and FITC scan to record location of gene carriers with respect to nucleus.
8. Obtain movies at desired spatiotemporal resolution, usually maximum.

3.7. Obtaining Displacements of Latex Beads and Gene Complexes

The process for obtaining particle trajectories from movies is similar despite various programs available to accomplish such tasks. In general, it involves identifying the first point of the trajectory for each particle, and the centroid of each particle as it moves across different frames is calculated based on algorithms that identify the outline and maximum intensity of fluorescence within a defined neighboring space ***(19)***. The protocol described below is specific to the Metamorph software 6.1.

1. Open the movie file in Metamorph. Confirm the number of frames imported matches the duration of the movie.
2. Create regions highlighting the area of interest. For particle tracking in cells, the region should encompass each cell; for particle tracking in glycerol or other extracellular fluids, four equal size regions should be used to evenly divide the tracking area.
3. Select a region, zoom-in 400%, then select Edit → Duplicate → Stack with Zoom (*see* **Note 10**).
4. Open the Track Objects Application. Use the Template Match algorithm. Select to only output object number, frame number, x and y positions.
5. Select Track, then highlight particles to be tracked. Ensure the inner box highlights only the fluorescent pixels from each particle. The outer box denotes the area where the algorithm will search for the particle in the next frame; as a general rule of thumb, the dimensions of the outer box should be roughly three times larger than the inner box. After selecting all particles to be tracked, hit OK to proceed.
6. Export the outputted information (Object number, frame number, x, y) into either an Excel or a text file.

3.8. Translating x, y Positional Data Over Time for Calculation of Mean Squared Displacements and Effective Diffusivities

The actual implementation of programming software to calculate displacements is both program and user dependent, and the steps below only outline the calculations needed. A distribution of the C++ implementation, without support, is available upon request to hanes@jhu.edu.

1. Import the particle frames and position data.
2. Calculate the geometric average of mean squared displacements for all particles across all time scales. This is a key and perhaps most confusing step in the analysis. The MSD at the shortest time scale of each particle is strictly the arithmetic mean of squared displacements for that particle across every frame, $\text{MSD} = (x_{n+1} - x_n)^2 + (y_{n+1} - y_n)^2$, where n represents the frame number. For the ensemble averaged MSD across the entire particle population, a geometric mean over all particle MSD at each time scale should be used to minimize effects of

outliers. When calculating MSD at other time scales, it is important to note that frame number and time scale are independent of each other. For example, MSD at a temporal resolution four times greater than the shortest temporal resolution does not imply calculating displacements between every fourth frame (i.e., 1→5, 5→9, 9→13…). Rather, it involves calculating the displacements between all frames that are a defined temporal resolution apart. Thus, MSD at a temporal resolution four times greater than the shortest temporal resolution is an arithmetic average of displacements between frame 1→5, 2→6, 3→7, etc.

3. Calculating the effective diffusivities (D_{eff}). Once the geometric MSD across all time scales is calculated, the $D_{eff}(\tau)$ is simply $D_{eff}(\tau) = MSD(\tau)/(4\tau)$.

3.9. Measuring the Tracking Resolution

There are two common methods to calculate the resolution of the microscope; fluorescently labeled particles can either be glued to the glass chamber and the thermal motions measured ***(15)***, or the particles can be suspended in a homogeneous viscous fluid such as glycerol ***(1)***. The latter offers the advantage of accounting for errors arising from implementation of algorithms used in locating centroids of particles from frame to frame.

1. Following the steps outlined in **Subheadings 3.2, 3.7**, and **3.8**, obtain a log-log plot of MSD vs. time scale τ.
2. Using any graphing software (e.g., Microsoft Excel), fit the data to the equation $MSD = 4D_o\tau + 4\sigma^2$, where σ is the tracking resolution. The resolution for well-aligned microscopes typically varies between 5 and 10 nm.

3.10. Characterizing the Mechanism of Transport

As discussed in the introduction, gene vectors in biological environments may undergo episodes of hindered transport, unobstructed diffusion, and facilitated active transport. Despite the ambiguity, it is possible to distinguish particles that move substantially faster or slower than diffusive particles by analyzing the Relative Change in effective diffusivity (RC). This method was first presented for the analysis of transport of particles microinjected in the cytoplasm of live carcinoma cells ***(9)*** and for comparing the transport of viral and nonviral gene vectors in live primary neurons ***(18)***. RC values are defined as $RC = D_{eff}(\tau^{comp})/D_{eff}(\tau^{ref})$, where τ^{comp} is the comparison time scale and τ^{ref} represents the reference time scale. For a population of purely Brownian particles, the RC values have a Gaussian distribution around 1. This classification scheme thus differentiates mechanisms of particle movements by considering the deviation of RC values of individual particles from Brownian behavior at short and long time scales. The analysis involves three major steps: (1) establishing the range of RC values describing diffusive transport at short and long time scales, (2) calculating the RC values of individual particles at short and long time scales, and (3) comparing the RC values of particles to the RC range to determine the mechanism of transport. These steps are illustrated in **Fig. 2**. Comparison of the mode of particle transport at short and long time scales affords an improved understanding of the dominant mechanism across different time scales. The overall mechanism of transport over time scale is classified based on the scheme presented in **Fig. 2**. It is important to note that this classification method strictly determines the mechanism of transport. All purely diffusive motions, regardless of the bulk fluid viscosity, would be similarly classified as diffusive.

An RC curve represents the range of relative effective diffusivities across time scale where a particle exhibits random walk behavior (*see* **Fig. 2**. Data of purely Brownian diffusion can be created through Monte Carlo simulation of random walks or from particle tracking in glycerol (*see* **Note 11**). The factors affecting the RC curve are time scale and the confidence interval with which

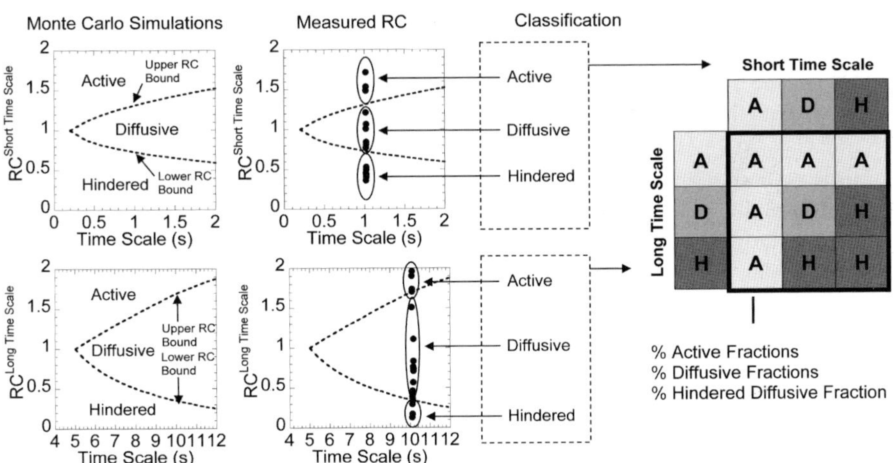

Fig. 2. Flow chart depicting the process of classifying overall transport mode of particle motions into hindered-diffusive, diffusive, and active based on the mechanism of transport at short and long time scale. On the basis of Monte Carlo simulations, an RC master curve mapping the range of RC values characterizing 95% of Brownian motions as diffusive over short and long time scale is established. Next, the experimentally obtained RC values are compared to the RC master curve. RC values above the upper RC bound are classified as active, within the RC bounds as diffusive, and below the lower RC bound as hindered-diffusive. Using this strategy, the transport modes over respective short and long time scale are obtained. Finally, the overall mechanism of gene vector transport across all time scales is determined based on the above classification scheme.

a particle is to be classified as Brownian. We have conveniently set this value at 95%, where the fastest 2.5% of particles may be mis-classified as actively transported and the slowest 2.5% as undergoing hindered diffusion. Monte Carlo simulations are preferred, as the sample size is not only much greater (e.g., 100,000 random walks are typically simulated compared to 100s of trajectories from particle tracking in glycerol) but also free of experimental error.

1. Monte Carlo simulation of random walks. The easiest method is to simulate sequential displacements of fixed length in random directions for a defined number of steps ($x_{n+1} = x_n + d\cos(\theta)$. $y_{n+1} = y_n + d\sin(\theta)$, where d is the fixed length (typically set as 1) and θ is generated from a random number generator). Then, perform the analysis discussed in **Subheading 3.8**. The unit length of motion is irrelevant as we are only calculating for the relative change in effective diffusivities. At least $n = 100{,}000$ trajectories should be simulated.
2. Calculate the RC values of simulated random walk using the equation RC = $D_{eff}(\tau^{comp})/D_{eff}(\tau^{ref})$, and sort the RC values in increasing order. The RC value dividing the lowest 2.5% of particles from the remaining 97.5% is considered lower RC range; the RC value dividing the highest 2.5% is considered the upper RC range. (Note: The choice of time scale considered as short or long depends on the fluid and particles used and is best decided by an expert. Typically, the short time scale transport mode is established at RC = $D_{eff}(5T)/D_{eff}(T)$, where T represents the smallest time scale. The long time scale transport mode is calculated at RC = $D_{eff}(50T)/D_{eff}(25T)$. It is important to confirm that a particular choice of time scale has little effect on the overall classification of particle transport. The temporal difference between the comparison and reference time scales should not be too large, as the ability to distinguish hindered motions is reduced at long time scales.
3. Calculate the RC values of experimental data in the same fashion as **step 2**.
4. Compare the RC values at both short and long time scales from experimental data to the RC range established in **step 2**. Particle RC value below the lower RC range is considered hindered diffusive and above the upper RC range is active transport. The predominant transport mechanism for each particle is classified from the mechanism at short and long time scales following **Fig. 2**.
5. Calculate the number of particles exhibiting an overall transport mechanism of hindered, diffusive, and active (*see* **Note 12**).

4. Notes

1. Movement of particles out of the plane of focus is rare for short movies in a viscoelastic environment, such as the cell cytoplasm or mucus. However, even if particles do move away from the plane of focus, the apparent size increase of particles should be of no consequence in microscopes that are well aligned as these out-of-plane movements should not translate into artificial *x*-*y* movements of the centroid of the particles.

2. To achieve simultaneous imaging with fluorescently labeled organelles through plasmids expressing fluorescently labeled proteins (e.g., EEA1-GFP fusion protein for labeling of early sorting endosomes), cells may be pre-transfected using electroporation. In such scenarios, electroporated cells should be plated >24 h before microscopy with a seeding density of 150,000 cells per 2 ml FBS-supplemented media.
3. This step ensures that the fluorescence detected during microscopy is not due to fluorescence of free polymers but of the actual gene vectors. From our experience, the intracellular transport of common gene vectors following either polymer fluorescence or DNA fluorescence is similar, so this purification process may not be necessary in some cases.
4. If maintaining the same concentration of complexes is important, the concentration of complexes in solution can be measured on a UV spectrophotometer for absorbance at 260 nm. Conjugation of DNA with cationic polymers, such as PEI, has no effect on A_{260} readings.
5. The exact volume of mucus is often difficult to determine because of its highly viscoelastic nature. However, estimation using a density of 1 g/ml is generally sufficient. The volume of particles to add is up to 3% v/v in such a scenario.
6. If using an 8-well glass chamber slide, the volume can be reduced to 200–300 μl. It may be difficult to transfer highly viscoelastic mucus using a typical 1 ml pipet tip. In such cases, a 1-ml syringe is recommended for transfer of mucus to glass chamber slides.
7. To achieve simultaneous imaging with fluorescently labeled organelles, ex vivo dyes (e.g., Lysotracker, FM 4-64, etc.) may be added 30 min to 1 h before microscopy.
8. Microtubules are typically involved in the intracellular facilitated active transport observed for most viruses and nonviral gene vectors *(1)*. To test whether the intracellular transport of specific gene vectors is dependent on microtubules, add 10 μg/ml nocodazole in the media 1 h before microscopy to depolymerize the microtubule network.
9. An additional round of acid wash before PBS washes may be included as well. Acid washes have been shown effective at removing membrane-bound objects. To confirm viability of live cells in the presence of gene vectors, a wash step with 0.4% w/v of Trypan blue (1 ml) for 30 s followed by three more washes with PBS can be added. Trypan blue is excluded from the plasma membrane of live cells; the viability of cells can be directly observed under a light microscope. Do not add Hoechst dyes for nucleus labeling if performing Trypan blue wash step. Trypan blue has the added advantage of reducing green fluorescence and can reduce green fluorescence signals of particles adhering to surface of cells or in dead cells.
10. This step increases the number of pixels in the tracking movie by 4× in each dimension, with intensity of new pixels an extrapolation of surrounding pixel intensities. While no new information is introduced, this step aids in achieving

better resolution of the particle tracking analysis by reducing the discreet pixel size.

11. We have analyzed the motion of hundreds of particles in glycerol using the classification scheme from Monte Carlo simulations and found correct classification of transport modes within the confidence interval.
12. To confirm the accuracy of the transport mode classification, plots of geometric MSD vs. time scale should be plotted for each group. The slope of the data on such plot should be <1 for hindered diffusion, ~1 for diffusive, and >1 for active.

Acknowledgments

The authors gratefully acknowledge financial support from the National Institute of Health (NIH IROIEB00358-01) the National Science Foundation (NSF BES0346716) and a post-graduate scholarship (PGSD) from the Natural Science and Engineering Research Council of Canada (SKL). The authors thank Sudhir Khetan, Jung Soo Suk, and Kaoru Hida for critical review.

References

1. Suh, J., Dawson, M., and Hanes, J. (2005) Real-time multiple-particle tracking: applications to drug and gene delivery. *Adv Drug Deliv Rev* **57,** 63–78.
2. Riordan, J. R., Rommens, J. M., Kerem, B., Alon, N., Rozmahel, R., Grzelczak, Z., et al. (1989) Identification of the cystic fibrosis gene: cloning and characterization of complementary DNA. *Science* **245,** 1066–73.
3. Hanes, J., Dawson, M., Har-el, Y., Suh, J., and Fiegel, J. (2003) Gene Delivery to the Lung. *Pharmaceutical Inhalation Aerosol Technology*, A.J. Hickey, Editor. Marcel Dekker Inc., New York, pp. 489–539.
4. Powell, K., and Zeitlin, P. L. (2002) Therapeutic approaches to repair defects in DeltaF508 CFTR folding and cellular targeting. *Adv Drug Deliv Rev* **54,** 1395–408.
5. Dawson, M., Wirtz, D., and Hanes, J. (2003) Enhanced viscoelasticity of human cystic fibrotic sputum correlates with increasing microheterogeneity in particle transport. *J Biol Chem* **278,** 50393–401.
6. Lai, S. K., O'Hanlon, E. D., Harrold, S., Man, S. T., Wang, Y. Y., Cone, R. A., et al. Rapid transport of large polymeric nanoparticles in fresh undiluted human mucus. *Proc Natl Acad Sci USA* **104,** 1482–7.
7. Dawson, M., Krauland, E., Wirtz, D., and Hanes, J. (2004) Transport of polymeric nanoparticle gene carriers in gastric mucus. *Biotechnol Prog* **20,** 851–7.
8. Suh, J., Wirtz, D., and Hanes, J. (2003) Efficient active transport of gene nanocarriers to the cell nucleus. *Proc Natl Acad Sci USA* **100,** 3878–82.
9. Suh, J., Choy, K. L., Lai, S. K., Suk, J. S., Tang, B., Prabhu, S., et al. PEGylation of nanoparticles improves their cytoplasmic transport. *Int J Nanomed* **2**(4), 1–7.
10. Suh, J., Wirtz, D., and Hanes, J. (2004) Real-time intracellular transport of gene nanocarriers studied by multiple particle tracking. *Biotechnol Prog* **20,** 598–602.

11. Lukacs, G. L., Haggie, P., Seksek, O., Lechardeur, D., Freedman, N., and Verkman, A. S. (2000) Size-dependent DNA mobility in cytoplasm and nucleus. *J Biol Chem* **275,** 1625–9.
12. Kusumi, A., Sako, Y., and Yamamoto, M. (1993) Confined lateral diffusion of membrane receptors as studied by single particle tracking (nanovid microscopy). Effects of calcium-induced differentiation in cultured epithelial cells. *Biophys J* **65,** 2021–40.
13. Anderson, C. M., Georgiou, G. N., Morrison, I. E., Stevenson, G. V., and Cherry, R. J. (1992) Tracking of cell surface receptors by fluorescence digital imaging microscopy using a charge-coupled device camera. Low-density lipoprotein and influenza virus receptor mobility at 4 degrees C. *J Cell Sci* **101**(Pt 2), 415–25.
14. Seisenberger, G., Ried, M. U., Endress, T., Buning, H., Hallek, M., and Brauchle, C. (2001) Real-time single-molecule imaging of the infection pathway of an adeno-associated virus. *Science* **294,** 1929–32.
15. Apgar, J., Tseng, Y., Fedorov, E., Herwig, M. B., Almo, S. C., and Wirtz, D. (2000) Multiple-particle tracking measurements of heterogeneities in solutions of actin filaments and actin bundles. *Biophys J* **79,** 1095–106.
16. Saxton, M. J. (1996) Anomalous diffusion due to binding: a Monte Carlo study. *Biophys J* **70,** 1250–62.
17. Saxton, M. J. (1994) Anomalous diffusion due to obstacles: a Monte Carlo study. *Biophys J* **66,** 394–401.
18. Suk, J., Suh, J., Lai, S. K., and Hanes, J. Quantifying the Intracellular transport of viral and nonviral gene vectors in primary neurons. *Exp Biol Med* **232**(3), 461–9.
19. Savin, T., and Doyle, P. S. (2005) Static and dynamic errors in particle tracking microrheology. *Biophys J* **88,** 623–38.

7

Production of Lentiviruses Displaying "Early-Acting" Cytokines for Selective Gene Transfer into Hematopoietic Stem Cells

E. Verhoeyen, D. Nègre, and F. L. Cosset

Summary

A major limitation of current lentiviral vectors (LVs) is their inability to govern efficient gene transfer into quiescent cells, such as human CD34+ cells that reside in the G0 phase of the cell cycle and that are highly enriched in hematopoietic stem cells. This hampers their application for gene therapy of hematopoietic cells. We describe here novel LVs that overcome this restriction by displaying early-acting cytokines on their surface. Display of thrombopoietin, stem cell factor or both cytokines on LV surface allows high transfer into quiescent cord blood CD34+ cells. Moreover, these surface-engineered LVs preferentially transduce and promote survival of resting CD34+ cells rather than cycling cells. These novel LVs allow superior gene transfer in the most immature CD34+ cells compared to conventional LVs, even in the presence of recombinant cytokines. This is demonstrated by their capacity to promote selective transduction in long-term culture initiating cell colonies (LTC-ICs) and of long-term non-obese diabetic/severe combined immunodeficient (NOD/SCID) repopulating cells (SRCs). Here we describe the production of these "early acting cytokine" displaying vectors and the methodology to confirm the capacity of these vectors to promote selective transduction of HSCs.

Key Words: Hematopoietic stem cells; NOD/SCID mice; SCF; TPO; lentiviral vector; gene therapy; targeted gene transfer.

1. Introduction

For the correction of many inherited or acquired defects of the hematopoietic system, the therapeutic gene must be delivered to cells that are able to both self-renew and differentiate into all hematopoietic lineages, that is, a single

From: *Methods in Molecular Biology, vol. 434: Volume 2: Design and Characterization of Gene Transfer Vectors*
Edited by: J. M. Le Doux © Humana Press, Totowa, NJ

HSC can generate erythroid, myeloid, megakaryotic, and lymphoid lineages. As such, these gene therapies must be targeted to the "right" cell, "the human hematopoietic stem cell" (HSC), without modifying its properties.

It is, however, difficult to isolate a HSC because no single antibody has been found that recognizes a HSC exclusively. Many studies suggest that HSC reside in a population that expresses the CD34+ antigen ***(1)***, which is expressed in 1–4% of the human bone marrow cells, 1% of the mononuclear cells isolated from cord blood, and 0.2 % from peripheral blood ***(2–5)***. One characteristic of the HSC is that it does not actively divide in vivo. This quiescent state provides an obstacle to the transduction of the HSC. Murine retroviral vectors need the target cells to proliferate for efficient gene transfer, which can only be achieved by using hematopoietic growth factors such as thrombopoietin (TPO), stem cell factor (SCF), and Flk-3 ***(6)***. However, strong cytokine stimulation induces proliferation often with an unwanted side-effect: primitive cells start to differentiate, which has a negative affect on ex vivo CD34+ cell gene therapy using MLV vectors.

In particular, the emerging class of lentiviral vectors (LVs), represent the only means to transduce these quiescent cells permanently. LVs can integrate into the chromatin of non-dividing cells such as neurons, fibroblasts, and hepatocytes ***(7–10)***, while resting lymphocytes are refractory to LV-mediated gene transfer ***(11–15)***. We and others have shown that stimulation into the G1b phase of the cell cycle is sufficient to allow productive transduction of quiescent T-cells with LVs, while these conditions are insufficient for transduction with murine leukemia virus (MLV)-based retroviral vectors ***(11–15)***. Whether HIV or the LVs can integrate into the genome of a truly quiescent cell, in particular an early progenitor HSC in G0, remains controversial ***(14,16–22)***. Compared to their MLV counterparts, LVs do not need extended cytokine stimulation in order to transduce HSCs, avoiding strong cell proliferation, which might lead to loss of stem cell potential ***(24)***. While the LV-mediated transduction of HSCs within the CD34+ cell population seems possible without cytokine stimulation, a clear boosting effect of strong cytokine cocktails on HSCs transduction was demonstrated ***(16–19,22,24–26)***. Additionally, in combination with extended cytokine stimulation, protocols often rely on high vector doses per target cell, multiple vector hits, or an increase in virus/target cell contact by centrifugation or use of fibronectin fragments. Such maneuvers are likely to induce activation and differentiation of stem cells and to promote multiple vector integrations.

Our targeting strategy for gene transfer consists of an interaction of a ligand displayed on the surface of the vector with its specific receptor thereby inducing signaling and stimulation of the target cells without changing their phenotype. As a consequence of the specific stimulation, gene transfer into the target cell is significantly enhanced. This concept was first shown with amphotropic MLV retroviral vectors whose glycoproteins were engineered so as to display the hepatocyte growth factor (HGF) or the interleukin-2 (IL-2). Such vectors could infect primary hepatocytes ***(27)*** or IL-2-dependent cells ***(28)***, respectively, that were refrac-tory to gene transfer by vectors carrying unmodified envelope glycoproteins.

As mentioned above, LV can transduce non-proliferating cells, but important primary targets of gene therapy such as resting T-cells and HSCs remain refractory to LV-mediated gene transfer. But by engineering LV particles displaying T-cell activating single chain antibody peptides, resting T cells could be efficiently transduced ***(12)***. We further improved this method by developing HIV-1 vector particles displaying IL-7 to increase transduction of resting T cells without inducing a naive-to-memory phenotypic switch ***(11)***.

Here we describe an analogous approach for selective gene transfer into HSCs. We have engineered recombinant membrane proteins and incorporated them into LV particles to display "early acting" cytokines on their surface ***(29)***. We postulated that the new surface-modified LVs would selectively and minimally stimulate HSCs in the CD34+ bulk population during gene transfer with the specific aim to promote high levels of transduction, indispensable for clinical application, in these targets. A recombinant membrane protein consisting of the transmembrane influenza hemagglutinin glycoprotein was fused to truncated forms of TPO, such that the TPO is displayed on the exterior of the vector particle. Similarly, the SCF cDNA was fused to the amphotropic Moloney MLV env glycoprotein, transfected into the packaging cells, and displayed on the LV particles. As the receptors for TPO and SCF, c-mpl and c-kit, respectively, are expressed on HSCs, vector particles displaying SCF and TPO can bind these receptors on HSCs. In addition, these vectors bind the receptors for VSV-G, which are phospholipids in the HSC membrane (*see* **Fig.** 1). With these engineered vectors displaying TPO, SCF, or both cytokines, transduction efficiency of quiescent CD34+ cells is increased several fold ***(29)***. These surface-engineered LVs preferentially transduced and promoted survival of resting CD34+ cells rather than cycling cells. Importantly, these novel LVs allowed superior gene transfer in the most immature CD34+ cells as compared with conventional LVs, even when the latter vectors were used to transduce cells in the presence of recombinant cytokines. Still once the CD34+ cell population is successfully transduced, a critical point is to document that the cells can give rise to all hematopoietic lineages. Our "early acting cytokine" displaying vectors are able to selectively transduce the HSCs in the CD34+ population with high efficacy and without disturbing their cell function or their capacity to differentiate. This conservation of HSC character of transduced cells is demonstrated by their capacity to differentiate in vitro into derived long-term culture initiating cell colonies (LTC-ICs) and to repopulate immunocompromised non-obese diabetic/severe combined immunodeficient (NOD/SCID) mice, defined as a measure for long-term human repopulating cells. In addition, HSCs treated

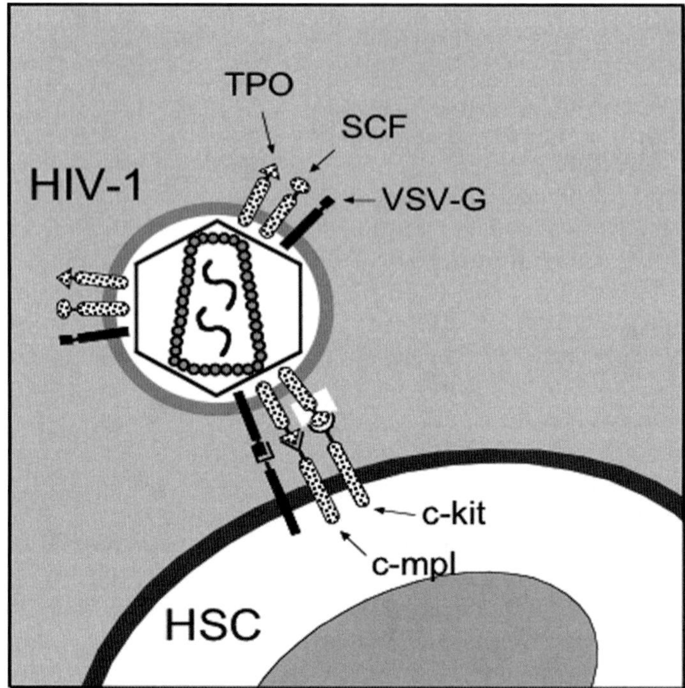

Fig. 1. Lentiviral vectors displaying "early-acting-cytokines." HIV-1 derived lentiviral particles display recombinant envelope glycoproteins N-terminally fused to stem cell factor (SCF) or thrombopoietin (TPO), two cytokines that can specifically target the vector particles to hematopoietic stem cells (HSCs) because they express the receptor for SCF, c-kit and the TPO receptor, c-mpl. In addition, the vector particles contain the vesicular stomatitis virus G (VSV-G) glycoproteins, which allow vector-cell fusion.

with these vectors do not need recombinant growth factors for survival because these are supplied by the vectors particles themselves. Here we describe the production of these "early acting cytokine" displaying vectors and the methodology to confirm the capacity of these vectors to promote selective transduction of HSCs.

2. Materials

2.1. Buffers and Solutions

1. 2× Hepes-buffered saline (HBS) and 2 M $CaCl_2$: Calphos Mammalian Transfection Kit (Clontech–Takara BioEurope, Saint Germain-en-Laye, France).
2. Phosphate-buffered saline (PBS) without calcium and magnesium, without sodium bicarbonate, sterile.

3. Trypsin–ethylenediaminetetraacetric acid (EDTA) 1× Hank's balanced salt solution without calcium and magnesium, sterile.
4. Ficol-Paque Plus, sterile.
5. Pyronine-Y (PY) staining buffer: Hanks balanced salt solution, 20 mM Hepes (*N*-2-hydroxyethylpiperazine-*N'*-2-ethanesulfonic acid), 1g/l glucose, 10% FCS, 1 μg/ml PY.

2.2. Media

1. Fetal calf serum (FCS), sterile.
2. Dulbecco's modified Eagle medium (DMEM) with 0.11 g/l sodium pyridoxine and pyridoxine. DMEM is supplemented with 10% FCS, 100 μg/l streptomycin, 100 U/ml penicillin (stored at 4°C).
3. Cellgro medium (serum-free medium; Cellgenix, Freiburg, Germany).
4. DMEM-α supplemented with 10% FCS.
5. Medium for LTC-IC culture: α-MEM supplemented with 12.5% FCS, 12.5 % horse serum and 10^{-4} M β-mercaptoethanol.
6. Methocult medium (Stem cell technologies) for CFC detection: 1% methylcellulose in Iscove' MDM supplemented with 30% FCS, 1% BSA, 10^{-4} M 2-mercapthoethanol, 2 mM L-glutamine, rhSCF (50 ng/ml), Epo (3U/ml), IL-3 (10 ng/ml), GM-CSF (10 ng/ml).

2.3. Nucleic Acids and Oligonucleotides

1. LV DNA encoding for an HIV-1 derived self-inactivating vector with the internal EF1-alpha promoter driving the reporter gene, green fluorescent protein (GFP) (*see* **Fig. 2A**).
2. Envelope glycoprotein expressing plasmids:
   a. *Fusion glycoprotein:* stomatitis virus G glycoprotein (VSV-G).
   b. *Activating and targeting glycoproteins* for HSCs (*see* **Fig. 2B**): (1) TPO-Env (TPO fused to hemaglutin envelope glycoprotein) and (2) SCF-Env (SCF fused to MLV envelope glycoprotein).
3. Virus structural protein (gag-pol) expressing plasmid (pCMV8.91).
4. Neuraminidase expressing plasmid pCMV-NA.

2.4. Cells and Tissue

1. 293T cells.
2. Hela cells.
3. MS5 stroma cells.
4. Source of HSCs: fresh neonatal (cord) blood (*see* **Note 1**).

2.5. Animals

1. NOD/SCID mice housed in sterile environment.

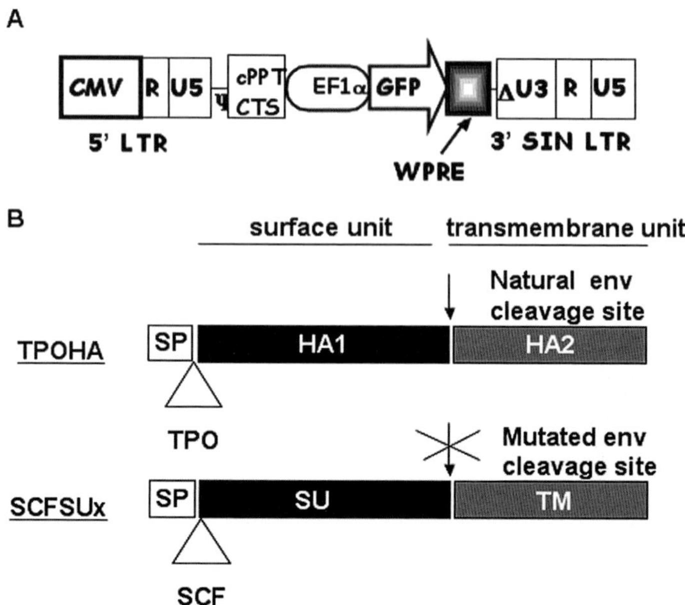

Fig. 2. Schematic representation of lentiviral vector (**A**) and cytokine-displaying envelope glycoproteins (**B**). (**A**) Lentiviral vector derived from HIV-1 with a deletion in the 3′ U3 region (delta U3) resulting in a self-inactivating (SIN) vector, which once integrated into the host genome only allows GFP expression from the internal "elongation factor 1 alpha" EF1-alpha promoter. LTR, long-terminal repeat; CMV, human cytomegalovirus early promoter; PBS, primer binding site; Ψ, packaging sequence; cPPT/CTS, central polypurine track and central termination sequence; WPRE, woodchuck post-transcriptional responsive element. (**B**) Design of the chimeric cytokine-displaying vectors. Thrombopoietin (TPO) was inserted N-terminally of the surface subunit HA1 from the influenza heamaglutinin glycoprotein. After cleavage the HA1 subunit is connected to HA2 by a non-covalent binding. Stem cell factor (SCF) was inserted N-terminally of the surface subunit SU from the murine leukemia virus (MLV) glycoprotein. In this chimeric envelope the natural cleavage site between SU and the transmembrane subunit was mutated to inhibit SU-TM cleavage. SP: signal peptide.

2.6. Special Equipment

1. Magnetic separation columns (Miltenyi Biotec France, Paris, France).
2. MAC magnetic device (Miltenyi).

2.7. Additional Reagents

1. CD34+ cell separation kit from Miltenyi containing anti-human CD34 Microbeads and blocking reagent.

2. 48-well and 24-well cell culture coated tissue culture plates.
3. 35-mm uncoated cell culture plates.
4. 0.45-μm filter.
5. Mouse monoclonal antibodies: anti-hCD45-phoecoerythrin (PE)-Cyanin 5 (Cy5), anti-hCD19-PE, anti-hCD15-PE, anti-hCD14-PE, anti-hCD56-PE, anti-hCD13-PE, and anti-hCD34-PE and corresponding PE-conjugated mouse IgG controls.

3. Methods

3.1. Production of Lentiviral Vectors Displaying Activating Polypeptides

1. Day 0: 2.5 × 10^6 (10^6 293T cells are seeded the day before transfection in 10-cm plates in a final volume of 10 ml DMEM).
2. Day 1: cotransfection of HIV packaging construct (8.6 μg) with the lentiviral gene transfer vector (8.6 μg) and the two or three glycoproteins: (1) VSV-G (1.5 μg) and (2) TPOHA or /and (3) SCFSUx (1.5 μg) using the Clontech calcium-phosphate transfection system (*see* **Note 2**).
3. Day 2: 15 h after transfection, the medium is replaced with 6 ml of fresh Cellgro medium (Celgenix; *see* **Note 3**).
4. Day 3: 36 h after transfection, vectors are harvested, filtrated through 0.45 μm pore-sized membrane and stored at –80 °C for 2–3 months.

3.2. Immunoselection of Human CD34+ Cells

1. Dilute cord blood 1:1 with PBS and gently layer 35 ml of this diluted product on 15-ml Ficoll in a 50-ml tube.
2. Centrifuge the cells at 850 g for 30 min, 20 °C without brake and collect the layer containing mononuclear cells.
3. Wash the collected mononuclear cell interface in PBS/2% FCS at 850 g, 20 °C for 10 min.
4. Resuspend the cells at 1–2 × 10^8/ml, add the anti-hCD34+ microbeads according to manufacturer's indication (Miltenyi), and incubate for 30 min while rocking at 4 °C.
5. Wash cells to remove the unbound antibody and resuspend in PBS/2% FCS.
6. Pre-incubate a MAC separation column with PBS/2% FCS. Then let the CD34+ labeled cells pass through a first column put on the MAC magnetic device.
7. Wash the column once with PBS/2% FCS and then remove the column from the magnet.
8. Put 1 ml of PBS/2% FCS on the column and flush out the CD34+ labeled cells.
9. Repeat **steps 6–8** once more, and after the cells passed through the second column, they are pelleted by centrifugation at 500 g and resuspended in Cellgro medium. The purity of the CD34+ cells is routinely 90–95%.

3.3. Titer Determination

1. Day –1: HeLa cells are seeded in DMEM at a density of 2×10^5 cells per well in 6-well plates in a final volume of 2 ml.
2. Day 0: Serial dilutions (*see* **Note 4**) of vector preparations were added to HeLa cells and incubated O/N.
3. Day 1: Medium on the cells is replaced with 2 ml fresh DMEM, and cells are incubated for 72 h.
4. Day 3: Cells are trypsinized and transferred to fluorescence activated cell sorter (FACS) tubes. The percentage of GFP-positive cells is determined by FACS analysis.

3.4. Analysis of Transduction and Titer

1. *Transduction* efficiency is usually determined as the percentage of GFP-positive cells after transduction of 3×10^5 target cells with 1 ml of viral supernatant.
2. *Infectious* titers are provided as transducing units (TU)/ml and can be calculated by using the formula: Titer = $\%\text{inf} \times (3 \times 10^5/100) \times d$; where "d" is the dilution factor of the viral supernatant and "%inf" is the percentage of GFP-positive cells as determined by FACS analysis using dilutions of the viral supernatant that transduce between 5 and 10% of GFP-positive cells.
3. *Multiplicities of infection (MOI)*: ratio between infectious particles and target cells that are required to optimally transduce target cells of interest, which are generally much less permissive to transduction than the cells used for titrations.

3.5. Cell-Cycle Fractionation by Viable Pyronin Y Staining (see Fig. 3)

1. RNA staining with Pyronin Y (PY) is performed: CD34+ cells were resuspended at concentration of 2×10^6 cells/ml in PY-staining buffer, and cells were incubated for 45 min at 37 °C.
2. Cells were washed once, resuspended in the same chilled buffer, analyzed, and sorted on a FACstar (BD Biosciences, Le Pont de Claix, France).
3. The living CD34+ cells were gated on morphology, and in this gate, cells in G0 were identified by their minimal RNA content, whereas cells in G1/S/G2+M phase were defined as those with high or maximal PY staining, thus allowing isolation of viable CD34+ cells in G0 or G1/S/G2+M. The two sorting gates were well separated to avoid contamination of each population.

3.6. Transduction of Human CD34+ Cells

1. 5×10^4 CD34+ CB cells are seeded in Cellgro medium in 48-well plates and were transduced with fresh LV supernatant at MOIs of 20 or 4.
2. Cells were transduced for 72 h or after 24 h transduction cells were washed and resuspended in Cellgro medium for a further 48 h before transduction efficiency was determined by flow cytometry.

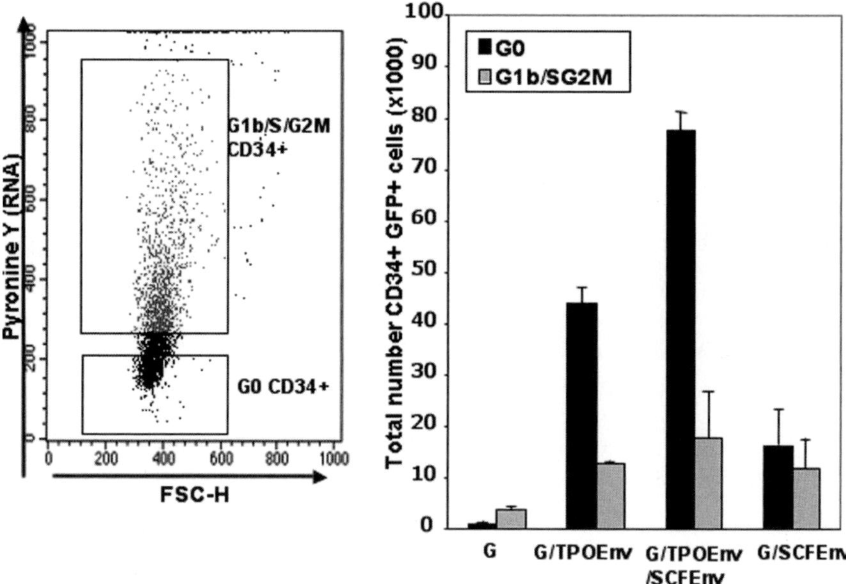

Fig. 3. TPO/SCF-displaying lentiviral vectors promote transduction of immature quiescent G0 CD34+ cells. (**A**) FACS analysis of CD34+ CB cells stained with Pyronin Y. Quiescent cells residing in the G0 phase of the cell cycle (G0 CD34+) have minimal RNA content, as indicated by low PY staining, whereas cells residing into G1b, S, G2+M phase (G1b/S/G2M) have a high RNA content. Sort windows to collect G0 and G1b/S/G2M cells are shown in the dot blot (**A**) as R1 and R2, respectively. R1 and R2 were kept well separated to avoid contamination of the two-cell categories. In (**B**), the G0 and cycling (G1b/S/G2M) CD34+ cell populations were transduced for 72 h with TPO-, SCF-, SCF/TPO-displaying vectors or with VSV-G-pseudotyped vectors. The total number of GFP+ CD34+ marked cells [= number of cells at start of infection (5×10^4) × cell expansion × % cell transduction × % cell survival] for the G0 and G1b/S/G2M populations are shown in (**B**). Data are shown as means ± SD, $n = 3$.

3.7. LTC-IC Determination

1. Day 0: Seed 60,000 MS5 cells/24-well in 1 ml DMEM-α/10% FCS.
2. Day 1: Seed 5000 transduced CD34+ cells/24-well on the MS5 stroma layer in 1 ml LTC-IC culture medium.
3. Subsequently, remove every week 500 μl of LTC-IC (*see* **Note 5**) medium and replace it for fresh LTC-IC medium up to 5 weeks after seeding the CD34+ cells.
4. After 5 weeks of culture, residual cells are harvested; 500 cells are distributed into 35-mm uncoated culture dishes containing CFC-methocult medium in duplicate. (This is done in duplicate as mentioned so 2 × 500 cells with 500 cells per dish.)

LTC-IC colonies (BFU: erythrocyte colonies; G-CFC: granulocyte colonies; M-CFC: macrophage colonies and mixed colonies; *see* **Note 6**) are scored after 14 days of culture in a moist chamber for GFP expression by a fluorescence microscope using a 488-nm excitation filter.

3.8. Evaluation of Targeted HSC Transduction by NOD/SCID Repopulating Assay (see **Fig. 4***)*

1. After a 24-h (MOI of 4) transduction with TPOHA-, SCFSUx-displaying LVs, CD34+ CB cells were injected by tail vain injection into sublethally irradiated (3.5 Gy) NOD/SCID mice without in vivo administration of cytokines.
2. 6–8 weeks post-transplantation, the bone marrow (BM) from femurs was harvested for staining of different human hematopoietic cell lineages. BM was isolated from the femurs by first piercing through the femur bone with a needle. Then about 1 ml of PBS is used to flush out the BM from the femur by means of syringe with needle. About 5–7 × 10^6 BM cells are normally harvested 5–7 × 10^6 BM cells

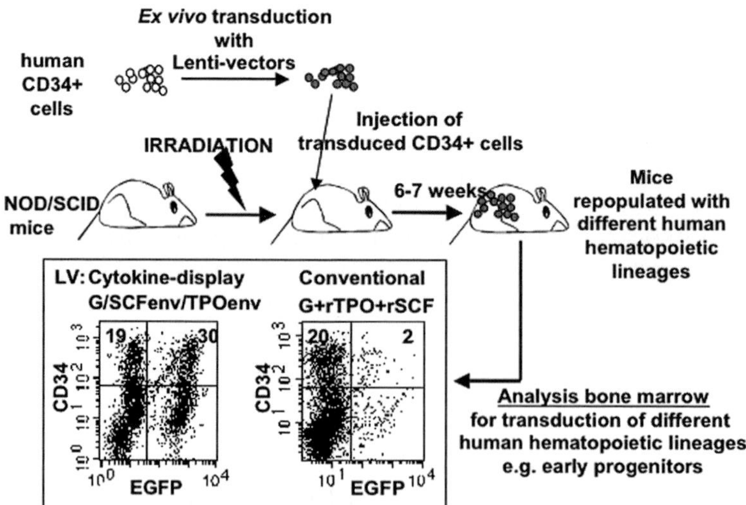

Fig. 4. Preferential transduction of NOD/SCID repopulating cells by TPO- and SCF- or TPO/SCF-displaying lentiviral vectors. NOD/SCID mice received 2 × 10^5 CD34+ cord blood cells, transduced for 24 h at an MOI of 4 with TPO- (G/TPOEnv), SCF- (G/SCFEnv), TPO/SCF- (G/TPOEnv/SCFEnv) or VSV-G pseudotyped vectors in the absence (–) or presence of the counterpart cytokines. At 7 weeks post-transplantation engraftment of transduced (GFP+) early human progenitor cells (hCD34+) in the femur BM of the mice was compared for conventional VSV-G pseudotyped and cytokine displaying vectors. The upper right quadrant shows the GFP+ cells within theCD34+ cells in the human graft (hCD45+ cells).

were incubated for 15 min with mouse IgG (1 μg/1 × 10^6cells) in PBS/2%FCS to block non-specific binding sites.

3. Then 5–7 × 10^6BM cells were incubated with 10 μl of anti-hCD45-PECy5 antibody (BD Pharmingen, Le pont de Claix, France) in 500 μl of PBS/2%FCS for 20 min to stain the population of human cells present in the mouse BM.
4. Cells were washed once with 4 ml of PBS/2% FCS, cells are pelleted down by centrifugation at 500 g.
5. The hCD45-labeled BM cells were divided into aliquots of 5 × 10^5cells in 100 μl of PBS/2%FCS and stained for different human hematopoietic lineages using anti-hCD19-, anti-hCD14-, anti-hCD15-, anti-hCD56, anti-hCD13-, and anti-hCD34-PE antibodies (3 μl of antibody/aliquot of cells). In all cases, corresponding PE-conjugated mouse IgG controls need to be used to evaluate specific labeling.
6. GFP+ cells are detected in the different human cell lineages by three-color flow cytometry analysis.

4. Notes

1. Many studies suggest that HSCs reside in a cell population expressing CD34+ antigen. Mononuclear cells from cord blood contain about 1% of CD34+ cells.
2. Together with TPOHA, the plasmid encoding neuraminidase is co-transfected to allow efficient release of virus from the producer cell. The HA envelope otherwise binds the vector particles to the producer cells through binding to its receptor sialic acid expressed on the producer cell ***(30)***.
3. Cellgro medium is adapted for culture of CD34+ cells and permits the maintenance of the functionality of the cytokines displayed on the vector surface, even after freezing.
4. Serial dilutions of 10 of the vector supernatant (e.g., 1, 10, 100 μl) are added to HeLa cells to identify the vector doses that will give no more than 10% transduced Hela cells, a dose that corresponds to only 1 vector genome integration per cell. Only this vector doses can be used for correct calculation of the titer.
5. This cell cycle fractionation is a variant of the PY (RNA)/Hoechst (DNA) staining method ***(31)***. As the DNA staining did not provide additional information, we omitted this step and shortened the procedure, which resulted in much higher cell survival.
6. Burst forming unit (BFU) colonies start appearing from 8 to 10 days of culture on and can be easily distinguished because they are red erythrocyte colonies that form huge grapes. G-CFC are granulocytes that are colonies consisting of round small white cells that are dense in the middle and more dispersed at the surroundings, whereas M-CFCs are macrophage colonies that consist of wide dispersed big white cells. Mixed colonies consist of one or two of the abovementioned cell types; e.g., GM-CFC is a colony consisting of granulocytes and macrophages. A high score of mixed colonies indicates presence of early progenitors because one cell gave rise to different cell lineages.

7. LTC-IC medium is a minimal medium that permits long-term survival of very early progenitors and eliminates the majority of lineage-committed cells.

Acknowledgments

This work was supported by the Agence Nationale pour la Recherche contre le SIDA (ANRS), the European community (contract LSHB-CT-2004-005242, "Consert"), Association Française contre les Myopathies (AFM), and INSERM. We acknowledge the contributions of our colleagues Chantal Bella, Caroline Costa, and Clémentine Schiltz to these studies.

References

1. Baumheter S, Singer MS, Henzel W, Hemmerich S, Renz M, Rosen SD, Lasky LA. (1993) Binding of L-selectin to the vascular sialomucin CD34. *Science* 262(5132):436–8.
2. Bernstein ID, Singer JW, Smith FO, Andrews RG, Flowers DA, Petersens J, Steinmann L, Najfeld V, Savage D, Fruchtman S, et al. (1992) Differences in the frequency of normal and clonal precursors of colony-forming cells in chronic myelogenous leukemia and acute myelogenous leukemia. *Blood* 79(7):1811–6.
3. Stewart AK, Imrie K, Keating A, Anania S, Nayar R, Sutherland DR. (1995) Optimizing the CD34+ and CD34+Thy-1+ stem cell content of peripheral blood collections. *Exp Hematol* 23(14):1619–27.
4. Fritsch G, Stimpfl M, Buchinger P, Printz D, Sliutz G, Wagner T, Agis H, Valent P, Gadner H. (1994) Does cord blood contain enough progenitor cells for transplantation? *J Hematother* 3(4):291–8
5. Bender JG, Unverzagt KL, Walker DE, Lee W, Van Epps DE, Smith DH, Stewart CC, To LB. (1991). Identification and comparison of CD34-positive cells and their subpopulations from normal peripheral blood and bone marrow using multicolor flow cytometry. *Blood* 77(12):2591–6.
6. Miller DG, Adam MA, Miller AD. (1990) Gene transfer by retrovirus vectors occurs only in cells that are actively replicating at the time of infection. *Mol Cell Biol* 10(8):4239–42.
7. Blomer U, Naldini L, Kafri T, Trono D, Verma IM, Gage FH. (1997) Highly efficient and sustained gene transfer in adult neurons with a lentivirus vector. *J Virol* 71(9):6641–9.
8. Naldini L, Blomer U, Gallay P, Ory D, Mulligan R, Gage FH, Verma IM, Trono D. (1996) In vivo gene delivery and stable transduction of nondividing cells by a lentiviral vector. *Science* 272(5259):263–7.
9. Reiser J, Harmison G, Kluepfel-Stahl S, Brady RO, Karlsson S, Schubert M. (1996) Transduction of nondividing cells using pseudotyped defective high-titer HIV type 1 particles. *Proc Natl Acad Sci USA* 93(26):15266–71.

10. Kafri T, Blomer U, Peterson DA, Gage FH, Verma IM. (1997) Sustained expression of genes delivered directly into liver and muscle by lentiviral vectors. *Nat Genet* 17(3):314–7.
11. Verhoeyen E, Dardalhon V, Ducrey-Rundquist O, Trono D, Taylor N, Cosset FL. (2003) IL-7 surface-engineered lentiviral vectors promote survival and efficient gene transfer in resting primary T lymphocytes. *Blood* 101(6):2167–74.
12. Maurice M, Verhoeyen E, Salmon P, Trono D, Russell SJ, Cosset FL. (2002) Efficient gene transfer into human primary blood lymphocytes by surface-engineered lentiviral vectors that display a T cell-activating polypeptide. *Blood* 99(7):2342–50.
13. Ducrey-Rundquist O, Guyader M, Trono D. (2002) Modalities of interleukin-7 induced human immunodeficiency virus permissiveness in quiescent T lymphocytes. *J Virol* 76(18):9103–11.
14. Korin YD, Zack JA. Progression to the G1b phase of the cell cycle is required for completion of human immunodeficiency virus type 1 reverse transcription in T cells. *J Virol* 72(4):3161–8.
15. Dardalhon V, Jaleco S, Kinet S, Herpers B, Steinberg M, Ferrand C, Froger D, Leveau C, Tiberghien P, Charneau P, Noraz N, Taylor N. (2001) IL-7 differentially regulates cell cycle progression and HIV-1-based vector infection in neonatal and adult CD4+ T cells. *Proc Natl Acad Sci USA* 98(16):9277–82.
16. Case SS, Price MA, Jordan CT, Yu XJ, Wang L, Bauer G, Haas DL, Xu D, Stripecke R, Naldini L, Kohn DB, Crooks GM. (1999) Stable transduction of quiescent CD34(+)CD38(-) human hematopoietic cells by HIV-1-based lentiviral vectors. *Proc Natl Acad Sci USA* 96(6):2988–93.
17. Miyoshi H, Smith KA, Mosier DE, Verma IM, Torbett BE. (1999) Transduction of human CD34+ cells that mediate long-term engraftment of NOD/SCID mice by HIV vectors. *Science* 283(5402):682–6.
18. Guenechea G, Gan OI, Inamitsu T, Dorrell C, Pereira DS, Kelly M, Naldini L, Dick JE. (2000) Transduction of human CD34+ CD38- bone marrow and cord blood-derived SCID-repopulating cells with third-generation lentiviral vectors. *Mol Ther* 1(6):566–73.
19. Uchida N, Sutton RE, Friera AM, He D, Reitsma MJ, Chang WC, Veres G, Scollay R, Weissman IL. (1998) HIV, but not murine leukemia virus, vectors mediate high efficiency gene transfer into freshly isolated G0/G1 human hematopoietic stem cells. *Proc Natl Acad Sci USA* 95(20):11939–44.
20. Follenzi A, Ailles LE, Bakovic S, Geuna M, Naldini L. (2000) Gene transfer by lentiviral vectors is limited by nuclear translocation and rescued by HIV-1 pol sequences. *Nat Genet* 25(2):217–22.
21. Zielske SP, Gerson SL. (2003) Cytokines, including stem cell factor alone, enhance lentiviral transduction in nondividing human LTCIC and NOD/SCID repopulating cells. *Mol Ther* 7(3):325–33.
22. Sutton RE, Reitsma MJ, Uchida N, Brown PO. (1999) Transduction of human progenitor hematopoietic stem cells by human immunodeficiency virus type 1-based vectors is cell cycle dependent. *J Virol* 73(5):3649–60.

23. Kittler EL, Peters SO, Crittenden RB, Debatis ME, Ramshaw HS, Stewart FM, Quesenberry PJ. (1997) Cytokine-facilitated transduction leads to low-level engraftment in nonablated hosts. *Blood* 90(2):865–72.
24. Ailles L, Schmidt M, Santoni de Sio FR, Glimm H, Cavalieri S, Bruno S, Piacibello W, Von Kalle C, Naldini L. (2002) Molecular evidence of lentiviral vector-mediated gene transfer into human self-renewing, multi-potent, long-term NOD/SCID repopulating hematopoietic cells. *Mol Ther* 6(5):615–26.
25. Robert D, Mahon FX, Richard E, Etienne G, de Verneuil H, Moreau-Gaudry F. (2003) A SIN lentiviral vector containing PIGA cDNA allows long-term phenotypic correction of CD34+-derived cells from patients with paroxysmal nocturnal hemoglobinuria. *Mol Ther* 7(3):304–16.
26. Amsellem S, Ravet E, Fichelson S, Pflumio F, Dubart-Kupperschmitt A. (2002) Maximal lentivirus-mediated gene transfer and sustained transgene expression in human hematopoietic primitive cells and their progeny. *Mol Ther* 6(5):673–7.
27. Nguyen TH, Pages JC, Farge D, Briand P, Weber A. (1998) Amphotropic retroviral vectors displaying hepatocyte growth factor-envelope fusion proteins improve transduction efficiency of primary hepatocytes. *Hum Gene Ther* 9(17):2469–79.
28. Maurice M, Mazur S, Bullough FJ, Salvetti A, Collins MK, Russell SJ, Cosset FL. (1999) Efficient gene delivery to quiescent interleukin-2 (IL-2)-dependent cells by murine leukemia virus-derived vectors harboring IL-2 chimeric envelope glycoproteins. *Blood* 94(2):401–10.
29. Verhoeyen E, Wiznerowicz M, Olivier D, Izac B, Trono D, Dubart-Kupperschmitt A, Cosset FL. (2005) Novel lentiviral vectors displaying "early-acting cytokines" selectively promote survival and transduction of NOD/SCID repopulating human hematopoietic stem cells. *Blood* 106(10):3386–95.
30. Sandrin V, Boson B, Salmon P, Gay W, Negre D, Le Grand R, Trono D, Cosset FL. (2002) Lentiviral vectors pseudotyped with a modified RD114 envelope glycoprotein show increased stability in sera and augmented transduction of primary lymphocytes and CD34+ cells derived from human and nonhuman primates. *Blood* 100(3):823–32.
31. Wilpshaar J, Falkenburg JH, Tong X, Noort WA, Breese R, Heilman D, Kanhai H, Orschell-Traycoff CM, Srour EF. (2000) Similar repopulating capacity of mitotically active and resting umbilical cord blood CD34(+) cells in NOD/SCID mice. *Blood* 96(6):2100–7.

8

Fiber-modified Adenoviruses for Targeted Gene Therapy

Hongju Wu and David T. Curiel

Summary

Human adenovirus serotype 5 (Ad5) has been widely explored as a gene delivery vector. To achieve highly efficient and specific gene delivery, it is often necessary to re-direct Ad5 tropism. Because the capsid protein fiber plays an essential role in directing Ad5 infection, our laboratory attempted to re-target Ad5 through fiber modification. We have developed two strategies in this regard. One is a bi-specific adaptor protein strategy, in which the adaptor protein is designed to bind both the Ad5 fiber and an alternative cell-surface receptor. Another is genetic modification, in which alternative targeting motifs are genetically incorporated into the fiber knob domain so that the Ad5 vectors can infect cells through the alternative receptors. In this chapter, we will focus on the genetic fiber modification strategy and provide a detailed protocol for generation of fiber-modified Ad5 vectors. A series of techniques/procedures used in our laboratory will be described, which include the generation of fiber-modified Ad5 genome by homologous recombination in a bacterial system, rescuing the modified Ad5 viruses, virus amplification and purification, and virus titration.

Key Words: Gene therapy; adenovirus; fiber modification; homologous recombination; rescuing adenoviral vectors; adenovirus amplification; CsCl gradient purification.

1. Introduction

Adenoviruses have several advantages as gene delivery vectors, which include minor pathological effect, efficient infection of both dividing and non-dividing cells, easy manipulation and production in vitro, and high capacity of transgene insert. Current adenovirus-mediated gene therapy protocols are mostly based on human subgroup C adenoviruses, in particular, the human adenovirus serotype 5 (Ad5).

From: *Methods in Molecular Biology, vol. 434: Volume 2: Design and Characterization of Gene Transfer Vectors*
Edited by: J. M. Le Doux © Humana Press, Totowa, NJ

Ad5 particles have an icosahedral morphology, in which the viral genome is packaged in a capsid composed of major capsid protein hexon (II), penton base (III), and fiber (IV), and several minor proteins including protein IIIa, VI, VIII, and IX (*see* **Fig. 1**). The Ad5 genome is a linear, double-stranded DNA with approximately 36 kilobases (kb) in size and is stabilized by its interaction with core proteins V, VII, and mu. Ad5 infection is primarily mediated by binding of its fiber protein to the coxsackie virus-adenovirus receptor (CAR) expressed on cell surface ***(1–3)***. Binding of fiber to CAR leads to subsequent interaction between the RGD epitope of penton base and αV integrins, which in turn triggers internalization of Ad5 viruses ***(4–6)***. Clearly, fiber is essential in determining the cells that Ad5 infect. The Ad5 native receptor, CAR, is expressed in various cell types at various levels. This is beneficial when widespread gene delivery is required. However, for therapeutic purposes, very often specific gene delivery into the target/diseased cells is needed. Targeted gene delivery is thus one of the most challenging tasks in the development of gene therapy protocols. Because Ad5 fiber is critical in directing Ad5 infection, fiber modification is considered an effective way to re-direct Ad5 tropism ***(7–9)***.

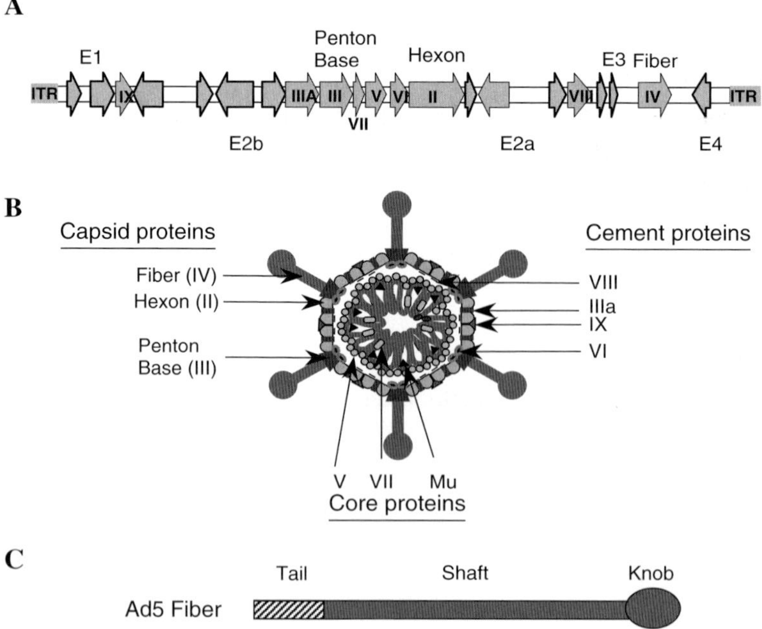

Fig. 1. Structural diagram of adenovirus serotype 5 (Ad5). (**A**) Ad5 genome structure; (**B**) Ad5 virion and structural components; (**C**) Diagram of Ad5 fiber protein.

Thus far, two strategies have been developed to re-direct Ad5 tropism. One is bi-specific adaptor protein strategy, in which the adaptor protein is designed to bind both the Ad5 vectors and an alternative cell-surface receptor ***(10,11)***. Typically, the adaptor protein is composed of a CAR domain that binds to the knob domain of Ad5 fiber and a ligand that binds to a cell-surface receptor that is highly enriched in the target cells. Another strategy is genetic modification, in which alternative targeting motifs are genetically incorporated into the fiber knob domain so that the Ad5 vectors can infect cells through the alternative receptors ***(12–18)***. This chapter is focused on the genetic modification of Ad5 fiber.

Ad5 fiber protein is composed of an N-terminal tail domain that associates with penton base, a shaft domain consisting of 22 repeats of a 15-amino acid residue motif, and a C-terminal knob domain that binds to the cell-surface receptor CAR (*see* **Fig. 1C**). The knob domain is responsible for receptor binding, and thus has been the major focus of genetic fiber modification. Two locales in fiber have been explored for genetic incorporation of foreign targeting motifs—the HI loop and the C-terminal end of fiber ***(12–18)***. The size of foreign peptides that can be genetically incorporated into the HI loop has been investigated, and it has been demonstrated that polypeptides as large as 83 amino acid residues can be incorporated into the HI loop with only marginal negative consequences on the key properties of the Ad5 vectors ***(17)***. The size limit for peptide incorporation at the C-terminal end of fiber has not been defined thus far.

In this chapter, we provide a detailed protocol for generating fiber-modified Ad5 vectors. A series of techniques/procedures used in our laboratory will be described, which include the generation of fiber-modified Ad5 genome by homologous recombination in a bacterial system, rescuing the modified Ad5 viruses, virus amplification and purification, and virus titration.

2. Materials

2.1. Cell Culture, Virus Amplification, and Purification

1. 293 cells expressing Ad-E1 genes are obtained from American Type Culture Collection (ATCC).
2. Dulbecco's modified Eagle's medium (DMEM)/F12 (Sigma, St. Louis, MO, USA) complete culture medium: 10% fetal bovine serum (FBS), 2 mM L-glutamine, and a mixture of penicillin (100 IU/ml) and streptomycin (100 μg/ml).
3. 2% DMEM/F12 medium: 2% FBS, 2 mM L-glutamine, and a mixture of penicillin (100 IU/ml) and streptomycin (100 μg/ml).
4. 2× DMEM medium: 20% FBS, 4 mM L-glutamine and a mixture of penicillin (200 IU/ml) and streptomycin (200 μg/ml).

5. Phosphate-buffered saline (PBS): 0.144 g/l KH_2PO_4, 9.00 g/l NaCl, and 0.795 g/l Na_2HPO_4, pH 7.4.
6. 0.05% Trypsin–ethylenediaminetetraacetic acid (EDTA).
7. Lipofectamine™ 2000 (Invitrogen, Carlsbad, CA, USA).
8. CsCl 1.33: 454.2 mg CsCl per ml in 5 mM HEPES, pH 7.8. Make 250 ml and sterilize with filters.
9. CsCl 1.45: 609.0 mg CsCl per ml in 5 mM HEPES, pH 7.8. Make 250 ml and sterilize with filter.
10. Dialysis buffer: 10% glycerol in PBS. Filter sterilization (optional).
11. Virion lysis buffer (VLB): 0.1% sodium dodecyl sulfate (SDS), 10 mM Tris–HCl (pH 7.4), 1 mM EDTA.
12. SeaPlaque agarose (low-melting temperature agarose).

2.2. Molecular Biology

1. Restriction enzymes including Swa I, Pac I, EcoR I, Kpn I, Nde I, and their buffers.
2. T4 DNA Ligase and buffer.
3. DNA preparation kits including minipreps, maxipreps, and gel purification.
4. Bacterial strains BJ5183 (Stratagene, La Jolla, CA, USA), DH10B (Invitrogen), and XL-1 blue (Stratagene).
5. Plasmid pNEB193 (New England Biolabs, Ipswich, MA, USA).
6. Phenol : chloroform : isoamyl alcohol (25:24:1) saturated with 10 mM Tris, pH 8.0, 1 mM EDTA.
7. LB liquid media: 10 g/l Tryptone, 5 g/l sodium chloride, and 5 g/l yeast extract. Sterilize.
8. 6× DNA-loading buffer: 0.25% bromophenol blue, 0.25% xylene cyanol FF, and 30% glycerol in water.
9. Ethidium bromide solution, 10 mg/ml.
10. Agarose.

3. Methods

3.1. Generation of Fiber-Modified Ad5 Genome

The intact Ad5 genome is about 36 kb in size. Several versions of Ad5 genome in plasmid format have been developed (such as pTG3602 and pAdeasy-1). This is a major advancement for Ad5-based study because it has greatly simplified the manipulation of Ad5 in vitro. It allows the Ad5 genome to be amplified in bacteria, purified using plasmid preparation methods, and manipulated using molecular biology techniques. The plasmids containing Ad5 genome usually range from 35 to 40 kb, which is too large for direct modification using common digestion and ligation techniques. Our laboratory has adopted a bacteria-based homologous recombination strategy in generating fiber-modified Ad5 genome (*see* **Fig. 2**). Two plasmid vectors are needed to accomplish the genetic fiber modification. One is a large plasmid (backbone plasmid) containing Ad5 genome (that is, pAd5 backbone). To facilitate desired recombination, the backbone needs to be linearized in the position of fiber gene. Another vector is a small plasmid (shuttle vector, named pFiberShuttle) containing fiber gene and its flanking sequences. The fiber gene in the shuttle vector can be modified using conventional molecular techniques, and the

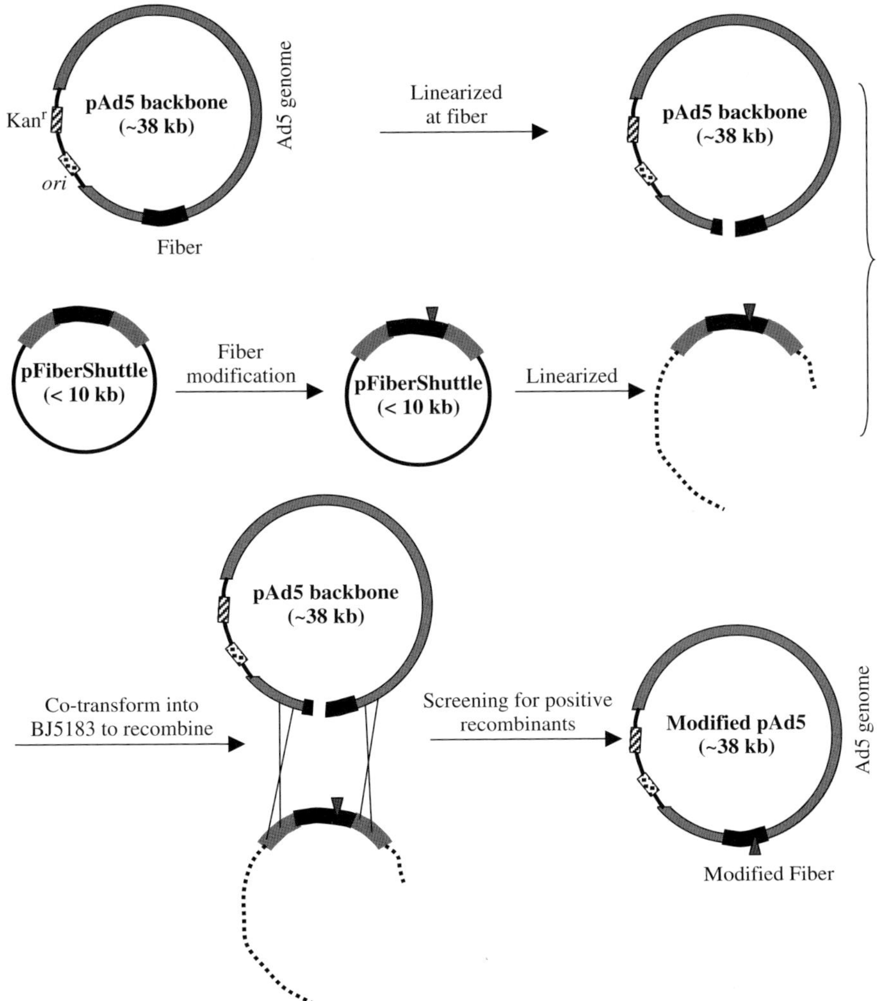

Fig. 2. Diagram illustrating homologous recombination-based genetic fiber modification strategy.

fiber-flanking sequences can be used for homologous recombination with the Ad5 backbone plasmid so that the modified fiber gene can be incorporated into the Ad5 genome. In addition, a bacteria strain that supports homologous recombination is needed. BJ5183 has been widely used in this regard, and it has been working well in our laboratory.

3.1.1. Generation of Ad5 Backbone for Fiber Modification

In this protocol, we will describe the procedure of generating a fiber-modified Ad5 vector using the Ad5 backbone plasmid lacking the E1 region. The E1-deleted vectors (also called the first generation of Ad5 vector) are replication deficient because the E1 early genes are essential in regulating viral gene expression and replication. In complementary cell lines, for example, human embryonic kidney 293 cells that stably express the E1 genes, Ad5 viruses can form and replicate to high titers. For experimental and therapeutic studies, the reporter genes or therapeutic genes can be incorporated into the E1 region. Incorporation of transgenes into the E1 region is described in commercially available kits and the manufacturers' protocols, and will not be discussed here. This protocol can be applied either before or after a transgene is incorporated into the E1 region. Here, as an example, we will describe a fiber-modification strategy using a pTG3602-derived backbone, in which the E1 region was replaced with CMV-promoter-driven reporter genes GFP and Luciferase—referred to as pAd5 herein.

To linearize the backbone at the position of fiber gene, a unique restriction enzyme site is needed. As a large plasmid, there are very few enzyme sites in the pAd5 backbone plasmids that cut uniquely. In fact, there is none in fiber that can be uniquely cut by any commercially available enzymes. Therefore, we attempted to introduce one into pAd5 fiber gene. Swa I, a commercially available, 8-cutter enzyme, was chosen to be incorporated into the fiber gene because its recognition site, ATTTAAAT, is not in pAd5 and is rarely found in other genes. In addition, we noticed there are only two sites for the restriction enzyme Nde I in the Ad5 genome, one of which is located at the 5′ end of the fiber gene. This is about the best site we could use to introduce Swa I site in. It should be noted that, although there are only two Nde I sites in Ad5 genome, there may be more sites in different vectors because of the non-Ad5 genome portion of the plasmid, and/or the transgenes incorporated into the Ad5 E1 region. For example, there are totally two Nde I sites in the original pTG3602, three Nde I sites in pAdeasy-1, and four Nde I sites in pAd5 we are using here (two more in the CMV-promoters that we introduced into the E1 region to drive the reporter gene expression).

A protocol for the generation of pAd5 containing a unique Swa I site in the fiber gene is provided below. It should be noted that other strategies may also be employed for this purpose. For example, polymerase chain reaction (PCR)-based, three-piece ligation strategy similar to what we used in hexon modification may also be designed for fiber modification ***(19)***.

1. Design and order an Nde I-Swa I linker—that is, a pair of oligonucleotides that contain a Swa I site and the sticky ends suitable for ligation into the Nde I-digested ends. For example, the following pair of oligos can be used for this purpose:

   5′ TACGCATTTAAATCGG   3′
   3′   GCGTAAATTTAGCCAT 5′

2. Partial digestion of pAd5 with Nde I. Set the reaction as following:

   Purified plasmid pAd5: 5 μg
   10× NEBuffer 4:5 μl
   Nde I (10 units/ul): 1 μl
   Add dd.H_2O to 50 μl

   Mix the reagents and briefly centrifuge the tube so that the liquid is collected at the bottom of the microcentrifuge tube. Incubate the tube at 37 °C for 30 min (*see* **Note 1**).
3. Add 10 μl of 6× loading dye into the reaction tube and load the sample on a pre-prepared 1.0% agarose gel containing ethidium bromide. Run gel electrophoresis for 1.0–1.5 h at 80 V (constant voltage) (*see* **Note 2**).
4. Cut out the largest linearized band under a UV transilluminator.
5. Purify the fragment with commercially available large-construct purification kits according to the manufacturer's protocols. Alternatively, the conventional phenol extraction method may be used.
6. Prepare the linker:
   a) Take 1 μl of each oligo stock (usually 100 pmol/ul) and dilute to a concentration of 1 pmol/ul.
   b) Mix 10 μl of each diluted oligo in a microcentrifuge tube (the final concentration for the linker is thus 0.5 pmol/μl).
   c) Incubate at 70°C for 5 min to denature the oligos.
   d) Turn off the heater and allow the tube to cool down to room temperature naturally. In this process, the two oligos will anneal.
7. Ligate the linker with the Nde I-digested and purified pAd5 backbone fragment. The molar ratio of insert: vector works better around 3:1, but may vary because of variable yield of purification and estimation of fragment concentration. Therefore, it is necessary to use various ratios. If the pAd5 backbone is finally eluted/dissolved in 40 μl of dd.H_2O, the concentration for the backbone should be approximately 0.01 pmol/μl. Two reactions can be set as following:

	a)	b)
pAd5/Nde I digested:	10 μl (=0.1 pmol)	10 μl (=0.1 pmol)
Linker:	0.5 μl (=0.25 pmol)	1 μl (=0.5 pmol)
10× ligation buffer	2 μl	2 μl
T4 DNA Ligase	1 μl	1 μl
dd.H_2O	6.5 μl	6.0 μl

Mix well and incubate at 14 °C overnight.

8. Transformation into DH10B by electroporation method according to manufacturer's protocol (*see* **Notes 3–5**).
9. To screen for positive clones, first pick 6–12 colonies and make minipreps using plasmid miniprep kits according to manufacturer's protocols.
10. Next, perform Swa I digestion. Check the digestion using agarose gel as described above. The potential right clones should show only one large linearized band (~38 kb) resulting from one cut by Swa I digestion.
11. Send the DNA of the potential right clones to a sequencing facility for sequencing analysis. It would be easier to get reliable sequencing data if the sequencing primers are 50–200 bases upstream of the modification site.
12. Pick one or two positive clones that are confirmed by sequencing analysis for maxiprep using maxiprep kits according to manufacturer's protocol. The resultant backbone is named pAd5_Swa I herein (*see* **Note 6**).

3.1.2. Generation of Fiber Shuttle Vector

As mentioned above, the required elements for a fiber shuttle vector include the fiber gene and its flanking sequences. The size of the flanking sequences may vary. We recommend it ranges from 1.0 to 3.0 kb. The reason is simple: the longer the homologous regions, the better chance for homologous recombination. However, if the homologous regions are too long, multiple recombinations may occur, which reduces the possibility of desired recombination. Following is an example of how a fiber shuttle vector may be created.

1. Cut out the 3.5 kb EcoR I–Kpn I fragment that contains the fiber gene and its flanking sequences (about 1.0 kb on each side) from the pAd5 backbone plasmid (pTG3602 based). For pAdeasy-1 (E3 deleted) vector, this strategy should work as well. The only difference is in pAdeasy-1; the E3 region containing the above-mentioned EcoR I site was deleted. However, the EcoR I site upstream of the E3 region can now be used, and similarly, the EcoR I–Kpn I fragment can be used to generate the fiber shuttle vector. To isolate the right EcoR I–Kpn I fragment, it is necessary to perform sequential digestions (for example, EcoR I digestion first, isolate the fragment including fiber, then Kpn I digestion, and isolate the right fragment).

2. Ligate the EcoR I–Kpn I fragment-containing fiber and its flanking sequences into pNEB193 vector digested with EcoR I and Kpn I (*see* **Note 7**).
3. Transform into bacterial strain XL-1 blue either by electroporation or by chemical transformation, and plate on Ampcillin-containing LB-agar plates.
4. To screen for positive clones, pick six colonies for minipreps, and perform EcoR I + Kpn I digestion. Run an agarose gel to see whether they contain the right sizes of insert and vector. Finally, send a couple of positive clones for sequencing to confirm. The resultant vector is named pFiberShuttle herein.
5. Make plasmid maxipreparation for pFiberShuttle using manufacturer's protocols.
6. Perform any desired fiber modification on this fiber shuttle vector using conventional molecular biology techniques.

3.1.3. Homologous Recombination in Bacterial System

1. Digest the backbone pAd5_Swa I with Swa I and purify the linearized fragment as described above (*see* **Note 5**).
2. Digest the modified shuttle vector with EcoR I + Kpn I, and purify the fragment-containing fiber and its flanking sequences. Alternatively, the shuttle vector may be linearized with any enzyme that cuts outside the fragment-containing fiber and its flanking sequences. The insert does not have to be cut out, but the plasmid needs to be linearized to facilitate homologous recombination.
3. Co-transform the pAd5 backbone and fiber shuttle fragments obtained above into bacterial strain BJ5183 using electroporation method (*see* **Note 4**). A molar ratio of 10:1 for shuttle vector : pAd5 backbone is recommended for efficient homologous recombination. Plate the co-transformed bacteria in an LB + Agar plate containing appropriate antibiotics (should be the same antibiotics as for the pAd5 backbone).
4. Incubate at 37 °C for 24 h.
5. Screen for clones containing the right modifications using PCR method directly on the bacterial colonies.

    a) PCR primers: design primers flanking the modification region. Take the size of the amplified region into consideration. The region should be easily amplified and can be easily distinguished from unmodified counterparts by electrophoresis on an agarose gel. Usually, a region of approximately 500 bp works well in this matter.
    b) Pick a colony from the recombination plate using a pipet tip; first, inoculate it on a new plate (simply touch the tip on the surface of a new plate) and mark the touched place with the colony number. Then, soak the tip in a microcentrifuge tube containing 20 μl of 0.1% SDS. Do the same for a few more colonies. The new plate is called a master plate for the colonies. Incubate the master plate at 37 °C overnight to allow the colonies to grow.
    c) Perform PCR using 1–2 μl of the bacterial colonies dissolved in 0.1% SDS as template.

   d) Run agarose gel for the PCR product and determine what colonies contain the right size of PCR product. If the PCR product is about 500 bp and the modification gives rise a >50 bp difference, on a 1.5% or higher agarose gel, the products can be easily distinguished.
6. Pick the colonies containing the right size of PCR product from the master plate and inoculate for minipreps (*see* **Note 8**).
7. Digest the miniprep DNA with Pac I to see whether they contain appropriate sizes of DNA fragment [right clones should contain a large fragment (~35 kb, Ad5 genome) and a small fragment (2.8 kb, prokaryotic plasmid part, which may be different in other backbones)].
8. Electroporate bacterial strain DH10B with 0.5–1.0 μl of the miniprep DNA of potential right recombinants (2–4 clones may be processed). Plate them on appropriate LB+ agar plates (*see* **Note 9**).
9. Pick two colonies from each plate and inoculate for minipreps (commercial kits may be used for this step).
10. Re-check the clones by using PCR and digestions as described above. The positive clones should be further confirmed by sequencing analysis.
11. Make maxiprep DNA for the identified right clones.

3.2. Rescuing Ad5 Viral Vectors in Mammalian Cells

3.2.1. Prepare Ad5 Genome for Virus Rescue

1. Digest the modified Ad5 vector with Pac I. Set reaction as following:

   Purified plasmid (the Ad5 vector obtained above): 20 μg
   10× NEB 1: 20 μl
   BSA: 2ul
   Pac I (10 units/ul): 5 μl
   Add dd.H_2O to 200 μl

   Mix the reagents and briefly centrifuge the tube so that the liquid is collected at the bottom of the microcentrifuge tube. Incubate the tube at 37 °C for 3 h.
2. Purify the digested DNA with phenol extraction method (*see* **Note 10**).

   a) For each reaction (200 μl), add 200 μl of buffered phenol, mix well (vortex is not recommended because of large Ad5 genome. Vortex may break DNA);
   b) Centrifuge for 1 min at 15,000 × *g* on a table-top microcentrifuge. Transfer the supernatant (top layer, containing DNA) to a new tube;
   c) Repeat **steps a)** and **b)** one more time;
   d) Add 20 μl of 3 M NaAc into the tube, mix well; then add 2.5 vol (~500 μl) of 100% ethanol, mix well; Incubate at –80 °C for 15 min;
   e) Centrifuge for 10 min at 15,000 × *g* on the table-top microcentrifuge;
   f) Remove the supernatant and add 500 μl of 70% ethanol to wash the pellet once. The washing step includes resuspending the pellet, centrifuging at 15,000 × *g* for 5 min, and removing the supernatant;

g) Air-dry the pellet (takes about 10 min);
h) Dissolve the pellet in 100 μl of dd.H_2O or buffer EB (Qiagen, Valencia, CA, USA).

3.2.2. Transfection Into 293 Cells (for E1-Deleted Vectors)

Plate 293 cells in 60-mm dishes or T25 flasks at a density of 7.5×10^5 cells per dish or flask. When the cells reach 50–70% of confluency, perform transfections as following (for each dish or flask):

1. Mix Lipofectamine™ 2000 gently before use and dilute 20 μl of Lipofectamine™ 2000 in 500 μl Opti-MEM® I (or DMEM/F12 medium without serum). Incubate for 5 min at room temperature.
2. Dilute 5 μg of Pac I-digested and phenol-purified DNA (obtained from 3.2.1) in 500 μl opti-MEM. Mix gently.
3. Combine the diluted DNA with diluted Lipofectamine™ 2000 (total volume 1 ml). Mix gently and incubate for 20 min at room temperature.
4. Replace the culture media in each dish or flask containing the 293 cells with 5 ml fresh complete culture media and add the DNA-lipofectamine complex (1 ml) into each dish. Mix gently by rocking the dish or flask back and forth several times.
5. Incubate the cells at 37 °C in a humidified CO2 (5%) incubator.

3.2.3. Overlay the Transfected Cells with Low-Melting Agarose (Optional)

This procedure is designed for 60-mm dishes and is required if plaque isolation is needed.

1. Prepare 1.8% SeaPlaque agarose: add 1.8 g SeaPlaque agarose into 100 ml sterile H_2O, melt the agarose in a microwave, and cool it to 37 °C in a waterbath.
2. Prepare complete 2× DMEM, warm up to 37 °C.
3. Mix 10 ml of the 1.8% of SeaPlaque agarose with 10 ml of complete 2× DMEM media.
4. Aspirate the media from the transfected cells and add 3 ml of the SeaPlaque–DMEM mixture (~37 °C) to each 60-mm dish. The mixture should be added carefully from the side of the dish so that it would not disturb the cell layer.
5. Incubate the cells at 37 °C in a humidified 5% CO_2 incubator overnight.
6. Add 3 ml of complete DMEM/F12 media into each dish containing the agarose-coated cells and continue cell culture in the humidified 37 °C, 5% CO_2 incubator.

3.2.4. Plaque Formation and Collection

1. Refresh the transfected cells every 3–4 days. For agarose-coated cells, aspirate the liquid media from the dishes. Be careful not to disrupt the agarose layer. Add 3 ml of fresh complete DMEM media into each dish carefully. Return the cells to the incubator.

2. Examine the cells using a light microscope regularly (daily). 293 cells infected by adenoviruses will round up and form grape-like clusters because of toxicity from massive virus amplification, which is so-called cytopathic effect (CPE). At first, individual cells that are transfected with intact Ad5 genome start to produce Ad5 virions. When the amount of viruses becomes overwhelming, these cells will break down and release the virus content, which subsequently infect the surrounding cells. When CPE spreads over the surrounding cells, clear plaques will appear (*see* **Fig. 3**). Usually, plaques will become visible under a light microscope 7–10 days after transfection. For some modifications, however, it may take 2–3 weeks.
3. Plaque collection: when most of the cells in a dish or flask become CPE, collect the cells. First scrape off the cells, then transfer the cells with media into a 15-ml sterile polypropylene centrifuge tube and centrifuge for 5 min at approximately 200 × *g*, 4 °C in Beckman GS-GR Centrifuge or equivalent. Remove the media and resuspend the cells in 1–2 ml of 2% media. For agarose-coated dishes, the agarose needs to be removed before collecting the cells with viruses. To do so, first aspirate the liquid media, then move a 200-μl pipet tip around the dish to loose the edge of the agarose, and slightly tilt the dish so that the agarose coat can slide down (to a waste dish). Add 1–2 ml of 2% media into the dish with cells and scrape off the cells with a cell scraper.
4. Perform four times of freeze–thaw cycle. Use dry ice to freeze and 37 °C waterbath to thaw. Vortex for 10 s each time after thawing.

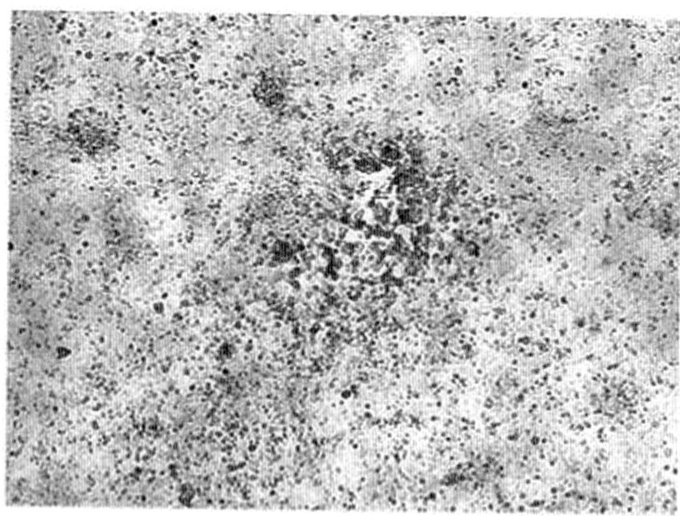

Fig. 3. An example of plaque formation. The plaque shown here is about 10 days after transfection of the Ad5 genome into 293 cells. A clear plaque is observed under phase contrast microscopy.

5. Centrifuge for 20 min at 4000 × *g*, 4 °C in Beckman GS-GR Centrifuge or equivalent. Transfer the supernatant to a microcentrifuge tube and store at –80 °C until use. Discard the pellet (cell debris). Alternatively, the viruses can be stored at –80 °C after the fourth freeze. They can be thawed and centrifuged right before use.

3.3. Virus Amplification and Purification

3.3.1. First Amplification: From Plaque Mixtures to 2× T75 Flasks

1. Grow 293 cells to approximately 80% confluence in two 75-cm^2 (T75) flasks in DMEM/F12 complete culture media.
2. Dilute viral stock (the plaque mixtures obtained above) in 10 ml of 2% DMEM/F12 media. Aspirate culture media from the T75 flasks containing 293 cells and add 5 ml of diluted viruses into each flask (*see* **Note 11**).
3. Incubate the cells at 37 °C, 5% CO_2 for 2–3 h.
4. Add 10 ml of complete DMEM/F12 media into each flask.
5. Incubate at 37 °C, 5% CO_2 until complete CPE is reached in the flasks.
6. Harvest cells and media from the flasks into 50-ml capped centrifuge tubes.
7. Centrifuge at 4 °C, approximately 200 × *g* for 5 min in Beckman GS-GR Centrifuge or equivalent.
8. Aspirate off media and resuspend the cell pellet in 2 ml of 2% DMEM/F12 media (*see* **Note 12**).
9. Perform four times of freeze–thaw cycle, and centrifuge at 4 °C, 4,000 g for 20 min to get rid of the cell debris. Store the supernatant at –80 °C.

3.3.2. Large Scale Viral Preparation

1. Grow 293 cells to approximately 80% confluence in twelve 185-cm^2 (T185) flasks in DMEM/F12 complete media.
2. Dilute viral stock of first amplification in 120 ml of 2% DMEM/F12 media. Aspirate culture media from the T185 flasks and add 10 ml into each flask (*see* **Note 11**).
3. Incubate the cells at 37 °C, 5% CO_2 for 2–3 h.
4. Add 20 ml of complete DMEM/F12 media into each T185 flask containing the infected 293 cells.
5. Incubate the cells at 37 °C, 5% CO2 until complete CPE is reached in each flask (48–72 h).
6. Harvest cells and media from the T185 flasks in 50-ml capped centrifuge tubes. Several 50-ml tubes will be needed. They should be combined after centrifugation and resuspension in 2% media (following).
7. Centrifuge at 4 °C, approximately 200 × *g* for 10 min in Beckman GS-GR Centrifuge or equivalent.
8. Aspirate off media and resuspend cell pellets in 2% DMEM/F12 media. Use a combined volume of approximately 8 ml for all the infected cells.
9. Perform four times of freeze–thaw cycle. Store the sample at –80 °C following the fourth freezing cycle until purification.

3.3.3. CsCl Purification of Ad5 Vectors

1. Thaw the frozen virus sample (obtained above) at 37 °C, vortex for 10–20 s, and centrifuge at 4 °C, approximately 4000 × *g* for 20 min in Beckman GS-GR Centrifuge or equivalent, and harvest the supernatant. Store the sample at 4 °C while setting up CsCl gradients.
2. Set up CsCl gradient in two Beckman ultra-clear centrifuge tubes for SW41 rotor as following: first add 4 ml CsCl 1.33 in the tube; then, carefully add 4 ml CsCl 1.45 from the bottom by inserting the tip through the CsCl 1.33 layer and slowly let out CsCl 1.45. The CsCl 1.33 layer will smoothly move up. A clear interface between the two layers should be visible. Finally, add 4 ml virus samples on the top of the gradient in each tube. If the two tubes do not appear to be balanced, add 5 mM HEPES to balance them. Be careful not to disturb the gradients.
3. Ultracentrifuge at 56,000 × *g*, 4 °C for 3 h.
4. Harvest the lowest band using a 3-ml syringe with 23-G3/4 needle. There are two methods to do it: side or bottom extraction. If using the side extraction method, the insertion of the needle should be approximately 0.5 cm below the desired band. Extract slowly while keeping the needle at the lower surface of the band. Stop extracting from the gradient when the virus band becomes invisible. Move the ultracentrifuge tube into a 50-ml centrifuge tube and take the syringe out from the side (this way the solution leaking from the needle hole goes into the 50-ml tube). For the bottom extraction: puncture the ultracentrifuge tube with a 23-G3/4 needle, allow the solution to drop into a waste tube; when the desired band settles at the bottom, use a new microcentrifuge tube to collect the drop-through until the band is gone.
5. Combine the viruses from the two ultracentrifuge tubes (~2 ml). Dilute with 5 mM HEPES, pH.7.8, to a total of approximately 4 ml.
6. Set up CsCl gradient in two new tubes as above, and load the viruses from above to one tube, and 5 mM HEPES to the other tube (as balance).
7. Ultracentrifuge at 107,000 × *g*, 4 °C overnight (>18 h).
8. Harvest the lower band as described above.
9. Dialyze the viruses using a dialysis cassette (Pierce, Rockford, IL, USA) in 1 liter of dialysis buffer. Change dialysis buffer every 2 h (three times).
10. Aliquot the viruses and store them at –80 °C.

3.4. Virus Titration

3.4.1. Viral Particle Titration

This method determines the titer of a virus suspension using viral particles (VP) as unit. It is simply based on the correlation between the number of VP and the DNA content. DNA content can be easily measured by absorbance at 260 nm (OD260) using a spectrophotometer. Each OD260 unit represents approximately 1.1×10^{12} adenovirus particles. This method can only be used after CsCl purification step because unpurified viruses contain culture medium that interferes with OD readings. In addition, the supernatant (culture media) unavoidably contains contents released from the dead cells (such as DNA, protein, lipids, and membrane and nuclei debris), and they interfere with OD260 reading. Also, CsCl purification allows concentration of the virus to a high-enough level for detection with a spectrophotometer as well as removal of defective particles. To be accurate, the OD260 readings should lie between 0.1 and 1.0 because only at this range the OD260 readings can accurately reflect the amount of DNA for most spectrophotometers.

1. Prepare VLB.
2. Take 70 μl of purified viruses obtained above, dilute as following in microcentrifuge tubes (*see* **Note 13**):

   1:3 dilution: 33.3 μl viruses + 66.7 μl of VLB
   1:5 dilution: 20 μl viruses + 80 μl of VLB
   1:10 dilution: 10 μl viruses + 90 μl of VLB

3. Mix briefly by vortex and incubate at 56 °C for at least 10 min with shaking.
4. Briefly centrifuge the tubes and allow the samples cool down to room temperature.
5. Measure OD260 in a spectrophotometer for each sample. Use 100 μl of VLB as blank.
6. Calculate the VP titer for each dilution using the formula: $(OD_{260}) \times$ (dilution factor) $\times 1.1 \times 10^{12}$ = ? VP/ml. Eliminate the OD260 readings out of the 0.1–1.0 range.
7. Average the VP titers obtained from the dilutions. This is the VP titer for the virus sample.

3.4.2. Plaque Formation Unit

This method measures the plaque formation activity of a virus sample (that is, how many plaques each milliliter of the virus sample can form). It involves infection of permissive 293 cells with the viruses, completion of an infection cycle, and re-infection of neighboring cells to generate a clear plaque. It is the most challenging and the most variable titration method described herein because many factors such as quality of the cells and plaque observations at different time or by different person may affect the results.

1. Plate 293 cells in a 6-well plate at a density of 5×10^5 cells/well in 3 ml complete culture media the day before infection.
2. Dilute the virus. A 5-μl aliquot of the virus should be more than enough for this experiment. Make dilutions in 2% medium as following. Mix well after each dilution (*see* **Note 13**).

10^{-2} dilution: 4 μl virus + 396 μl 2% medium (=400 μl)
10^{-4} dilution: 10 μl of 10^{-2} dilution + 990 μl 2% medium (=1000 μl)
10^{-6} dilution: 10 μl of 10^{-4} dilution + 990 μl 2% medium (= 1000 μl)
10^{-8} dilution: 10 μl of 10^{-6} dilution + 990 μl 2% medium (=1000 μl)
10^{-9} dilution: 300 μl of 10^{-8} dilution + 2700 μl 2% medium (=3000 μl)
10^{-10} dilution: 300 μl of 10^{-9} dilution + 2700 μl 2% medium (=3000 μl)
10^{-11} dilution: 300 μl of 10^{-10} dilution + 2700 μl 2% medium (=3000 μl)

3. Infect the cells in the 6-well plate with 1 ml of 10^{-9}, 10^{-10}, and 10^{-11} diluted viruses. Use two wells for each dilution.
4. Add 2 ml of complete culture media 2 h later. Incubate the cells at 37 °C, 5% CO_2 overnight (12–16 h).
5. Coat the wells with SeaPlaque agarose + 2× DMEM mixture, 3 ml for each well, as described in 3.2.3. The only difference is, do not add liquid media to the top of the agarose coat.
6. Refresh the cells with 1 ml of SeaPlaque + 2× DMEM mixture every 4 days. Check plaque formation from time to time.
7. Count plaques. Wait for at least 14 days to allow all potential plaques become readily visible under microscope. The wells that contain well-isolated plaques should be counted (*see* **Note 14**).
8. Calculate the titer (pfu/ml). For each well, the titer (pfu/ml) = the plaque count × the dilution factor. Take average of the titers obtained from each well as the PFU titer of the virus sample.

3.4.3. Tissue Culture Infectious Dose 50 ($TCID_{50}$) Method

This method is based on the development of CPE in 293 cells using end-point dilutions. It is becoming more widely used than PFU method because it is also an infection-based method, but takes less time, and is less variable (more repeatable). The following protocol is modified from the manual of Quantum Adeasy™ vector system ***(20)*** (*see* **Note 15**).

1. Plate 293 cells in a 96-well plate, with a density of 10^4 cells/well in 100 μl of 10% complete culture media, the day before infection.
2. Dilute viruses as following to make 10^{-2}, 10^{-4}, 10^{-5}, … , 10^{-12} dilutions. Mix well after each dilution. The range can change according to the virus sample concentration, but make sure in the highest dilution, no well develops CPE. And in the lowest dilution, all wells develop CPE (*see* **Note 13**).

10^{-2} dilution: 4 μl virus + 396 μl 2% medium (=400 μl)
10^{-4} dilution: 20 μl of 10^{-2} dilution + 1980 μl 2% medium (=2000 μl)
10^{-5} dilution: 200 μl of 10^{-4} dilution + 1800 μl 2% medium (=2000 μl)
....
10^{-12} dilution: 200 μl of 10^{-11} dilution + 1800 μl 2% medium (=2000 μl)

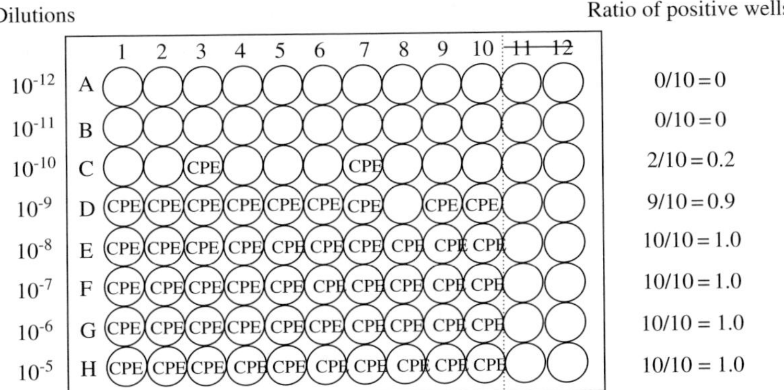

Fig. 4. An example of TCID50 titration result.

3. In each row of the 96-well plate containing 293 cells, add 100 μl viruses of one dilution into each well of the wells 1–10 (10 wells per dilution). Use the eight highest dilutions (that is 10^{-5}–10^{-12} dilutions here) for the eight rows of the 96-well plate. When distributing the dilutions, always start with the highest dilution in the top row. Wells in columns 11 and 12 can serve as negative control.
4. Incubate the plate at 37 °C, 5% CO2 in a humidified incubator for 10 days.
5. After 10 days, count the wells with visible CPE per row under an inverted light microscope. A well should be counted as positive even if only a small spot showing CPE. If in doubt, observe again in 2–3 days.
6. Determine the ratio of positive wells per row. If all wells are positive, the ratio is 10/10 = 1.0; all wells are negative means the ratio is 0/10 = 0.0, and if some wells are positive and some wells negative, the ratio is n/10, in which n represents the number of positive wells (For example, two wells positive in one row, the ratio would be 2/10 = 0.2). See example in **Fig. 4**.
7. Calculate the TCID50 titer using the KÄRBER statistical method. For 100 μl of dilution, the titer is: $T = 10^{1+d(s-0.5)}$, of which, $d = 1$ when using 10-fold dilution (as used here), and s = the sum of the ratios starting with 10^{-1} dilution. Although 10^{-1}–10^{-4} are omitted (as in **Fig. 4**), they need to be included as ratios of 1.0. In the example, s = 1.0 + 1.0 + 1.0 + 1.0 + 1.0 + 1.0 + 1.0 + 1.0 + 0.9 + 0.2 + 0.0 + 0.0 = 9.1.
   Titer: $T = 10^{1+1(9.1-0.5)} = 10^{9.6}$ (for 100 μl of virus)
   $T = 10 \times 10^{9.6}$ TCID50/ml $= 10^{10.6}$ TCID50/ml $= 3.98 \times 10^{10}$ TCID50/ml

4. Notes

1. The enzyme dose and incubation time for partial digestion may need to be adjusted. A good partial digestion in this case should result in most DNA as linearized (one cut), very few molecules were uncut, or cut twice or more.

2. For electrophoresis of large DNA fragments, in general, lower voltage running for longer time in a higher percentage gel allows better separation of the DNA fragments. Therefore, gel-running condition illustrated in this protocol may be adjusted for individual needs.
3. For large plasmids, avoid using fast-growing bacterial strains such as XL-1 blue, TOP10, etc. Other than DH10B, DH5α and STBL can also be used.
4. For large plasmid transformation, electroporation is preferred over chemical transformation because chemical transformation may cause partial transformation of the intact plasmid.
5. For large plasmid and DNA fragments, avoid vortex as it may break the DNA.
6. The pAd5_Swa I backbone may be further improved by completely eliminating the native fiber gene (which can increase the possibility of desired recombination). This may be achieved by completely eliminating the fiber gene in the shuttle vector while incorporating a unique site such as Swa I in the fiber position, then recombining with the pAd5_SwaI backbone.
7. Any other small plasmids with the proper restriction sites can be used instead of pNEB193 in generating fiber shuttle vector.
8. Commercial kits are not recommended for minipreps from BJ5183 because BJ5183 do not make high quality or amount of plasmids. Simply add P1, P2, and P3, and centrifuge as recommended by the kit's protocol, then precipitate DNA with 2.5 vol of 100% ethanol. And centrifuge again for 10 min at 15,000 × *g*. Dry the pellet and dissolve in 30 μl of dd.H_2O.
9. Because BJ5183 supports homologous recombination, plasmids in BJ5183 are not stable. Therefore, BJ5183 cannot be used to make high quality, large amount of plasmids. The recombinant plasmid DNA obtained by miniprep from BJ5183 should be transformed into DH10B (or DH5a, STBL, etc.) to make stable, high quality of plasmid DNA.
10. It is not necessary to run an agarose gel to separate the Ad5 genome and the plasmid portion when preparing the fragment for rescuing viruses. Commercially available large fragment purification/desalting kits, as well as phenol extraction method, can be used to desalt the digested fragments so that they are ready for transfection into 293 cells.
11. During amplification steps, adjust the amount of viruses or cells used so that complete CPE can be reached in each flask around 48 h after infection. Usually, the yield is acceptable if the complete CPE reaches as early as 24 h or as late as 72 h after infection. If it took too long to reach complete CPE, a second (in five T75 flasks) or third amplification step (in 10 T75 flasks) will be necessary.
12. When collecting the rescued viruses or first amplified viruses, you can collect both the cells and the media. This apparently gives you more viruses, but the large volume is more difficult to store and to perform subsequent freeze/thaw cycles. Our experience suggests that cell-associated viruses are enough for subsequent steps. Thus, to simplify the subsequent procedures, we usually only collect the cell-associated viruses.

13. The dilution strategy is designed to minimize the use of the viruses, to minimize the sampling errors, and for easy manipulations. The highest desired dilution varies with different virus concentration, thus may need to be adjusted individually. The dilutions outlined in this protocol usually work well for viral samples in 10^{12} VP/ml range.
14. When counting plaques in PFU titration method, it is usually easier and more accurate to count the wells containing 3–30 plaques. Therefore, it is only necessary to count the wells with the highest one or two dilutions of viruses. In addition, to minimize the observation errors, try to count at different time points such as 10, 14, 18, and 21 days. You will see how different the results could be. Take results from the longest time point with clearly separated plaques.
15. It has been validated that the titer measured by TCID50 is about 0.7 log higher than the titer by standard plaque assay ***(20)***. Therefore, TCID50/ml can be transferred into pfu/ml. For example, $T = 10^{10.6}$ TCID50/ml = $10^{10.6-0.7}$ pfu/ml $= 10^{9.9}$ pfu/ml = 7.94×10^{9} pfu/ml.

Acknowledgments

This work is supported in part by NIH grants (to D.T.C.) 1P01CA104177-01A2, 1P01HL076540, R01CA083821 and 1R01CA111569-01A1. H.W. is supported by NIH brain SPORE grant P50 CA097247 and Juvenile Diabetes Research Foundation grant 1-2005-71.

References

1. Louis, N., Fender, P., Barge, A., Kitts, P., and Chroboczek, J. (1994) Cell-binding domain of adenovirus serotype 2 fiber. *J Virol* **68,** 4104–6.
2. Bergelson, J. M., Cunningham, J. A., Droguett, G., Kurt-Jones, E. A., Krithivas, A., Hong, J. S., *et al.* (1997). Isolation of a common receptor for Coxsackie B viruses and adenoviruses 2 and 5. *Science* **275,** 1320–3.
3. Santis, G., Legrand, V., Hong, S. S., Davison, E., Kirby, I., Imler, J. L., *et al.* (1999) Molecular determinants of adenovirus serotype 5 fibre binding to its cellular receptor CAR. *J Gen Virol* **80 (Pt 6),** 1519–27.
4. Bai, M., Harfe, B., and Freimuth, P. (1993) Mutations that alter an Arg-Gly-Asp (RGD) sequence in the adenovirus type 2 penton base protein abolish its cell-rounding activity and delay virus reproduction in flat cells. *J Virol* **67,** 5198–205.
5. Wickham, T. J., Mathias, P., Cheresh, D. A., and Nemerow, G. R. (1993) Integrins alpha v beta 3 and alpha v beta 5 promote adenovirus internalization but not virus attachment. *Cell* **73,** 309–19.
6. Li, E., Brown, S. L., Stupack, D. G., Puente, X. S., Cheresh, D. A., and Nemerow, G. R. (2001) Integrin alpha(v)beta1 is an adenovirus coreceptor. *J Virol* **75,** 5405–9.
7. Curiel, D. T. (1999) Strategies to adapt adenoviral vectors for targeted delivery. *Ann N Y Acad Sci* **886,** 158–71.

8. Silman, N. J., and Fooks, A. R. (2000) Biophysical targeting of adenovirus vectors for gene therapy. *Curr Opin Mol Ther* **2,** 524–31.
9. Mizuguchi, H., and Hayakawa, T. (2004) Targeted adenovirus vectors. *Hum Gene Ther* **15,** 1034–44.
10. Dmitriev, I., Kashentseva, E., Rogers, B. E., Krasnykh, V., and Curiel, D. T. (2000) Ectodomain of coxsackievirus and adenovirus receptor genetically fused to epidermal growth factor mediates adenovirus targeting to epidermal growth factor receptor-positive cells. *J Virol* **74,** 6875–84.
11. Zhu, Z. B., Makhija, S. K., Lu, B., Wang, M., Rivera, A. A., Preuss, M., *et al.* (2004) Transport across a polarized monolayer of Caco-2 cells by transferrin receptor-mediated adenovirus transcytosis. *Virology* **325,** 116–28.
12. Krasnykh, V. N., Mikheeva, G. V., Douglas, J. T., and Curiel, D. T. (1996) Generation of recombinant adenovirus vectors with modified fibers for altering viral tropism. *J Virol* **70,** 6839–46.
13. Wickham, T. J., Tzeng, E., Shears, L. L., 2nd, Roelvink, P. W., Li, Y., Lee, G. M., *et al.* (1997) Increased *in vitro* and *in vivo* gene transfer by adenovirus vectors containing chimeric fiber proteins. *J Virol* **71,** 8221–9.
14. Dmitriev, I., Krasnykh, V., Miller, C. R., Wang, M., Kashentseva, E., Mikheeva, G., *et al.* (1998) An adenovirus vector with genetically modified fibers demonstrates expanded tropism via utilization of a coxsackievirus and adenovirus receptor-independent cell entry mechanism. *J Virol* **72,** 9706–13.
15. Wu, H., Seki, T., Dmitriev, I., Uil, T., Kashentseva, E., Han, T., and Curiel, D. T. (2002) Double modification of adenovirus fiber with RGD and polylysine motifs improves coxsackievirus-adenovirus receptor-independent gene transfer efficiency. *Hum Gene Ther* **13,** 1647–53.
16. Krasnykh, V., Dmitriev, I., Mikheeva, G., Miller, C. R., Belousova, N., and Curiel., D. T. (1998) Characterization of an adenovirus vector containing a heterologous peptide epitope in the HI loop of the fiber knob. *J Virol* **72,** 1844–52.
17. Belousova, N., Krendelchtchikova, V., Curiel, D. T., and Krasnykh, V. (2002) Modulation of adenovirus vector tropism via incorporation of polypeptide ligands into the fiber protein. *J Virol* **76,** 8621–31.
18. Xia, H., Anderson, B., Mao, Q., and Davidson, B. L. (2000) Recombinant human adenovirus: targeting to the human transferrin receptor improves gene transfer to brain microcapillary endothelium. *J Virol* **74,** 11359–66.
19. Wu, H., Dmitriev, I., Kashentseva, E., Seki, T., Wang, M., and Curiel, D. T. (2002) Construction and characterization of adenovirus serotype 5 packaged by serotype 3 hexon. *J Virol* **76,** 12775–82.
20. Quantum Adeasy^{TM} Vector System Application Manual. (2000) Quantum Biotechnologies, Montreal, Canada.

9

PEGylated Adenovirus for Targeted Gene Therapy

Catherine R. O'Riordan and Antonius Song

Summary

Bifunctional polyethylene glycol (PEG) molecules provide a novel approach to retargeting viral vectors without the need to genetically modify the vector. Modification of the surface of adenovirus with heterofunctional PEG allows further modification of the capsid with ligands. In addition, heterofunctional PEG modification ablates the normal tropism of the virus and reduces transduction of non-target tissues in vivo. Moreover, the addition of PEG chains to the surface of the virus shields antigen-binding sites, significantly reducing the susceptibility of the virus to antibody neutralization. Finally, T cell subsets from mice exposed to the PEGylated vector demonstrate a marked decrease in Th1 and Th2 responses, suggesting that PEG modification may help reduce the immune response to the vector.

Key Words: Adenovirus; PEGylation; targeting; FGF2; innate immunity; adaptive immunity; antibody neutralization.

1. Introduction

Adenoviral vectors can efficiently transduce a broad range of cell types and have been used extensively in pre-clinical and clinical studies for gene delivery applications. There are two potential hurdles for using adenoviruses as gene therapy vectors. First, target tissues may not express adequate levels of adenovirus receptors to permit efficient adenovirus entry into these cells. Second, expression of native receptors on healthy tissue may lead to vector or transgene-related toxicity through the entry of vector into non-target cells, in particular antigen-presenting cells (APCs) of the immune system and Kupffer cells of the liver. These potential disadvantages highlight the importance of

From: *Methods in Molecular Biology, vol. 434: Volume 2: Design and Characterization of Gene Transfer Vectors*
Edited by: J. M. Le Doux © Humana Press, Totowa, NJ

developing targeted adenovirus vectors with improved efficiency and specificity of gene transfer.

The initial phase of Ad infection involves at least two sequential steps. The first is attachment of the virus to the cell surface through binding of the knob domain of the fiber to CAR ***(1,2)*** (*see* **Fig. 1**). After attachment, Interaction between the RGD motif of the penton base with secondary host cell receptors α_v integrins facilitates internalization through receptor-mediated endocytosis ***(3,4)***. Furthermore, interaction between the KKTK motif on the fiber shaft with heparin sulfate proteoglycans promotes accumulation of Ad vector in liver following systemic administration ***(5–8)***. It is well accepted in the field of adenoviral gene therapy that higher adenoviral vector doses invariably lead to hepatotoxicity and acute inflammatory responses because of activation of innate immunity. Innate immune response is activated following transduction of macrophages and dendritic cells resulting in the activation of NFkB pathways that augment expression of several proinflammatory cytokines and chemokines ***(9–12)***, (*see* **Fig. 2**). Multiple inflammatory cytokines including tumor necrosis factor (TNF)-α interleukin (IL)-6, IL-8, IL-12, MIP-2, RANTES, are expressed following Ad vector administration in a dose-dependent manner ***(13,14)***. This first phase of the innate immune response, which occurs as early as 1 h after systemic administration of the virus, is also accompanied by elevated liver enzymes. Kupffer cells of the liver, macrophages, and dendritic cells in the spleen and peritoneum all release cytokines into the circulation, which leads to the development of the second phase of the immune response to Ad vectors. Data suggest that virus capsid proteins alone are responsible for induction of this first phase of the immune response ***(15)***.

The second phase of the immune response begins 5–7 days after administration of vector. Adenovirus-specific major histocompatibility complex (MHC) class I-restricted $CD8^+$ T cells directed against cells expressing viral genes and transgene products are generated ***(16,17)***. MHC class II helper $CD4^+$ T cells largely stimulated by the presence of adenovirus capsid proteins are also generated ***(18)*** (*see* **Fig. 2**). $CD4^+$T cell–dependent humoral immunity is responsible for the production of anti-adenoviral neutralizing antibodies, which rapidly clear the virus from the circulation and prevent successful gene transfer upon readministration (*see* **Fig. 2**).

Fig. 1. (A) Binding of adenovirus to cells via the knob domain of the fiber to CAR (B) Adenovirus following PEGylation with bifunctional PEG molecules (C) Conjugation of FGF2 to Ad/PEG allows virus binding to the targeted FGF receptor on the cell surface.

Retargeting Adenovirus

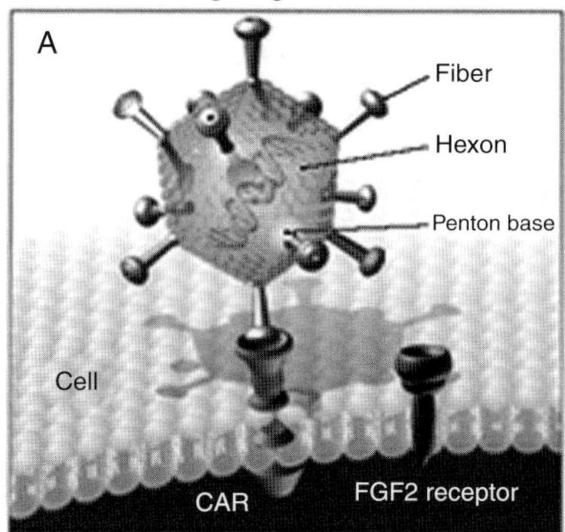

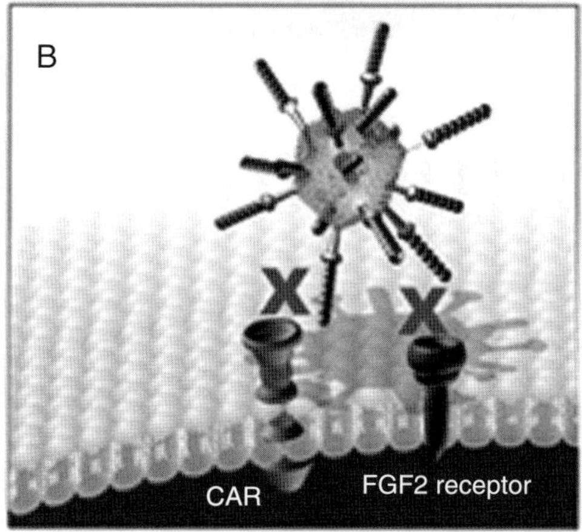

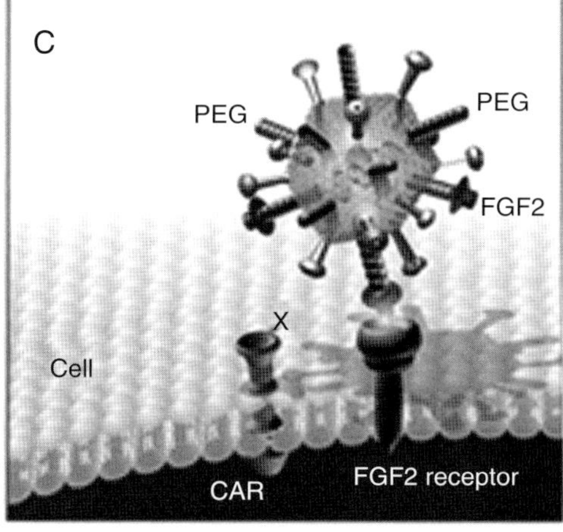

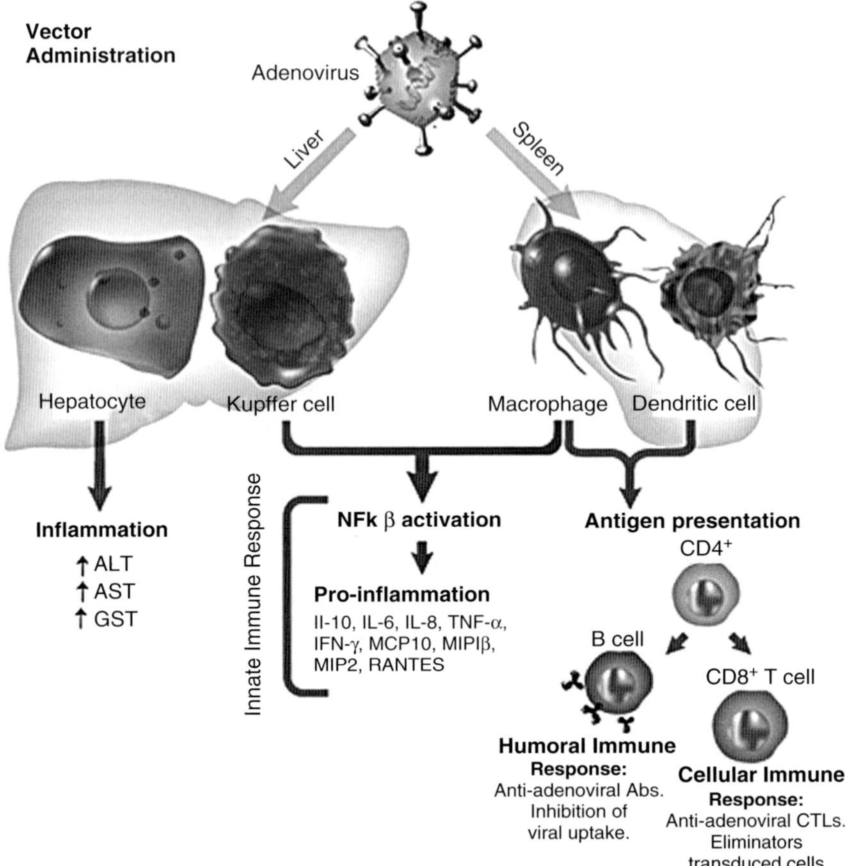

Fig. 2. Diagrammatic representation of the development of both the innate and adaptive immune response to an adenoviral vector.

In conclusion, strategies to retarget adenoviral vectors to target tissues should include ablation of the native tropism of the virus to reduce transduction of non-target tissues. This should help overcome some of the other limitations associated with use of this vector, in particular immunogenicity and toxicity.

Strategies to eliminate natural Ad tropism, based on genetic modification of particular viral capsid proteins such as fiber ***(19–24)*** and penton base, have been reported ***(8,21,22,24)***. These genetically modified vectors have helped elucidate the contribution of fiber and penton base to adenoviral cell entry and provide an important platform for evaluating the targeting potential of ligands. However, drawbacks to the clinical use of these modified vectors include poor propagation and limitations on the size of the retargeting ligand that can be incorporated into proteins such as fiber or hexon.

Alternative retargeting approaches include the use of chemical linkers to attach targeting ligands to the surface of the virus. Ligands for cell-surface receptors have been linked to capsid proteins of adenoviral vectors by bifunctional antibodies ***(25–42)*** or by polymers such as polyethylene glycol (PEG) ***(43–47)*** and poly-[*N*-(2-hydroxypropyl) methacrylamide] (HPMA) ***(48)***. Using these approaches, the native tropism of the virus is ablated either by the addition of polymer to fiber knob, penton base and fiber shaft (*see* **Fig. 1B**) or by the use of an anti-fiber neutralizing antibody in the context of a bifunctional conjugate ***(25–28)***. Whereas specific targeting for viral gene delivery vehicles has been achieved, the changes to the viral structure can potentially yield varied efficiencies. The efficacy of the targeted viral construct is highly dependent on the moiety used for targeting, the method of attachment onto the viral surface, and the targeted cell type. Ideally, the targeting ligand should be covalently attached to the surface of the virus so as to minimize dissociation following in vivo delivery. The use of PEG as a linker has many advantages. Modification of adenovirus with both monofunctional and heterofunctional PEGs protects the vector from neutralizing antibodies both in vitro and in vivo ***(44,48–50)***. Furthermore, PEGylation reduces the innate immune response to adenovirus ***(50)***, in addition to reducing Th1- and Th2-cell responses ***(47,50)*** while also reducing liver toxicity ***(51)***. The use of heterofunctional PEG molecules (*see* **Fig. 3**) allows for the modification of adenoviruses with ligands ***(43–46)***. Suitable ligands include peptides ***(43,45)*** and growth factors ***(47)***. In this chapter, we will focus on methods to redirect adenovirus to the FGF2 receptor. Methods to characterize the modified vector in vivo and in vitro will be described.

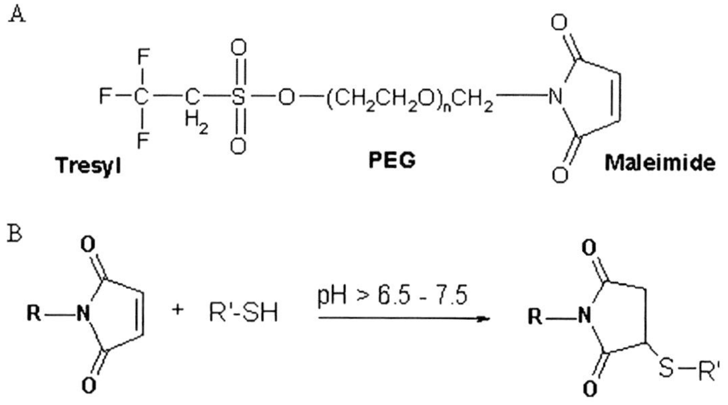

Fig. 3. (A) Chemical structure of Tresyl-PEG-maleimide. (B) Reaction between the maleimide group and reactive thiol groups which results in the formation of a stable thioether bond.

1.1. Redirecting Adenovirus to Receptors on Target Cells

1.1.1. Fibroblast Growth Factor Receptors

Adenoviral retargeting by fibroblast growth factor (FGF) family ligands can achieve targeted gene therapy to FGF-receptor-expressing cells, such as those within tumors and healing wounds. The receptor for FGF2 (basic FGF) is upregulated in many tumor cell lines including the human ovarian cancer cell line SKOV3.ip1, making it an attractive receptor for targeting ***(28,30)***. Increasing the efficiency of transduction by redirecting adenovirus to receptors that are highly expressed on the target cells reduces the number of adenoviral particles required for optimal levels of gene transfer. This in turn should permit high levels of transduction at doses of vector below those associated with acute toxicity. In addition, ablation of native viral tropism by the addition of PEG polymer and ligand reduces the affinity of the retargeted Ad vector for its native receptor and in turn its affinity for non-target tissues such as liver and spleen ***(47)***. The generation of Ad/PEG/FGF2 complexes occurs in two steps. First, PEG-maleimide (*see* **Fig. 3**) is added to the surface of the virion (PEGylation), then FGF2 ligand is coupled to the maleimide group on the PEGylated vector.

1.2. PEGylation of Adenovirus Vectors

PEG is a water-soluble polymer that can covalently attach to proteins. Covalent attachment of PEG to proteins requires activation of the hydroxyl terminal group of the polymer with a suitable leaving group that can be displaced by nucleophilic attack of the ε-amino terminal of lysine residues. Many methods for the activation of PEG result in part of the activating group being incorporated into the final PEG-peptide, and although PEG itself is immunologically inert ***(53)***, these residual coupling moieties can themselves be immunogenic. Many of these problems have been overcome by the biological optimization of a PEGylated method using tresyl-MPEG (TMPEG) (*see* **Fig. 3A**) ***(54,55)***. This optimized PEGylation reaction has been used to attach both monofunctional ***(48)*** and heterofunctional PEG molecules ***(43–45)*** to capsid proteins without compromising the integrity of the virion. Heterofunctional PEG molecules, such as TMPEG-maleimide (*see* **Fig. 3A**), with two independent functional groups allow for attachment to lysine residues of capsid proteins and the additional modification of ligand coupling (*see* **Fig. 1 and 3B**). Following attachment of the tresyl group to lysine residues on capsid proteins, FGF2 is coupled to the Ad/PEG complex through the formation of a stable thioether bond between the maleimide group on the PEG and a surface thiol group on FGF2 (*see* **Fig. 1 and 3B**).

2. Materials

2.1. Quantitation of Ad/PEG Particles

2.1.1. Determination of Virus Particle Concentration for Unmodified Virus

1. 50% sucrose, sterile filtered.
2. Phosphate-buffered saline, (PBS) pH 7.0, sterile filtered.
3. Ad2/CMV β-gal 4 virus in PBS, pH 7, 5% sucrose. This is an adenovirus serotype 2 (Ad2)-based vector with most of the E1 region deleted and replaced with the β-galactosidase transgene; it contains the complete wild-type E4 region ***(56)***.
4. SDS (sodium dodecyl sulfate) 10%.

2.2. Preparation of Ad/PEG Complexes

2.2.1. PEGylation of Adenovirus

1. 130 mM sodium phosphate, pH 7.0, sterile filtered.
2. TMPEG-maleimide PEG, MW 5000 (Nektar, Huntsville, AL, USA).
3. 50% sucrose, sterile filtered.
4. PBS, pH 7.0, sterile filtered.

2.2.2. Removal of Uncoupled TMPEG and Purification of Ad/PEG-maleimide by Size Exclusion Chromatography

1. Fractogel BioSec size exclusion resin (EM Sciences, Gibbstown, NJ, USA) 150 ml (2.6 × 30 cm) in a XK 26/60 column (Pharmacia, Piscataway, NJ, USA).
2. PBS, pH 7.0.
3. Akta Purifier (Pharmacia).
4. Sucrose 50%, sterile filtered.

2.2.3. Generation of an Ad/PEG/FGF2 Complex

2.2.3.1. Preparation of FGF2 Mutein Protein for Coupling to AD/PEG

1. FGF2-mutein [FGF2 (Selective Genetics, San Diego, CA, USA) or FGF2 wild-type native protein (Sigma, St. Louis, MO, USA)].
2. Slide-a-lyzer cassette, 10,000 MW (Pierce, Rockford, IL, USA).
3. Dialysis Buffer A: 0.1 M sodium phosphate, 0.1 M sodium chloride, 1 mM ethylenediaminetetraacetic acid (EDTA) pH 7.5.
4. DTT dithiothreitol (Sigma Chemicals).
5. Centrispin 6 columns (Biorad, Hercules, CA, USA).

2.2.3.2. Purification of AD/PEG/FGF2 by Ion Exchange Chromatography

1. Fractogel DEAE resin, 40–90 μm (EM Sciences).
2. DEAE Buffer A: 10 mM potassium phosphate, pH 7.5 (1 M = 15 g dibasic, 2 g monobasic per L) containing 0.15 M NaCl, 0.01% Tween-80 (sterile filtered).

3. DEAE Buffer B: 10 mM potassium chloride, pH 7.5 containing 1 M KCl, 1 M NaCl, 0.01% Tween-80 (sterile filtered).
4. 50% sucrose solution (sterile filtered).
5. XK16/60 glass column (Pharmacia).

2.3. Determination of Particle Number of Ad/PEG/FGF2 Complexes

2.3.1. Quantitation of FGF2 on Ad/PEG/FGF2 Conjugates

1. Basic FGF (FGF2) ELISA kit (R&D systems).
2. OPD peroxidase substrate (Sigma): Make fresh in 20 ml water as required.
3. PBS, pH 7.

2.4. In vitro Transduction Assays

2.4.1. Assessment of Transduction Efficiency of Ad/PEG/FGF2 Conjugates

1. Human ovarian carcinoma cell line SKOV3.ip1. The SKOV3.ip1 cell line was established from ascites that developed in a *nu/nu* mouse given an intraperitoneal (i.p.) injection of parental SKOV3 cells. These SKOV3.ip1 variant cells are phenotypically more aggressive, more cancerous, than their parental SKOV3 cells, because of a higher expression of protein p185 encoded by the protooncogene c-erbB-2/neu ***(57)***.
2. F-12K Kaighn's modified media (Gibco BRL, Carlsbad, CA, USA) containing 10% fetal bovine serum (FBS), 100 units/ml penicillin G sodium, 100 μg/ml streptomycin sulfate.
3. FBS (JRH Biosciences, Lenexa, KS, USA).
4. Penicillin G sodium (Invitrogen, Carlsbad, CA, USA).
5. Streptomycin sulfate (Invitrogen).
6. 1450 MicroBeta Trilux Luminometer (Perkin Elmer, Boston, MA, USA).
7. Galacto-Star β-galactosidase detection kit (Tropix/Applied Biosystems, Foster City, CA, USA)

2.4.2. Specificity of Binding of Ad/PEG/FGF2 to the FGF Receptor: Competitive Binding Assays

1. SKOV3.ip1 cells. Human ovarian carcinoma cell line SKOV3.ip1. The SKOV3.ip1 cell line was established from ascites that developed in a *nu/nu* mouse given an i.p. injection of parental SKOV3 cells. These SKOV3.ip1 variant cells are phenotypically more aggressive, more cancerous, than their parental SKOV3 cells, because of a higher expression of protein p185 encoded by the protooncogene c-erbB-2/neu ***(57)***.
2. F-12K Kaighn's modified media (Gibco BRL) containing 10% FBS, 100 units/ml penicillin G sodium, 100 μg/ ml streptomycin sulfate.
3. FBS (JRH Biosciences).
4. Penicillin G sodium (Invitrogen).

5. Streptomycin sulfate (Invitrogen).
6. 1450 MicroBeta Trilux Luminometer (Perkin Elmer).
7. Galacto-Star β-galactosidase detection kit (Tropix/Applied Biosystems)
8. FGF2 mutein protein (Selective Genetics).
9. Ad2 fiber knob protein (gift from Dr J. Zabner, University of Iowa).

2.4.3. Endothelial Cell Proliferation Assay

1. Adult bovine endothelial cells (ATCC).
2. Dulbecco's modified essential medium (DMEM) (Invitrogen).
3. Modified DMEM: DMEM containing 4 mM L-glutamine adjusted to contain 1.5 g/l sodium bicarbonate and 4.5 g/l glucose, FBS 10%.
4. L-Glutamine (Invitrogen).
5. Sodium bicarbonate (Invitrogen).
6. Glucose (Invitrogen).
7. FBS (JRH Biosciences).
8. CellTiter 96® AQ_{ueous} cell proliferation assay (MTS) kit (Promega, Madison, WI, USA).
9. FGF2 mutein protein (Selective Genetics).

2.5. In vivo Infection of Intraperitoneal Tumors with Ad/PEG/FGF2

2.5.1. Preparation of SKOV3.ip1 Cells and Generation of a Human Ovarian Cancer Mouse Model for In Vivo Ad/PEG/FGF2 Targeting

1. SKOV3.ip1 cells. Human ovarian carcinoma cell line SKOV3.ip1. The SKOV3.ip1 cell line was established from ascites that developed in a *nu/nu* mouse given an i.p. injection of parental SKOV3 cells. These SKOV3.ip1 variant cells are phenotypically more aggressive, more cancerous, than their parental SKOV3 cells, because of a higher expression of protein p185 encoded by the protooncogene c-erbB-2/neu ***(57)***.
2. F-12K Kaighn's modified media (Gibco BRL) containing 10% FBS, 100 units/ml penicillin G sodium, 100 ug/ml streptomycin sulfate.
3. FBS (JRH Biosciences).
4. Penicillin G sodium (Invitrogen).
5. Streptomycin sulfate (Invitrogen).
6. PBS, pH 7 (sterile filtered).
7. Sucrose 50% (sterile filtered).
8. Tumor lysis buffer (Tropix lysis buffer with 1 mM Pefabloc, 5 ug/ml leupeptin, 0.5 uM DTT).
9. Dithiothreitol (Boehringer-Mannheim: Roche Diagnostics, Indianapolis, IN, USA).
10. Pefabloc (Boehringer-Mannheim: Roche Diagnostics).
11. Leupeptin (Boehringer-Mannheim: Roche Diagnostics).
12. Tissue Tearor homogenizer (Biospec, Bartlesville, OK, USA).
13. Trypsin–EDTA 0.05% (w/v) (Gibco BRL).

2.5.2. Measurement of Transgene Expression and Viral DNA in Tumor Samples

2.5.2.1. β-Galactosidase ELISA

1. β-Galactosidase ELISA kit (Boehringer-Mannheim: Roche Diagnostics).
2. BCA assay reagents (Pierce).
3. OPD peroxidase substrate (Sigma): Make fresh in 20 ml water as required.

2.5.2.2. TaqMan Analysis of hexon DNA

1. Zirconia beads (Biospec Products): autoclave in small glass vials.
2. DNA lysis buffer: 100 mM Tris-HCl, pH 8, 5 mM EDTA, 0.4% SDS, 200 mM NaCl.
3. 2-ml sterile screw cap tubes (Axygen, Union City, CA, USA).
4. DEPC water (water treated with diethylpyrocarbonate to eliminate RNAse/Dnase).
5. 1.5-ml no-stick Rnase-free microfuge tubes (Ambion, Austin, TX, USA).
6. Proteinase K solution (Roche).
7. Taqman Universal PCR mix (Applied Biosystems).
8. Polymerase chain reaction (PCR) mix:

Reagent	μl/reaction
DEPC water	18.6
Universal PCR mix	25
25 μm 5´ primer	0.6
25 μm 3´ primer	0.6
25 μm probe	0.2
TOTAL	45

9. ABI PRISM 7700 Sequence Detector (Applied Biosystems).
10. Primers and probe (IDT, Coralville, IA, USA): Reconstitute in TE to 1 mg/ml. To make 25 μM working concentrations, dilute stock primers in water, and stock probe in TE. Store all stocks and working primers at –20ºC. Store working probe solution at 4ºC in the dark for up to 1 month. PCR primers and probes were designed for the adenovirus hexon capsid protein. See **Subheading 3.5.2.2** for sequence.

2.6. Characterization of the Innate Immune Response to PEGylated Vectors

2.6.1. Measurement of Liver Enzymes and Cytokines

1. Aspartate aminotransferase (AST) and alanine aminotransferase (ALT) kits (Olympus/Essex, UK/Ireland).

2. Mouse IL-12, IL-6, interferon (IFN)-γ and TNF-α ELISA kits (R&D systems).
3. Balb/C mice (Jackson Labs, Bar Harbor, ME, USA).

2.7. Characterization of the Adaptive Immune Response to PEGylated Vectors

2.7.1. Evaluation of Th1- and Th2-Cell Responses

2.7.1.1. Cytokine Measurements

1. Balb/C mice (Jackson Labs).
2. FBS (JRH Biosciences).
3. Penicillin G sodium (Invitrogen).
4. Streptomycin sulfate (Invitrogen).
5. Ultraculture media (Cambrex, Charles City, IOWA USA) with 100 units/ml penicillin G sodium and 100 μg/ml streptomycin sulfate, L-glutamine and 10% FBS.
6. ELISA kits for IL-12, IL-10, IFN-γ and IL-4 (R&D Systems).

2.7.2. Protection from Neutralization Assay

1. Human ovarian carcinoma cell line SKOV3.ip1. The SKOV3.ip1 cell line was established from ascites that developed in a *nu/nu* mouse given an i.p. injection of parental SKOV3 cells. These SKOV3.ip1 variant cells are phenotypically more aggressive, more cancerous, than their parental SKOV3 cells, because of a higher expression of protein p185 encoded by the protooncogene c-erbB-2/neu ***(57)***.
2. F-12K Kaighn's modified media (Gibco BRL) containing 10% FBS, 100 units/ml penicillin G sodium, 100 μg/ml streptomycin sulfate.
3. FBS (JRH Biosciences).
4. Penicillin G sodium (Invitrogen).
5. Streptomycin sulfate (Invitrogen).
6. Human anti-adenovirus-neutralizing serum (Genzyme Corp, Cambridge, MA, USA).
7. 1450 MicroBeta Trilux Luminometer (Perkin Elmer).
8. Galacto-Star β-galactosidase detection kit (Tropix/Applied Biosystems).

3. Methods

3.1. Quantitation of Ad/PEG Particles

The concentration of unmodified virions is typically determined by optical absorbance of the virion DNA (*see* **Subheading 3.1.1.**, **ref.** ***58***). This method relies on the use of an extinction coefficient for the virion while optical absorbance is determined after disruption of the virion with SDS to release virion DNA. As the efficiency of disruption of the virion by SDS may be altered by PEG/ligand modification, the optimal method to determine concentration of Ad/PEG/ligand complexes is to use chromatographic methods that utilize integration of peak areas and retention time of the peaks on the resin. Modified virus peak is quantitated chromatographically at A_{260} by comparison to standard curves constructed by injecting 10^8–10^{12} CsCl-purified unmodified virions that had been characterized for total particles by A_{260} in 0.1% SDS (**Subheading 3.1.1.**, **ref. *58***). Peak area typically increases linearly with virus particle number. The viral concentration of modified virus is determined by using linear regression analysis of the standard curve. The volume of unmodified virus used to generate the standard curve should be equal to the volume of the test article, i.e. the modified virus.

3.1.1. Determination of Virus Particle Concentration for Unmodified Virus

1. Add 1 µl of 10% SDS to 100 µl of virus. This releases encapsidated viral DNA.
2. Prepare an equivalent sample in PBS/5% sucrose for blank.
3. Incubate about 30 min at room temperature.
4. Vortex.
5. Measure A_{260}; use extinction co-efficient of 1 optical density (OD) = 1.1×10^{12} particles/ml ***(58)***.
6. There are 6.02×10^{23} particles/L in a 1 M solution.

3.2. Preparation of Ad/PEG Complexes

3.2.1. PEGylation of Adenovirus

1. Dilute $CsCl_2$ purified Ad2/βgal4 vector (1×10^{12}particles/ ml) in PBS containing 5% sucrose), 1:1 (vol : vol) with 130 mM sodium phosphate pH 7.0, 5.0% sucrose (*see* **Notes 1–4**).
2. Add 1.0, 5.0, or 10% (wt/vol) addition of TMPEG-maleimide (TMPEG-mal) to virus solution. For 10% TMPEG reaction, add the TMPEG in 5% increments for 30 min each. This prevents precipitation of the virus.
3. Incubate mixture at room temperature; for a 1% PEGylation reaction incubate for 45 min, 5% PEGylation incubate for 30 min, and for 10% TMPEG, 2× 5% TMPEG addition for 30 min each.
4. Place mixture on ice to stop reaction. The tresyl group on TMPEG is not reactive at temperatures below 4°C (*see* **Note 5**).

3.2.2. Removal of Uncoupled TMPEG and Purification of Ad/PEG-Maleimide by Size Exclusion Chromatography

Fractogel Biosec size exclusion resin separates proteins on the basis of molecular size. The molecular weight of a virus particle can be estimated to be approximately 175×10^8kDa with a diameter of 90 nm ***(59,60)***. Ad/PEG-maleimide (referred to as Ad/PEG) is recovered in the void volume of this resin as a sharp peak whereas uncoupled TMPEG-maleimide (MW 5000) is recovered in the included volume generally as a broad peak. Ad/PEG elutes from the column in a sharp peak. Multiple peaks indicate that the structure of the virus is compromised and may suggest dissociation of the viral particle into its constituent polypeptides.

1. Equilibrate a BioSec resin 150 ml (2.6 × 30 cm) with PBS, pH 7.0 at a flow rate of 2.0 ml/min, before loading sample (*see* **Note 6**).
2. Load sample onto column. Sample volume should not exceed 4.0 ml for best separation (*see* **Notes 7 and 8**).
3. Monitor A_{260}/A_{280} nm while collecting void volume in 1-ml fractions. Ad/PEG elutes in the void volume; unreacted TMPEG elutes as a smaller peak in the included volume (*see* **Notes 9–10**).
4. Add sucrose to collected peak fractions to a final concentration of 5.0% (*see* **Note 11**).

3.2.3. Generation of an Ad/PEG/FGF2 Complex

For easy conjugation to the maleimide group on Ad/PEG-mal, a 155-amino acid FGF2 (bFGF) mutein, in which the cysteine at position 96 is mutagenized to serine ***(61)***, is used. Of the four cysteines in FGF2, two surface cysteines are predicted by structural studies to be the probable sites of conjugation ***(61)***. Mutagenizing one of these cysteines to a serine generates a monoreactive cysteine FGF2; this minimizes the formation of FGF2/virion aggregates. As an alternative native basic FGF (FGF2) can also be used; however, this may result in a more heterogeneous population of Ad/PEG/FGF2 conjugates because of the presence of two reactive sulfhydryls on the surface of the FGF2 molecule. If native basic FGF is being used, care must be taken to work under reducing conditions. This ensures that the surface sulfhydryls are available for coupling to the maleimide group on the Ad/PEG–mal complex.

The maleimide group is specific for sulfhydryl groups (*see* **Fig. 3B**) when the pH of the reaction is maintained between pH 6.5 and 7.5. At pH 7, the reaction of maleimides with sulfhydryls is 1000-fold faster than with amines. Above this pH, the reaction with amines becomes more significant; thus, it is important to work at physiological pH. EDTA is included in the coupling buffer to prevent oxidation of the sulfhydryls. A stable thioether linkage between the maleimide group and the reacted sulfhydryl cannot be cleaved under physiological conditions.

3.2.3.1. Preparation of FGF2 Mutein Protein for Coupling to Ad/PEG

Determine amount of FGF2 to be conjugated.

1 Assume 1000 PEG/virion for 1% Ad/PEG-mal, ***(50)***; [PEG] = 1000 × [virus], using a molecule: molecule calculation ***(49)***.

2 Dialyze FGF2 using a Slide-A-Lyzer cassette into Dialysis Buffer A. Dialyze overnight at 4ºC.
3 Add a 1000 molar excess of FGF2 mutein protein to purified Ad/PEG and incubate for 6 h at 4ºC with gentle inversion (*see* **Note 12**).
4 If using native basic FGF2, dialyze FGF2 into dH_2O at a final protein concentration of 500 µg/ml. DTT is added to a final concentration of 25 mM; incubate on ice for 30 min to reduce disulfide bonds between surface sufhydryls. Remove DTT immediately using Centrispin 6 columns (Biorad) equilibrated in PBS pH 7.2. Dialyze against PBS pH 7.2 for an additional 2 h and then proceed as directed from **step 3** of this section (*see* **Note 13**).
5 If using a protein that does not have available sulfhydryls for coupling, see **Note 14**.

3.2.3.2. Purification of Ad/PEG/FGF2 by Ion Exchange Chromatography

DEAE is an anion exchange resin thus, Ad/PEG/FGF2 and free FGF2 are separated on the basis of charge. FGF2 is a basic protein with a pI of 9.1 whereas the major coat protein on the Ad capsid, hexon, has a lower pI of 4. Under the salt and pH conditions used during chromatography, Ad/PEG/FGF2 binds to the DEAE resin whereas uncoupled FGF2 is recovered in the flow through.

1. Prepare 10.0 ml of Fractogel DEAE in a XK16/60 column (*see* **Notes 15 and 16**).
2. Prepare buffer A.
3. Prepare buffer B.
4. Equilibrate the DEAE resin with buffer A at 2 ml/min; this is twice the flow rate used in the actual run (*see* **Note 17**).
5. After loading sample, wash the column with 1 column volume (CV) of Buffer A before starting the ionic gradient actual run (*see* **Note 17**).
6. The gradient is increased to 12.5% buffer B for 2 CV; gradient is increased again to 100% buffer B over 4 CV.
7. Ad/PEG/FGF2 will be eluted at about 25% buffer B, equivalent to 0.5 M salt.
8. Monitor column flow through with an A_{260}/A_{280} nm wavelength detector; free FGF2 is recovered in the flow-through fraction.
9. When the absorbance increases, collect 1 ml fractions across the Ad/PEG/FGF2 peak.
10. These fractions are pooled and 50% sucrose is added to a final concentration of 5% (*see* **Notes 18 and 19**).

3.3. Determination of Particle Number of Ad/PEG/FGF2 Complexes

Particle concentration of Ad/PEG/FGF2 is determined as described in **Subheading 3.2.3.2**. The concentration of the Ad/PEG/FGF2 sample is estimated from a standard curve of unmodified virus. Peak areas from DEAE chromatography of unmodified virus in the range of 10^9–10^{12} viral particles comprise the standard curve. The viral concentration of modified virus is determined by using line regression analysis of the standard curve.

3.3.1. Quantitation of FGF2 on Ad/PEG/FGF2 Conjugates

1. Basic FGF (FGF2) ELISA kit. The ELISA kit used employs the quantitative sandwich enzyme immunoassay technique. Ad/PEG/FGF2 samples are serially diluted twofold from 1:25 in PBS and assayed according to kit instructions.
2. A FGF2-specific monoclonal antibody is immobilized on the plate and reacts with FGF2 conjugated to the virus and FGF2 standards.
3. A secondary monoclonal antibody linked to horseradish peroxide (HRP) enzyme reacts with the bound substances and can be detected by chemiluminescence.
4. Color development is proportional to the amount of FGF2 bound.
5. Regression analysis can be used to determine the FGF2 concentration using the standard curve of human recombinant FGF2.

3.4. *In vitro* Transduction Assays

The efficiency of gene transfer of Ad/PEG/FGF2 conjugates is compared with unmodified Ad by measuring transgene expression (β-galactosidase) in cells following infection with equal particles of modified and unmodified vector. Transduction assays are performed on SKOV3.ip1 cells that express FGF receptors (*see* **Notes 20–22**).

3.4.1. Assessment of Transduction Efficiency of Ad/PEG/FGF2 Complexes

1. SKOV3.ip1 cells are maintained in F-12K Kaighn's modified media in a 37°C/5% CO_2/95% humidity incubator.
2. 2.5 e x 10^4 cells/500 μl media per well are plated in a 12-well cell culture plate.
3. 24 h later, dilute Ad2/PEG/FGF2 in media to desired particle number/cell (range 50–500) and add to cells in triplicate. Place cells in 37°C incubator, infect for 1 h.
4. Aspirate infection media, rinse wells, and then incubate with 1 ml/well fresh media.
5. Lyse cells 48 h post-infection to assay for β-galactosidase expression.
6. Aspirate media and add 500 μl lysis buffer (provided in Tropix kit) per well. Plates are incubated at room temperature for 10 min.
7. Add 100 μl of a 2% substrate solution per well using a 96-well opaque plate.
8. Add 20 μl of supernatant from lysed cells to the substrate solution in the 96-well plate in quadruplicate and incubate for 1 h at room temperature before reading in a luminometer.

3.4.2. Specificity of Binding of Ad/PEG/FGF2 to the FGF Receptor: Competitive Binding Assays

It is important to demonstrate that the Ad2/PEG/FGF2 complex is binding to the target FGF2 receptor and not to its native receptor, CAR. To demonstrate this, transduction assays are performed in the presence of competitors such as (1) fiber knob protein and (2) excess FGF2 mutein protein.

1. SKOV3.ip1 cells maintained in F12K Kaighn's modified media, 5×10^4 cells/well, are seeded in a 12-well cell culture plate.
2. 24 h later, FGF2 protein (0.5 μg/ml) or excess Ad2 fiber knob protein (1 μg/ml) is added to the cells. A control group of cells with no protein added is also included.
3. Cells plus protein are incubated for 2 h at 4°C (*see* **Note 23**).
4. Equivalent particles (500) of Ad or Ad/PEG/FGF2 are added to the cells, and the cells are incubated with virus at 37°C for 4 h.
5. Following infection fresh media is added to the cells.
6. Cells are harvested 48 h later and assayed for β-galactosidase activity as described in **Subheading 3.4.1**.

3.4.3. Endothelial Cell Proliferation Assay

FGF2 can bind to two receptors in vivo and in vitro. The target receptor is the high-affinity FGF receptor, but FGF2 can also bind to low-affinity heparin receptors. It is important to demonstrate that the Ad/PEG/FGF2 complex is binding and activating the high-affinity receptor. To demonstrate this, an endothelial cell proliferation assay is performed, which distinguishes between binding to either the low- or high-affinity receptor. The Ad/PEG/FGF2 complex should bind to the high-affinity receptor on endothelial cells and cause cell proliferation. Failure to promote proliferation indicates that the Ad/PEG/FGF2 complex is not binding to the target high-affinity receptor.

1. Adult bovine endothelial cells (ATTC) are maintained in modified DMEM at 37°C/5% CO_2.
2. 2×10^3 cells/well are seeded in a 96-well flat-bottomed cell culture plate in 100 μl of media.
3. Incubate cells at 37°C/5% CO_2 for 24 h.
4. Add FGF2 protein as a serial dilution curve of 400–0.4 pM. Samples are done in triplicate.
5. In a separate set of wells, add Ad2/PEG/FGF2 to achieve this same FGF2 concentration range. The concentration of FGF2 on the Ad2/PEG/FGF2 is determined from the number of FGF2 molecules attached to the Ad2/PEG/FGF2 (determined as described in **Subheading 3.3.1**.). Samples are done in triplicate.
6. The following controls are also included: Ad/PEG plus FGF2, Ad/PEG alone. In the controls, all inputs are adjusted to achieve concentrations equivalent to those of the corresponding Ad/PEG/FGF2 groups. All controls are done in triplicate.

7. After 5 days in culture, cellular proliferation is determined using a cell proliferation kit [CellTiter 96® AQ_{ueous} cell proliferation assay (MTS) (Promega)].
8. Add 20 µl CellTiter reagent per well.
9. Incubate 37ºC/5% CO_2 for exact 4 h.
10. Read absorbance on plate reader at 490 nm.

3.5. In Vivo Infection of Intraperitoneal Tumors with Ad/PEG/FGF2

An animal model of human ovarian carcinoma in CB-17 SCID mice is used to measure the efficiency of transduction of Ad/PEG/FGF2 conjugates in vivo. Xenografting of the SKOV3.ip1 cell line has been shown to allow the establishment of a disease model with many of the characteristics of human ovarian carcinoma ***(57)***. In studies described here, an adenoviral vector with a β-galactosidase transgene is used however, for a therapeutic application of a FGF2-retargeted vector, others have used an adenoviral vector expressing the herpes simplex virus thymidine kinase gene. Expression of the viral thymidine kinase gene renders cells sensitive to the toxic effects of nucleoside analogs such as ganciclovir ***(30)***. FGF2-retargeted adenovirus-mediated delivery of the herpes simplex virus thymidine kinase gene results in augmented therapeutic benefit in a murine model of ovarian cancer ***(30)***.

3.5.1. Preparation of SKOV3.ip1 Cells and Generation of a Human Ovarian Cancer Mouse Model for In Vivo Ad/PEG/FGF2 Targeting

1. Determine the number of SKOV3.ip1cells required for experiments (2×10^7 cells/mouse × number of mice).
2. 4×10^6 cells are seeded into each T150 flask and grown to 50% confluency (1.5×10^7 cells/flask).
3. To harvest cells, rinse the cell layer with PBS then add 0.05% (w/v) trypsin-EDTA to flask.
4. Wait 5–15 min until cells are rounded.
5. Lightly tap flask to detach cells, then add complete F12K Kaighn's medium and pipette cells into conical tube.
6. Spin at 1200 × *g* for 5 min, remove media, and resuspend in PBS, 5% sucrose pH 7.
7. Inject 2×10^7 cells in 200 µl intraperitoneum (i.p.) per mouse. Tumors will grow to approximately 1–3 mm diameter in 14 days.
8. On day 14, inject 6×10^{10} particles of Ad, Ad/PEG, or Ad/PEG/FGF2 i.p. per mouse.
9. Kill mice 48 h post-injection.
10. Remove tumors from the peritoneal cavity and store on dry ice or frozen at –80ºC.
11. Remove liver (1/10 the total weight) and spleen (1/3 the total weight) for Taqman analysis. When removing tissue for PCR analysis, care should be taken not to

contaminate the samples. A different set of instruments should be used per animal and per tissue to prevent cross contamination. In addition, the instruments used to open the animal should be different from those used to remove the tissue. All tubes used for tissue storage should be RNAse and DNAse free.

12. Prepare tumor lysis buffer.
13. Thaw tumors on ice and add 0.5 ml tumor lysis buffer per 100 μg tissue.
14. Homogenize tumors on ice for 10–15 s at lowest setting.
15. Centrifuge homogenates at 10,000 x *g* for 10 min at 4°C.
16. Transfer supernatant to fresh tubes.
17. Store samples at –80°C until ready to assay.

3.5.2. Measurement of Transgene Expression and Viral DNA in Tumor Samples

3.5.2.1. β-GALACTOSIDASE ELISA

This assay allows quantitative measurement of β-gal expression in cells at picogram levels. In this sandwich ELISA, β-gal antibodies are pre-bound on the surface of a microtiter plate. Cell lysates from transduced cells expressing β-gal are added to the plate and are bound by the antibodies present. Digoxigenin-labeled antibody to β-gal is added to the plate, followed by a peroxidase-conjugated antibody to digoxigenin. In the next step, the peroxidase substrate is cleaved yielding a green color product; absorbance of product is read at A_{405} nm.

1. Thaw supernatants completely at room temperature.
2. Centrifuge at 10,000 × *g* for 10 min at 4°C.
3. Transfer supernatants to clean tubes and discard pellets.
4. Dilute samples 1:50 for the β-gal ELISA and BCA protein assays.

3.5.2.2. TAQMAN ANALYSIS OF HEXON DNA

This assay uses fluorescent Taqman methodology and an instrument capable of measuring fluorescence (ABI Prism 7700 Sequence Detector). The Taqman reaction requires a hybridization probe labeled with two different fluorescent dyes. One dye is a reporter dye (FAM) and the other is a quenching dye (TAMRA). During the PCR, probe degradation, by 5´-3´ exonuclease activity of the Taq polymerase, releases the FAM from the quenching influence of the TAMRA, leading to an increase in fluorescence. The measurement of emission-fluorescence is performed at each cycle and is normalized for each sample by the emission intensity of an internal reference dye, ROX (6-carboxy-X-rhodamine). The resultant value, ΔRn, is plotted against cycle number; the significant value is the *Ct* threshold cycle, which is the cycle number at which the ΔRn has reached a threshold ***(63)***. A standard curve is then calculated by plotting the threshold cycle *Ct* versus input copy number. PCR primers and probes were designed for the adenovirus hexon capsid protein; hence, this assay allows for the measurement of virus in tissue samples rather than the measurement of transgene expression (*see* **Notes 24–27**).

The forward primer sequence is CAGACCTGGGCCAAAACCTT and the reverse primer TCCACCTCAAAGTCATGTCTAGC and the probe sequence is TCTACGCCAACTCCGCCCACG.

1. Add 1/8′′-1/4′′ zirconia beads and 1 ml DNA lysis buffer.
2. Homogenize for 2–3 min in mini-bead beater.
3. Add 100 μl proteinase K and briefly vortex.
4. Incubate 12–18 h at 55–60ºC.
5. Before proceeding, make sure all tissue pieces are mostly digested.
6. If not, add more proteinase K and digest longer.
7. Centrifuge (16,000 x *g*, 10 min).
8. Transfer 500 μl lysate to tube containing 500 μl phenol : chloroform : IAA (isoamyl alcohol).
9. Vortex for 20 s and centrifuge (14000 × *g*, 15 min).
10. Transfer aqueous layer (~450 μl) to tube containing 1 ml absolute ethanol.
11. Invert tube several times and store up to 30 min at –20ºC.
12. Centrifuge (11,500 × *g*, 10 min).
13. Wash with 400 μl 75% ethanol.
14. Centrifuge (7,500 × *g*, 5 min).
15. Resuspend pellet in water.
16. Quantitate using OD_{260}/ OD_{280}.
17. Make working dilution at 100 ng/μl in water.
18. Add 45 μl PCR mix to each well.
19. Add 5 μl of every standard and DNA in duplicate.
20. Reserve four wells for negative water controls.

3.6. Characterization of the Innate Immune Response to PEGylated Vectors

To demonstrate that reduced localization of Ad/PEG/FGF2 to liver and spleen can help ablate cytokine responses and liver toxicity, an in vivo experiment must be performed. This includes systemic administration of Ad or Ad/PEG/FGF2 complexes to mice followed by serum analysis. Serum is analyzed 6 h post vector administration to monitor the early phase of the innate immune response (*see* **Subheading 1**). In addition, sera from later time points, 8 and 14 days, are also analyzed to monitor later effects such as liver toxicity.

3.6.1. Study Aim: To Compare Liver Toxicity and Cytokine Profile After Systemic Administration of Ad or Ad/PEG/FGF2 to Mice

1. Balb/C mice: 22 mice in total.
2. Test articles Ad, Ad/PEG/FGF2 at a dose of 1 × 10^{11} particles/200 μl (*see* **Note 28**).

3 Divide mice into the following groups: Ad (8 mice total), Ad/PEG/FGF2 (8 mice total), vehicle (PBS) (8 mice total).
3. All mice are given test article by tail vein injection, maximum volume recommended is 200 μl or by intraperitoneal injection (200 μl).
4. 6 h after vector administration, four animals in each group are killed and serum fraction is recovered for analysis.
5. The remaining four animals in each group are subjected to eye bleeds 8 days post vector administration and are killed 14 days post vector administration. Serum fraction is retained for analysis after killing.
6. The following analysis are performed on the collected sera:

Sera collected 6 h, 8 and 14 days post vector administration is analyzed for IL-12, IL- 6, IFN-γ, and TNF-α

Sera collected 8 and 14 days post vector administration is analyzed for AST and ALT using enzyme kits, according to manufacturer's instructions (*see* **Notes 28–29**).

3.7. Characterization of the Adaptive Immune Response to PEGylated Vectors

Activation of APCs and B cells (*see* **Fig. 2**) by input viral capsid proteins underlies the mechanism responsible for the production of the humoral immune response to Ad vectors. Activated $CD4^+$ T cells (helper T cells) stimulate B cells to proliferate and secrete neutralizing antibodies against the Ad vector. A direct correlation between neutralizing antibody and the block to readministration of vector has been established by passive transfer of immunity by sera from treated to naïve animals ***(64)***. In addition, $CD4^+$ T cells contribute to CTL-mediated clearance of Ad-transduced cells by stimulating $CD8^+$T cells (*see* **Fig. 2**). This constitutes the cellular immune response to Ad vectors. There are two functionally distinct subclasses of helper T cells that can be distinguished by the interleukins that they secrete upon activation. Th1 cells secrete IL-2 and γ-interferon and are concerned with helping CTLs. Th2 cells secrete IL-4, IL-5, and IL-10 and are concerned mainly with helping B cells. In this section, experiments are described to compare the Th1 and Th2 response to PEGylated vectors, in addition, experiments to measure the transduction efficiency of the modified vectors in the presence of anti-Ad neutralizing sera will be described.

3.7.1. Evaluation of Th1 and Th2 Cell Responses to PEGylated Vectors

3.7.1.1. Study Aim: To Characterize the Cytokines Released from in Vitro Stimulation of Splenocytes Isolated from Animals Injected with PEGylated or Unmodified Ad Vectors

1. Balb/C mice—24 total.
2. Test articles Ad, Ad/PEG/FGF2 at a dose of 2.5×10^{10} particles per 200 μl.

3. Divide mice into the following groups Ad (8 animals), Ad/PEG/FGF2 (8 animals), and vehicle control (8 animals). Vehicle control is PBS.
4. Vector is administered to the animals by tail vein injection or intraperitoneal injection.
5. 11 days past vector administration, animals are killed and the spleens are removed aseptically and placed in UltraCulture media.
6. Crush each spleen between frosted ends of sterile glass slides in a Petri dish with 10 ml of ultraculture media.
7. Let settle, then transfer supernatant to a 15-ml conical tube.
8. Spin cells for 5 min at 1200 × *g*.
9. Resuspend in 2 ml cold 0.83% NH_4Cl at room temperature for 2 min.
10. Transfer to 50-ml tube with 23 ml of media, count cells.
11. Spin 5 min at 1200 × *g* and resuspend to desired concentration (2×10^6 cells/ml) in modified Ultraculture media.
12. Aliquot 5 ml of the spleen cells /tube into sterile polystyrene tubes.
13. Add 2.5 μg of heat-inactivated (inactivated at 42ºC for 20 min) recombinant unmodified Ad2. Ad2 is at a concentration of 1×10^8 viral particles/ml.
14. Transfer cells (1×10^6) from tube to 24-well plate in 1 ml/well.
15. Incubate for 72 h at 37°C, 5% CO_2.
16. Analyze the cell supernatant with commercial ELISA kits for IL-12, IL-10, IFN-γ, IL-[4]

3.7.2. Protection from Neutralization Assay

Transduction assays are performed on SKOV3.ip1 cells that express FGF receptors. The efficiency of gene transfer of Ad/PEG/FGF2 conjugates in the presence of human anti-Ad-neutralizing serum is compared with gene transfer by unmodified Ad in the presence of anti-Ad-neutralizing serum.

1. SKOV3.ip1 cells are maintained in F-12K Kaighn's modified media in a 37ºC/5% CO_2/95% humidity incubator.
2. 5×10^4 cells /well are plated in a 12-well cell culture plate.
3. Ad or Ad2/PEG/FGF2 is added to serial dilutions of heat-inactivated human anti-adenovirus-neutralizing serum (1:400–1:1600) and incubated for 1 h.
4. Human serum is inactivated by heating at 56ºC for 30 min.
5. Following incubation with serum, equal numbers of particles (500 particles/cell) of Ad or Ad2/PEG/FGF2 are added to SKOV3.ip1 cells in triplicate.
6. Place cells in 37ºC incubator; infect for 1 h.
7. Aspirate infection media, rinse wells, and then incubate with 500 μl–1 ml/well fresh media.
8. Lyse cells 48 h post-infection for β-galactosidase expression as described below.
9. Aspirate media and add 500 μl lysis buffer (provided in Tropix kit) per well. Plates are incubated at room temperature for 10 min.
10. Add 100 μl of a 2% substrate solution per well using a 96-well opaque plate.

11. Add 20 μl of supernatant from lysed cells to the substrate solution in the 96-well plate in quadruplicate and incubate for 1 h at room temperature before reading in a luminometer.

4. Notes

1. Adenovirus should be diluted with a sodium phosphate buffer pH 7 before the addition of TMPEG. At high concentrations of TMPEG, contaminating TFSA and/or HCl in the TMPEG can lower the pH of the solution. Lowering the pH below 6.5 promotes dissociation of the Ad particle into its constituent polypeptides ***(65)***. The buffering capacity of PBS alone is insufficient to maintain pH during PEGylation at high concentrations of TMPEG.
2. Virus should be placed on ice for thawing and for short-term storage during experiments. For long-term storage, virus should be kept at –80°C.
3. Do not vortex solutions containing virus. Gently tap or invert to mix.
4. Flush all buffers with N_2 to remove dissolved O_2.
5. TMPEG should be stored at –80°C under nitrogen in a vacuum pouch with dessicant.
6. Equilibration of the Fractogel BioSec resin should be confirmed by conductivity and pH measurement of the equilibration buffer after passing through the resin.
7. The conductivity of the sample must not exceed that of the equilibration buffer. The pH of the sample should equal, within 5%, the pH of the buffers used.
8. Volume of load sample should be < % of CV.
9. Void volume represents 25–30% of the total CV.
10. For maximum separation of Ad/PEG from unreacted TMPEG, flow rates of 0.5–1.5 ml/cm^2 × min based on the cross-sectional area of the column should be used.
11. The Fractogel BioSec resin is cleaned with 0.5 M NaOH and stored in 20% ethanol, 0.15 M NaCl.
12. An excess of targeting ligand needs to be prepared to maximize the amount of ligand coupled to PEG. Aim to add approximately 1,000-fold excess FGF2 over assumed PEG concentration in the virus stock.
13. If native FGF2 protein is used, there is the possibility that oligomers, particularly dimers can form between Cys-78 and Cys 96 ***(61)***. If this is suspected, mild reduction of these native FGF2 protein dimers is suggested [i.e., 25 mM DTT (dithiothreitol) for 30 min on ice]. Rapid removal of DTT is suggested using a Bio-spin 6 GF column (Biorad) followed by extensive dialysis.
14. For proteins that do not have an available sulfhydryl for coupling, Traut's reagent (Pierce) can be used to introduce sulfhydryls. Quantitate introduced sulfhydryls on target protein using a thiol and sulfide quantitation kit (Molecular Probes, Eugene, OR, USA). Follow manufacturer's instructions except reduce all volumes fivefold to run on a microplate. Include a known concentration (20–80 μM) of Traut's reagent and unthiolated target protein as controls.

15. The Fractogel resin series from EM sciences has the preferred architecture for separation of adenoviral virions. It is a tentacle-like design that reduces trapping of the virus particle in the pores of the resin ***(66,67)***.
16. After packing the column, compress the column manually using the column adapter and plunger to reduce the CV by 30%. This compression will allow for the best separation of proteins.
17. Equilibration of resin should be confirmed by conductivity and pH measurement of equilibration buffer after it has flowed through the resin.
18. Fractogel DEAE column is cleaned with 1 CV 1 M sodium chloride and 0.01 M hydrochloric acid, then 4 CV of 0.5 M NaOH.
19. Fractogel DEAE resin is stored in 20% ethanol and 0.1 M sodium chloride.
20. To ensure maximum receptor density during infection, cultures must not exceed 50% confluency before infection. The FGF-receptor density decreases with increased cell culture confluency ***(30)***.
21. To ensure that the Ad/PEG/FGF2 vector no longer recognizes its native receptor CAR, transduction assays should be performed on 293 or HeLa cells, which are known to have a high density of the CAR receptor.
22. It is important that all transduction assays are performed on the basis of equal particle number for modified and unmodified vectors.
23. FGF2 and soluble fiber are pre-bound to their respective receptors by pre-incubation at 4ºC. These ligands are pre-bound at low temperature to prevent receptor-mediated endocytosis of the ligand.
24. Taqman is a very sensitive technique. A small room or a section of the laboratory should be dedicated entirely for Taqman analysis.
25. Change gloves often to avoid contamination. Disinfect bench and cover working areas.
26. All DNA work is performed in a hood.
27. Use the highest quality equipment and reagents (i.e., no DNAse, RNAse, or protease activity).
28. To measure the immediate cytokine response following systemic administration of adenovirus, a minimum dose of 1×10^{11} vector particles must be administered. Doses below this value do not give measurable cytokine responses.
29. Following systemic administration of Ad vectors (doses greater than 1×10^{11} particles per animals), there is a biphasic rise in liver enzymes that occurs around 8 days and then 14 days post vector administration. It is necessary to assay serum samples at these two time points for both aminotransferases AST and ALT (liver enzymes), which are markers of liver damage.

References

1. Bergelson JM, Cunningham JA, Droguett G, Kurt-Jones EA, Krithivas A, Hong JS, Horwitz MS, Crowell RL, Finberg RW. (1997) Isolation of a common receptor for Coxsackie B viruses and Adenoviruses 2 and 5. *Science* **275**(5304):1320–3.

2. Tomko RP, Xu R, Philipson L. (1997) HCAR and MCAR: the human and mouse cellular receptors for subgroup C adenoviruses and group B coxsackieviruses. *Proc Natl Acad Sci USA* **94**(7):3352–6.
3. Wickham TJ, Filardo EJ, Cheresh DA, Nemerow GR. (1994) Integrin alpha v beta 5 selectively promotes adenovirus mediated cell membrane permeabilization. *J Cell Biol* **127**(1):257–64.
4. Wickham TJ, Mathias P, Cheresh DA, Nemerow GR. (1993) Integrins alpha v beta 3 and alpha v beta 5 promote adenovirus internalization but not virus attachment. *Cell* **73**(2):309–19.
5. Nakamura T, Sato K, Hamada H. (2003) Reduction of natural adenovirus tropism to the liver by both ablation of fiber-coxsackievirus and adenovirus receptor interaction and use of replaceable short fiber. *J Virol* **77**(4):2512–21.
6. Smith TA, Idamakanti N, Rollence ML, Marshall-Neff J, Kim J, Mulgrew K, Nemerow GR, Kaleko M, Stevenson SC. (2003) Adenovirus serotype 5 fiber shaft influences in vivo gene transfer in mice. *Hum Gene Ther* **14**(8):777–87.
7. Smith TA, Idamakanti N, Marshall-Neff J, Rollence ML, Wright P, Kaloss M, King L, Mech C, Dinges L, Iverson WO, Sherer AD, Markovits JE, Lyons RM, Kaleko M, Stevenson SC. (2003) Receptor interactions involved in adenoviral-mediated gene delivery after systemic administration in non-human primates *Hum Gene Ther* **14**(17):1595–604.
8. Vigne E, Dedieu JF, Brie A, Gillardeaux A, Briot D, Benihoud K, Latta-Mahieu M, Saulnier P, Perricaudet M, Yeh P. (2003) Genetic manipulations of adenovirus type 5 fiber resulting in liver tropism attenuation. *Gene Ther* **10**(2): 153–62.
9. Bruder JT, Kovesdi I. (1997) Adenovirus infection stimulates the Raf/MAPK signaling pathway and induces interleukin-8 expression. *J Virol* **71**(1):398–404.
10. Lieber A, He CY, Meuse L, Schowalter D, Kirillova I, Winther B, Kay MA. (1997) The role of Kupffer cell activation and viral gene expression in early liver toxicity after infusion of recombinant adenovirus vectors. *J Virol* **71**(11):8798–807.
11. Muruve DA, Barnes MJ, Stillman IE, Libermann TA. (1999) Adenoviral gene therapy leads to rapid induction of multiple chemokines and acute neutrophil-dependent hepatic injury in vivo. *Hum Gene Ther* **10**(6):965–76.
12. Shifrin AL, Chirmule N, Gao GP, Wilson JM, Raper SE. (2005) Innate immune responses to adenoviral vector-mediated acute pancreatitis. *Pancreas* **30**(2): 122–9.
13. Elkon KB, Liu CC, Gall JG, Trevejo J, Marino MW, Abrahamsen KA, Song X, Zhou JL, Old LJ, Crystal RG, Falck-Pedersen E. (1997) Tumor necrosis factor alpha plays a central role in immune-mediated clearance of adenoviral vectors. *Proc Natl Acad Sci USA* **94**(18):9814–9.
14. Zaiss AK, Liu Q, Bowen GP, Wong NC, Bartlett JS, Muruve DA. (2002) Differential activation of innate immune responses by adenovirus and adeno-associated virus vectors. *J Virol* **76**(9):4580–90.
15. Liu Q, Muruve DA. (2003) Molecular basis of the inflammatory response to adenovirus vectors. *Gene Ther* **10**(11):935–40.

16. Jooss K, Ertl HC, Wilson JM. (1998) Cytotoxic T-lymphocyte target proteins and their major histocompatibility complex class I restriction in response to adenovirus vectors delivered to mouse liver. J Virol. **72**(4):2945–54.
17. Yang Y, Jooss KU, Su Q, Ertl HC, Wilson JM. (1996) Immune responses to viral antigens versus transgene product in the elimination of recombinant adenovirus-infected hepatocytes in vivo. Gene Ther. **3**(2):137–44.
18. Yang Y, Wilson JM. (1995) Clearance of adenovirus-infected hepatocytes by MHC class I-restricted CD4+ CTLs in vivo. J Immunol. **155**(5):2564–70.
19. Leissner P, Legrand V, Schlesinger Y, Hadji DA, van Raaij M, Cusack S, Pavirani A, Mehtali M. (2001) Influence of adenoviral fiber mutations on viral encapsidation, infectivity and in vivo tropism. Gene Ther. **8**(1):49–57.
20. Alemany R, Curiel DT. (2001) CAR-binding ablation does not change biodistribution and toxicity of adenoviral vectors.Gene Ther. **8**(17):1347–53.
21. Einfeld DA, Schroeder R, Roelvink PW, Lizonova A, King CR, Kovesdi I, Wickham TJ. (2001) Reducing the native tropism of adenovirus vectors requires removal of both CAR and integrin interactions. J Virol. **75**(23):11284–91.
22. Mizuguchi H, Koizumi N, Hosono T, Ishii-Watabe A, Uchida E, Utoguchi N, Watanabe Y, Hayakawa T. (2002) CAR- or alphav integrin-binding ablated adenovirus vectors, but not fiber-modified vectors containing RGD peptide, do not change the systemic gene transfer properties in mice. Gene Ther. **9**(12):769–76.
23. Smith T, Idamakanti N, Kylefjord H, Rollence M, King L, Kaloss M, Kaleko M, Stevenson SC. (2002) *In vivo* hepatic adenoviral gene delivery occurs independently of the coxsackievirus-adenovirus receptor. Mol Ther. **5**(6):770–9.
24. Koizumi N, Kawabata K, Sakurai F, Watanabe Y, Hayakawa T, Mizuguchi H. (2006) Modified adenoviral vectors ablated for coxsackievirus-adenovirus receptor, $alpha_v$ integrin, and heparan sulfate binding reduce in vivo tissue transduction and toxicity. Hum Gene Ther. **17**(3):264–79.
25. Wickham TJ, Segal DM, Roelvink PW, Carrion ME, Lizonova A, Lee GM, Kovesdi I. (1996) Targeted adenovirus gene transfer to endothelial and smooth muscle cells by using bispecific antibodies. J Virol. **70**(10):6831–8.
26. Douglas JT, Rogers BE, Rosenfeld ME, Michael SI, Feng M, Curiel DT. (1996) Targeted gene delivery by tropism-modified adenoviral vectors. Nat Biotechnol. **14**(11):1574–8
27. Wickham TJ, Lee GM, Titus JA, Sconocchia G, Bakacs T, Kovesdi I, Segal DM. (1997) Targeted adenovirus-mediated gene delivery to T cells via CD3. J Virol. **71**(10):7663–9.
28. Goldman CK, Rogers BE, Douglas JT, Sosnowski BA, Ying W, Siegal GP, Baird A, Campain JA, Curiel DT. (1997) Targeted gene delivery to Kaposi's sarcoma cells via the fibroblast growth factor receptor. Cancer Res. **57**(8):1447–51.
29. Rogers BE, Douglas JT, Ahlem C, Buchsbaum DJ, Frincke J, Curiel DT. (1997) Use of a novel cross-linking method to modify adenovirus tropism. Gene Ther. **4**(12):1387–92.
30. Rancourt C, Rogers BE, Sosnowski BA, Wang M, Piche A, Pierce GF, Alvarez RD, Siegal GP, Douglas JT, Curiel DT. (1998) Basic fibroblast growth

factor enhancement of adenovirus-mediated delivery of the herpes simplex virus thymidine kinase gene results in augmented therapeutic benefit in a murine model of ovarian cancer. Clin Cancer Res. **4**(10):2455–61.

31. Gu DL, Gonzalez AM, Printz MA, Doukas J, Ying W, D'Andrea M, Hoganson DK, Curiel DT, Douglas JT, Sosnowski BA, Baird A, Aukerman SL, Pierce GF. (1999) Fibroblast growth factor 2 retargeted adenovirus has redirected cellular tropism: evidence for reduced toxicity and enhanced antitumor activity in mice. Cancer Res. **59**(11):2608–14.
32. Blackwell JL, Miller CR, Douglas JT, Li H, Reynolds PN, Carroll WR, Peters GE, Strong TV, Curiel DT. (1999) Retargeting to EGFR enhances adenovirus infection efficiency of squamous cell carcinoma. Arch Otolaryngol Head Neck Surg. **125**(8):856–63.
33. Doukas J, Hoganson DK, Ong M, Ying W, Lacey DL, Baird A, Pierce GF, Sosnowski BA. (1999) Retargeted delivery of adenoviral vectors through fibroblast growth factor receptors involves unique cellular pathways. FASEB J. **13**(11): 1459–66.
34. Yoon SK, Mohr L, O'Riordan CR, Lachapelle A, Armentano D, Wands JR. (2000) Targeting a recombinant adenovirus vector to HCC cells using a bifunctional Fab-antibody conjugate. Biochem Biophys Res Commun. **272**(2):497–504.
35. Trepel M, Grifman M, Weitzman MD, Pasqualini R. (2000) Molecular adaptors for vascular-targeted adenoviral gene delivery. Hum Gene Ther. **11**(14): 1971–81.
36. Tillman BW, Hayes TL, DeGruijl TD, Douglas JT, Curiel DT. (2000) Adenoviral vectors targeted to CD40 enhance the efficacy of dendritic cell-based vaccination against human papillomavirus 16-induced tumor cells in a murine model. Cancer Res. **60**(19):5456–63.
37. Reynolds PN, Zinn KR, Gavrilyuk VD, Balyasnikova IV, Rogers BE, Buchsbaum DJ, Wang MH, Miletich DJ, Grizzle WE, Douglas JT, Danilov SM, Curiel DT. (2000) A targetable, injectable adenoviral vector for selective gene delivery to pulmonary endothelium in vivo. Mol Ther. **2**(6):562–78.
38. Ebbinghaus C, Al-Jaibaji A, Operschall E, Schoffel A, Peter I, Greber UF, Hemmi S. (2001) Functional and selective targeting of adenovirus to high-affinity Fc gamma receptor I-positive cells by using a bispecific hybrid adapter. J Virol. **75**(1):480–9.
39. Hoganson DK, Sosnowski BA, Pierce GF, Doukas J. (2001) Uptake of adenoviral vectors via fibroblast growth factor receptors involves intracellular pathways that differ from the targeting ligand. Mol Ther. **3**(1):105–12.
40. Grill J, Van Beusechem VW, Van Der Valk P, Dirven CM, Leonhart A, Pherai DS, Haisma HJ, Pinedo HM, Curiel DT, Gerritsen WR. (2001) Combined targeting of adenoviruses to integrins and epidermal growth factor receptors increases gene transfer into primary glioma cells and spheroids. Clin Cancer Res. **7**(3): 641–50.
41. Nettelbeck DM, Miller DW, Jerome V, Zuzarte M, Watkins SJ, Hawkins RE, Muller R, Kontermann RE. (2001) Targeting of adenovirus to endothelial cells by a

bispecific single-chain diabody directed against the adenovirus fiber knob domain and human endoglin (CD105). Mol Ther. **3**(6):882–91.
42. Israel BF, Pickles RJ, Segal DM, Gerard RD, Kenney SC. (2001) Enhancement of adenovirus vector entry into CD70-positive B-cell lines by using a bispecific CD70-adenovirus fiber antibody. J Virol. **75**(11):5215–21.
43. Smith JS, Keller JR, Lohrey NC, McCauslin CS, Ortiz M, Cowan K, Spence SE. (1999) Redirected infection of directly biotinylated recombinant adenovirus vectors through cell surface receptors and antigens. Proc Natl Acad Sci U S A. **96**(16):8855–60
44. Romanczuk H, Galer CE, Zabner J, Barsomian G, Wadsworth SC, O'Riordan CR. (1999) Modification of an adenoviral vector with biologically selected peptides: a novel strategy for gene delivery to cells of choice. Hum Gene Ther. **10**(16): 2615–26.
45. Romanczuk H, Galer CE, Zabner J, Barsomian G, Wadsworth SC, O'Riordan CR. (1999) Dressing up adenoviruses to modify their tropism. Hum Gene Ther. **10**(16):2575–6.
46. Drapkin PT, O'Riordan CR, Yi SM, Chiorini JA, Cardella J, Zabner J, Welsh MJ. (2000) Targeting the urokinase plasminogen activator receptor enhances gene transfer to human airway epithelia. J Clin Invest. **105**(5):589–96.
47. Lanciotti J, Song A, Doukas J, Sosnowski B, Pierce G, Gregory R, Wadsworth S, O'Riordan C. (2003) Targeting adenoviral vectors using heterofunctional polyethylene glycol FGF2 conjugates. Mol Ther. **8**(1):99–107
48. Fisher KD, Stallwood Y, Green NK, Ulbrich K, Mautner V, Seymour LW. (2001) Polymer-coated adenovirus permits efficient retargeting and evades neutralizing antibodies. Gene Ther. **8**(5):341–8
49. O'Riordan CR, Lachapelle A, Delgado C, Parkes V, Wadsworth SC, Smith AE, Francis GE. (1999) PEGylation of adenovirus with retention of infectivity and protection from neutralizing antibody in vitro and in vivo. Hum Gene Ther. **10**(8):1349–58.
50. Croyle MA, Chirmule N, Zhang Y, Wilson JM. (2001) "Stealth" adenoviruses blunt cell-mediated and humoral immune responses against the virus and allow for significant gene expression upon readministration in the lung. J Virol. **75**(10): 4792–801.
51. Croyle MA, Chirmule N, Zhang Y, Wilson JM. (2002) PEGylation of E1-deleted adenovirus vectors allows significant gene expression on readministration to liver. Hum Gene Ther. **13**(15):1887–900.
52. Croyle MA, Le HT, Linse KD, Cerullo V, Toietta G, Beaudet A, Pastore L. (2005) PEGylated helper-dependent adenoviral vectors: highly efficient vectors with an enhanced safety profile. Gene Ther. **12**(7):579–87.
53. Richter AW, Akerblom E. (1983) Antibodies against polyethylene glycol produced in animals by immunization with monomethoxy polyethylene glycol modified proteins. Int Arch Allergy Appl Immunol. **70**(2):124–31.

54. Delgado C, Patel JN, Francis GE, Fisher D. (1990) Coupling of poly(ethylene glycol) to albumin under very mild conditions by activation with tresyl chloride: characterization of the conjugate by partitioning in aqueous two-phase systems. Biotechnol Appl Biochem. **12**(2):119–28.
55. Francis GE, Fisher D, Delgado C, Malik F, Gardiner A, Neale D. (1998) PEGylation of cytokines and other therapeutic proteins and peptides: the importance of biological optimization of coupling techniques. Int J Hematol. **68**(1):1–18.
56. Armentano D, Zabner J, Sacks C, Sookdeo CC, Smith MP, St George JA, Wadsworth SC, Smith AE, Gregory RJ. (1997) Effect of the E4 region on the persistence of transgene expression from adenovirus vectors. J Virol. **71**(3): 2408–16.
57. Yu D, Wolf JK, Scanlon M, Price JE, Hung MC. (1993) Enhanced c-erbB-2/neu expression in human ovarian cancer cells correlates with more severe malignancy that can be suppressed by E1A. Cancer Res. **53**(4):891–8.
58. Maizel JV Jr, White DO, Scharff MD. (1968) The polypeptides of adenovirus. I. Evidence for multiple protein components in the virion and a comparison of types 2, 7A, and 12. Virology. **36**(1):115–25.
59. Green, M., Pina, M., Kimes, R., Wensik, P., Machattie, L., and Thomas Jr., C. (1967) Adenovirus DNA. Molecular weight and conformation. Proc.Natl. Acad. Sci. USA **57,** 1302–1309
60. van der Eb AJ, van Kesteren LW, van Bruggen EF. (1969) Structural properties of adenovirus DNA's. Biochim Biophys Acta. **182**(2):530–41
61. Lappi DA, Matsunami R, Martineau D, Baird A. (1993) Reducing the heterogeneity of chemically conjugated targeted toxins: homogeneous basic FGF-saporin. Anal. Biochem. **212**(2):446–51.
62. Zhang JD, Cousens LS, Barr PJ, Sprang SR. (1991) Three-dimensional structure of human basic fibroblast growth factor, a structural homolog of interleukin 1 beta. Proc Natl Acad Sci U S A. **88**(8):3446–50.
63. Heid CA, Stevens J, Livak KJ, Williams PM. (1996) Real time quantitative PCR. Genome Res. **6**(10):986–94.
64. Yang Y, Li Q, Ertl HC, Wilson JM. (1995) Cellular and humoral immune responses to viral antigens create barriers to lung-directed gene therapy with recombinant adenoviruses. J Virol. **69**(4):2004–15
65. Prage L, Pettersson U, Hoglund S, Lonberg-Holm K, Philipson L. (1970) Structural proteins of adenoviruses. IV. Sequential degradation of the adenovirus type 2 virion. Virology. **42**(2):341–58
66. Huyghe BG, Liu X, Sutjipto S, Sugarman BJ, Horn MT, Shepard HM, Scandella CJ, Shabram P. (1995) Purification of a type 5 recombinant adenovirus encoding human p53 by column chromatography. Hum Gene Ther. **6**(11):1403–16.
67. O'Riordan CR, Lachapelle AL, Vincent KA, Wadsworth SC. (2000) Scaleable chromatographic purification process for recombinant adeno-associated virus (rAAV). J Gene Med. **2**(6):444–54.

10

Transposon-Based Mutagenesis Generates Diverse Adeno-Associated Viral Libraries with Novel Gene Delivery Properties

James T. Koerber and David V. Schaffer

Summary

The engineering of novel properties and functions into viral vectors for improved gene delivery remains a barrier to the development of efficient, customized gene delivery vehicles. Rational methods for designing improved viral vectors are often experimentally challenging and laborious, particularly when knowledge of viral structure–function relationships is limited. As an alternative, high-throughput libraries may be rapidly and efficiently selected for viral variants with a desired function. Here we describe a transposon-based insertional mutagenesis approach to generate large diverse adeno-associated viral (AAV) libraries containing a randomly located peptide. Briefly, a selectable marker is randomly inserted throughout the AAV2 *cap* gene and the resulting "bookmarked" AAV *cap* gene is cloned into an AAV packaging vector. The selectable marker is then replaced with a defined oligonucleotide, and the final AAV library is used to package a diverse pool of AAV virions, which can used for functional selection.

Key Words: Adeno-associated virus; transposase; mutagenesis; gene therapy; protein engineering.

1. Introduction

Adeno-associated viral (AAV) vectors show tremendous potential for the safe and effective treatment of a range of genetic disorders, including Alzheimer's disease, hemophilia, and Parkinson's disease ***(1,2)***. AAV is a non-pathogenic parvovirus with a 4.7-kb single-stranded DNA genome that contains two viral genes: *rep* and *cap* ***(3,4)***. While *rep* encodes four proteins (Rep78, Rep68,

From: *Methods in Molecular Biology, vol. 434: Volume 2: Design and Characterization of Gene Transfer Vectors*
Edited by: J. M. Le Doux © Humana Press, Totowa, NJ

Rep52, and Rep40) essential for viral replication, *cap* encodes three structural proteins (VP1-3) that self-assemble as a 60-mer to form the viral protein shell or capsid. The extensive safety record coupled with the high efficiency of AAV vectors has furthered their use to deliver genes to various dividing and nondividing cell types in vivo ***(5–7)***. While natural evolution has generated numerous alternatives to the well-characterized and most clinically utilized AAV serotype 2 (AAV2) ***(8,9)***, numerous remaining challenges, such as a robust universal purification platform and engineering cell-specific tropism, limit the gene delivery potential of these viral vectors.

Genetic engineering has greatly extended our knowledge of AAV biology and enhanced its gene delivery properties. Site-directed mutagenesis studies have identified amino acids critical to AAV2 function ***(10–13)***, and the insertion of some peptides in defined positions in the AAV capsid, guided by sequence alignment with related parvoviruses ***(14,15)*** and crystal structures ***(12,16)***, has conferred recombinant AAV vectors with some cell-specific delivery properties. However, these rational design approaches can be laborious, and results are often highly variable, as evidenced by large differences in functional peptide display and viral infectivity for various peptides inserted into the same location ***(10,12)***. Alternative high-throughput library approaches have selected AAV vectors with novel cell-targeting peptides inserted at a defined capsid location ***(17,18)***, altered receptor binding properties ***(19)***, and the ability to evade antibody neutralization ***(19)***. However, the same insertion location is likely not optimal for displaying all functional peptide inserts, and directed evolution approaches do not typically involve the insertion of a peptide of defined function into a protein. Therefore, inserting known peptides or domains randomly throughout the entire primary amino acid sequence of the viral structural proteins may generate viral capsids with novel functions. Endonuclease methods have been used to engineer both bacterial and mammalian proteins ***(20,21)***, but such techniques fail to yield large diverse random libraries because of low DNA ligation efficiencies and biased insertions. Alternatively, transposases, enzymes capable of moving or copying DNA sequences randomly from one DNA template to another (*see* **Fig. 1A**), have greatly facilitated prokaryotic and eukaryotic evolution through random DNA insertions into an organism's genome ***(22)***. Insertional mutagenesis using such transposases has improved functional genomics studies of viral genomes ***(23–25)***, and we have recently built on this approach to identify novel peptide insertion sites within the vesicular stomatitis virus protein for retroviral and lentiviral vector engineering ***(26)***.

Transposon insertional mutagenesis relies on the transposase-facilitated transfer of a unique drug-resistance gene from a donor plasmid to a random location in an acceptor plasmid, containing the gene of interest, followed by

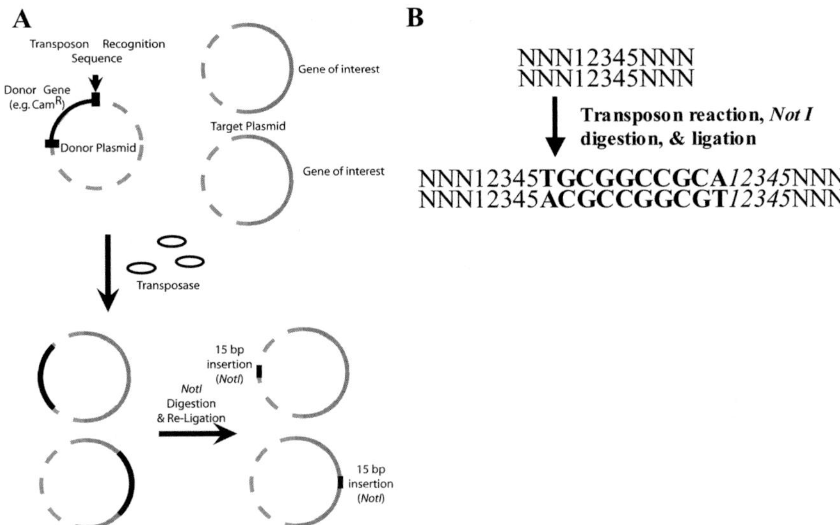

Fig. 1. Overview of transposon-based insertional mutagenesis reaction. (**A**) Donor plasmid containing the donor gene (black) and acceptor plasmids containing the gene of interest (gray) are mixed in the presence of a transposase, resulting in the transfer of the donor gene from the donor to the acceptor plasmid. The resulting pool of plasmids, which contains the donor gene randomly inserted throughout the acceptor plasmid, may be digested with *Not*I and re-ligated to yield a plasmid library containing a "bookmark" (i.e., *Not*I site) ideally covering every possible internucleotide position. (**B**) A random 5-bp sequence in the acceptor plasmid at the insertion site is duplicated (shown in italics) and placed after the inserted *Not*I site (shown in bold).

subsequent selection with the appropriate antibiotics. Subsequent replacement of the drug-resistance gene with an oligonucleotide encoding a desired peptide results in a diverse plasmid library, which can be used to produce virus containing the peptide randomly located within the viral capsid. Here we present a detailed protocol for employing a transposon-based system to randomly insert an oligonucleotide encoding for a peptide of interest [i.e., a hexahistidine (His_6) tag] throughout the entire AAV2 *cap* gene. Other motifs to modulate cell surface binding or other viral properties can also be inserted. Therefore, this general protocol is readily extended to other peptides, AAV serotypes, and viruses to generate customized viral gene delivery vectors.

2. Materials

1. 2.5 M calcium chloride ($CaCl_2$).
2. 2× HeBS: 1.5 mM Na_2HPO_4, 50 mM HEPES, 280 mM NaCl, pH 7.10.
3. AAV lysis buffer: 50 mM Tris–HCl, 150 mM NaCl, pH 8.5.

3. Methods

3.1. Construction of AAV cap Plasmid

1. To obtain AAV *cap* gene from viral genomic DNA or plasmid, mix in a PCR tube: 5 μl 10× Thermopol buffer, 5 μl dimethyl sulfoxide (DMSO), 1 μl 10 mM dNTP mix, ~0.1 pmol DNA template, 10 pmol of 5′ and 3′ primers containing suitable restriction sites for cloning, 5 U Vent DNA polymerase, and water to 50 μl. For example, to recover the AAV2 *cap* gene, use 5′-GCGG**AAGCTT**CGATCAACTACGC-3′ as the 5′ primer and 5′-GG**GGCGCGCC**GCAATTACAGATTACGAGTCAGGTATCTGGTG-3′ as the 3′ primer. These primers introduce *Hin*dIII and *Asc*I restriction sites (in bold), respectively, to facilitate cloning.
2. Using a QIAquick purification kit or similar spin column kit, purify the polymerase chain reaction (PCR).
3. The construction of a small plasmid vector (e.g., pBluescript) containing only the AAV *cap* gene ensures that more insertion events occur in the *cap* gene versus the vector backbone and thus reduces the final required library size, which should be significantly larger than the total number of base pairs in the template plasmid to ensure every possible position is well represented. However, the unique *Not*I site within pBluescript must first be eliminated because the transposon reaction introduces a *Not*I site. This may be accomplished by either mutagenesis or insertion of a small oligonucleotide linker at the *Not*I site to introduce a unique *Asc*I site (*see* **Note 1**).
4. Digest ~1 μg both the 2.6-kb AAV *cap* PCR product and modified pBluescript plasmid with *Hin*dIII and *Asc*I. Purify the digested products using a ~1% agarose gel and a commercial gel extraction kit, such as Qiagen QIAEX II Gel Extraction Kit.
5. Ligate 75–150 fmol of AAV *cap* PCR product and 25 fmol of pBluescript with 5 U of T4 DNA ligase in a 15 μl reaction.
6. Transform 10 μl of ligation into TOP10 bacteria and select for growth in the presence of ampicillin (100 μg/ml).
7. To screen for positive clones, digest plasmid DNA with *Hin*dIII and *Asc*I. Positive clones will yield two bands: 2.6 and 3.0 kb.

3.2. Transposon-Based Insertion

1. The use of commercially available transposon kits, such as the Mutation Generation System used here, permits efficient generation of a diverse library containing restriction site markers located randomly throughout the plasmid template (*see* **Note 2**).
2. To perform the transposition reaction, mix in a PCR tube: (40 ng × plasmid size in kilobases) plasmid template, 4 μl 5× reaction buffer for MuA Transposase, 1 μl Entraceposon (M1-Cam^R or M1-Kan^R) (*see* **Note 3**), 1 μl MuA Transposase, and water to 20 μl.

3. In a thermal cycler, incubate the reaction at 30°C for 1 h, followed by a 10 min incubation at 75°C. This results in either the *Cam*^{R} or *Kan*^{R} gene, flanked by *Not*I sites and with a 5 bp duplication, being randomly inserted throughout the plasmid template (*see* **Fig. 1** and **Note 4**).
4. Purify the DNA mixture by ethanol precipitation through addition of 2 µl 3M NaOAc, pH 5.2, and 50 µl 100% ethanol. We recommend adding glycogen (100 µg/ml final concentration) carrier as a pellet marker. Thoroughly wash the pellet with 70% ethanol and air dry. Resuspend the dried pellet in no more than 10 µl water.
5. Transform the purified ligation reaction into ElectroMAX DH10B bacteria through electroporation according to the manufacturer's instructions. Take a small fraction of the electroporation reaction and streak on a bacterial agar plate with ampicillin and chloroamphenicol. Estimate the initial plasmid diversity of the library from the number of bacterial colonies. Typical initial library diversity should be on the order of 10^6 independent bacterial clones.
6. Inoculate a 100 ml TB culture with the remaining reaction and shake culture at 250 rpm (1.3 × g) for 13 h at 37°C in the presence of 100 µg/ml ampicillin and 10 µg/ml chloroamphenicol.
7. Purify DNA from culture using a standard DNA purification method such as polyethylene glycol (PEG) precipitation or a commercial purification kit. Quantify purified DNA by measuring the absorbance at 260 nm using a UV-Vis spectrophotometer.

3.3. Construction of AAV Plasmid Library

1. To transfer the AAV *cap* gene containing the *Cam*^{R} gene, digest ~1 µg pBS *cap*-Cam^R and the appropriate modified AAV packing vector, such as pSub2 ***(19)*** with *Hin*dIII and *Asc*I. The pSub2 packaging vector contains the entire AAV2 genome except for the *cap* gene, which has been replaced with unique *Hin*dIII and *Not*I sites. Here, the *Not*I site in pSub2 has been replaced with an *Asc*I site through mutagenesis to create pSub2Asc. Purify the digested products using a 1% agarose gel. Gel extract the 3.7-kb band from the pBS *cap*-Cam^R sample and 5.7 kb linearized pSub2Asc plasmid.
2. Ligate 75–150 fmol of *cap*-Cam^R fragment and 25 fmol of pSub2Asc with 5 U of T4 DNA ligase in a 15 µl reaction (*see* **Note 5**).
3. Purify the ligation reaction as in **step 4**, **Subheading 3.2** and transform into ElectroMAX DH10B bacteria as in **step 5**, **Subheading 3.2**. Estimate the library diversity size as in **step 5**, **Subheading 3.2**.
4. Inoculate a large-scale TB culture in the presence of ampicillin and chloroamphenicol and purify as in **steps 6 and 7**, **Subheading 3.2**. A diagnostic restriction digest screen with *Age*I (which cuts near the start of the Cam^R gene) and *Asc*I (which cuts at end of *cap* gene) should yield a smear ranging from ~1.2 kb to 3.8 kb in size.
5. To replace the chloroamphenicol-resistance gene with a desired oligonucleotide, design oligonucleotides such that the sequences at the 5′ and 3′

ends are compatible with *Not*I sites (*see* **Fig. 1B**). For example, the following oligonucleotides were used for insertion of a His_6 tag (histidine codons shown in bold): 5′-GGCCGGT**CACCACCACCACCACCAC**TC-3′ and 5′-GGCCGA**GTGGTGGTGGTGGTGGTG**ACC-3′ (*see* **Note 6**). Mix equal amounts of single-stranded oligonucleotides in TE buffer (10 mM Tris, 1 mM EDTA, pH 8.0) supplemented with 50 mM NaCl, heat to 94°C, and gradually cool to the anneal the oligonucleotides. Phosphorylate the resulting double-stranded oligonucleotide by mixing 300 pmol DNA, 5 μl T4 DNA ligase buffer, 1 mM dATP, and 10 units T4 polynucleotide kinase. Incubate the reaction at 37 °C for 30 min and inactivate the enzyme by incubation at 65 °C for 20 min.

6. Digest ~1 μg pSub2Asc *cap*-Cam^R with *Not*I and gel extract the 8.3-kb DNA fragment.
7. Ligate 75–150 fmol of phosphorylated oligonucleotide and 25 fmol of pSub2Asc *cap* fragment with 5 U of T4 DNA ligase in a 15-μl reaction. Incubate at 14 °C for at least 6 h.
8. Purify the ligation reaction as in **step 4**, **Subheading 3.2** and transform into ElectroMAX DH10B bacteria as in **step 5**, **Subheading 3.2**. Estimate the library diversity size as in **step 5**, **Subheading 3.2**.
9. Inoculate a large-scale TB culture in the presence of ampicillin and purify as in **steps 6–7**, **Subheading 3.2**. A diagnostic restriction digest screen with *Eag*I (which cuts once in the oligo insert and once at end of the *cap* gene) should yield a smear ranging from 20 bp to 2.6 kb in size (*see* **Fig. 2**).

3.4. Production of Viral Library

1. Plate ~10^7 HEK 293 cells in 25 ml of DMEM onto a 15-cm tissue culture dish such that cells are ~75% confluent after 24 h.
2. After ~24 h, transfect cells by calcium phosphate precipitation ***(19)***. Briefly, mix 7 ng pSub2Asc library, 25 μg pBluescript, and 25 μg pHelper with 120 μl 2.5

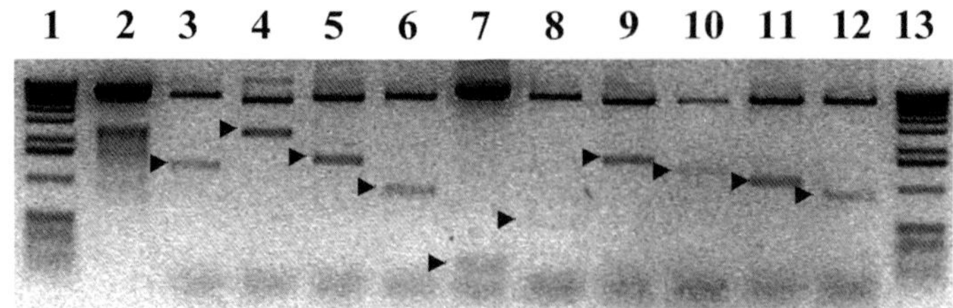

Fig. 2. Analysis of plasmid library diversity. The His_6 AAV insertion library and single clones contain two *Eag*I sites, one that cuts once in the His_6 insertion and one at 3′ end of the AAV *cap* gene. Successful insertions into *cap* will yield a large band ranging from 8.3 to 5.7 kb, along with a smaller band ranging up to 2.6-kb in size (indicated by arrow). Lanes: 1, 1-kb ladder; 2, His_6 AAV library; 3–12, 10 randomly selected single His_6 AAV clones; 13 1-kb ladder.

M $CaCl_2$ and water to 2.5 ml total volume. The pBluescript helps to maintain a constant DNA : calcium phosphate ratio, which we have found necessary for maintaining high efficiency DNA transfection and viral packaging. Add DNA/$CaCl_2$ solution dropwise to 2× HeBS solution. Mix once and add mixture dropwise to cells. Remove media after 6–8 h and replace with 25 ml DMEM. This $1{:}2 \times 10^{-4}$ molar ratio of plasmid DNA to pSub2Asc library was calculated such that >90% of cells received approximately one member of pSub2Asc library, assuming each cell receives ~50,000 total plasmids ***(27)***. This helps to ensure that most virions contain a viral genome with a *cap* gene encoding their capsid (*see* **Note 7**).

3. After 48 h, scrape cells from the plate and centrifuge at 1000 × *g* for 2 min. Aspirate medium and resuspend cell pellet in 1 ml PBS or AAV lysis buffer (50 mM Tris, 150 mM NaCl, pH 8.5).
4. Freeze/thaw three times using a dry ice/ethanol bath, or sonicate the cell suspension, to lyse the cells.
5. Centrifuge the lysate at 13,000 × *g* for 10 min to clarify, or pellet cell debris. The resulting supernatant contains the AAV viral library, which can be quantified by standard protocols such as dot blotting, ELISA, or quantitative PCR. If necessary, the library can be purified by density ultracentrifugation, such as with iodixanol or CsCl ***(19)***. A representative titer from two independent viral productions is shown in **Fig. 3A**. After this stage, the viral library can be selected for variants with a desired enhanced function (*see* **Note 8**). For example, to select for the functional display of His_6 tags on the viral surface, a mixture of 1 volume of cell lysate containing ~10^{11} viral particles, 0.5 volume binding buffer (10 mM Tris-HCl pH 8.0, 300 mM NaCl, and 20 mM imidazole), and 500 μl of 50%

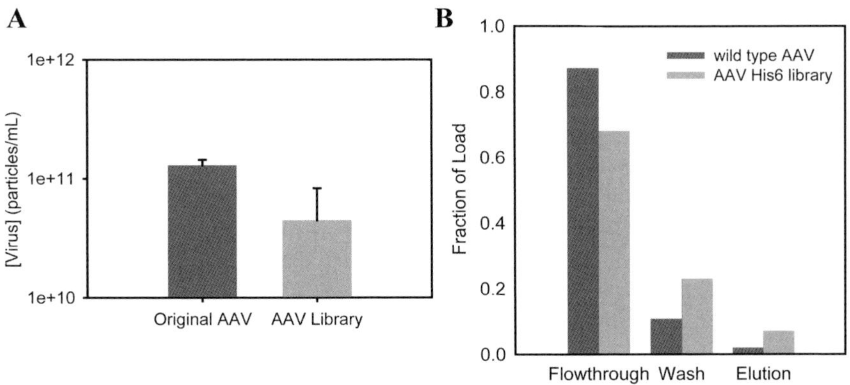

Fig. 3. Characterization of His_6 AAV Library. (**A**) Representative titers of both wildtype AAV2 and the His_6 AAV library from two independent preparations. (**B**) Chromatogram of viral binding fractions to Ni-NTA resin for both wildtype AAV2 and His_6 AAV library.

Ni-NTA agarose (Qiagen, Valencia, CA, USA) was agitated gently overnight at 4 °C. This slurry was then loaded onto a plastic column (Kontes, Vineland, NJ, USA) before washing with 3 ml of wash buffer (10 mM Tris-HCl pH 8.0, 50 mM imidazole) and eluting with 3 ml of elution buffer (10 mM Tris-HCl pH 8.0, 500 mM imidazole). Characterization of the binding profile of the viral library showed elevated levels of Ni-NTA binding relative to the wild type AAV2 control (*see* **Fig. 3B**).

4. Notes

1. Alternative restriction sites may be used for the construction of this plasmid, provided the sites do not occur within the inserted drug-resistance gene or the acceptor plasmid (map provided with transposon kit).
2. Other commercial kits may result in different final insertion sizes, alternate reaction conditions, or alternate antibiotic selections. Be sure to thoroughly review the transposon kit's instructions before use.
3. Choose a drug-resistance gene that differs from gene present in the plasmid template.
4. Include a control reaction without either MuA Transposase or Entraceposon to verify the transfer of the drug-resistance gene.
5. The ratio of the amount of the plasmid insert to plasmid backbone may be varied to identify the optimal cloning condition.
6. Alternate oligonucleotides or gene fragments may be used provided the ends contain restriction sites compatible with *Not*I overhangs. Alternatively, the *Not*I overhangs may be blunted or filled in with Klenow, and the oligonucleotides may be cloned into the modified site.
7. Alternative transfection methods such as electroporation or lipofection may be used. In all cases, high transfection efficiency and proper plasmid amounts are essential for producing a diverse viral library that can be easily functionally selected.
8. For selection protocols involving infection of a cell line, such as HEK293 ***(19)***, care should be taken to avoid infection with a large (>1000) number of virions per cell [i.e., multiplicity of infection (MOI)]. The presence of multiple AAV genomes with different *cap* genes in one cell leads to potential recombination between the *cap* genes or production of several different versions of the VP1-3 proteins, yielding chimeric or mosaic virions ***(28,29)***. Hence, isolation of the genomes from these mosaic virions will fail to recover the genotype responsible for the novel property or phenotype.

Acknowledgments

We thank Julie Yu for technical assistance. This work was supported by a NSF Graduate Fellowship (to J.T.K.). UC Discovery bio 05–10559, and NIH EB003007.

References

1. Manno, C. S., Chew, A. J., Hutchison, S., Larson, P. J., Herzog, R. W., Arruda, V. R., et al. (2003) AAV-mediated factor IX gene transfer to skeletal muscle in patients with severe hemophilia B. *Blood* **101,** 2963–72.
2. Kay, M. A., Manno, C. S., Ragni, M. V., Larson, P. J., Couto, L. B., McClelland, A., et al. (2000) Evidence for gene transfer and expression of factor IX in haemophilia B patients treated with an AAV vector. *Nat Genet* **24,** 257–61.
3. Fields, B. N., Knipe, D. M., Howley, P. M., and Griffin, D. E. (2001) *Fields Virology*, Lippincott Williams & Wilkins, Philadelphia.
4. Srivastava, A., Lusby, E. W., and Berns, K. I. (1983) Nucleotide sequence and organization of the adeno-associated virus 2 genome. *J Virol* **45,** 555–64.
5. Flotte, T. R., Afione, S. A., Conrad, C., McGrath, S. A., Solow, R., Oka, H., et al. (1993) Stable in vivo expression of the cystic fibrosis transmembrane conductance regulator with an adeno-associated virus vector. *Proc Natl Acad Sci USA* **90,** 10613–7.
6. Kaplitt, M. G., Leone, P., Samulski, R. J., Xiao, X., Pfaff, D. W., O'Malley, K. L., et al. (1994) Long-term gene expression and phenotypic correction using adeno-associated virus vectors in the mammalian brain. *Nat Genet* **8,** 148–54.
7. Kaspar, B. K., Llado, J., Sherkat, N., Rothstein, J. D., and Gage, F. H. (2003) Retrograde viral delivery of IGF-1 prolongs survival in a mouse ALS model. *Science* **301,** 839–42.
8. Gao, G., Vandenberghe, L. H., Alvira, M. R., Lu, Y., Calcedo, R., Zhou, X., et al. (2004) Clades of Adeno-associated viruses are widely disseminated in human tissues. *J Virol* **78,** 6381–8.
9. Chiorini, J. A., Kim, F., Yang, L., and Kotin, R. M. (1999) Cloning and characterization of adeno-associated virus type 5. *J Virol* **73,** 1309–19.
10. Wu, P., Xiao, W., Conlon, T., Hughes, J., Agbandje-McKenna, M., Ferkol, T., et al. (2000) Mutational analysis of the adeno-associated virus type 2 (AAV2) capsid gene and construction of AAV2 vectors with altered tropism. *J Virol* **74,** 8635–47.
11. Rabinowitz, J. E., Xiao, W., and Samulski, R. J. (1999) Insertional mutagenesis of AAV2 capsid and the production of recombinant virus. *Virology* **265,** 274–85.
12. Shi, W., Arnold, G. S., and Bartlett, J. S. (2001) Insertional mutagenesis of the adeno-associated virus type 2 (AAV2) capsid gene and generation of AAV2 vectors targeted to alternative cell-surface receptors. *Hum Gene Ther* **12,** 1697–711.
13. Lochrie, M. A., Tatsuno, G. P., Christie, B., McDonnell, J. W., Zhou, S., Surosky, R., et al. (2006) Mutations on the external surfaces of adeno-associated virus type 2 capsids that affect transduction and neutralization. *J Virol* **80,** 821–34.
14. Grifman, M., Trepel, M., Speece, P., Gilbert, L. B., Arap, W., Pasqualini, R., et al. (2001) Incorporation of tumor-targeting peptides into recombinant adeno-associated virus capsids. *Mol Ther* **3,** 964–75.
15. Girod, A., Ried, M., Wobus, C., Lahm, H., Leike, K., Kleinschmidt, J., et al. (1999) Genetic capsid modifications allow efficient re-targeting of adeno-associated virus type 2. *Nat Med* **5,** 1052–6.

16. Xie, Q., Bu, W., Bhatia, S., Hare, J., Somasundaram, T., Azzi, A., et al. (2002) The atomic structure of adeno-associated virus (AAV-2), a vector for human gene therapy. *Proc Natl Acad Sci USA* **99,** 10405–10.
17. Perabo, L., Buning, H., Kofler, D. M., Ried, M. U., Girod, A., Wendtner, C. M., et al. (2003) In vitro selection of viral vectors with modified tropism: the adeno-associated virus display. *Mol Ther* **8,** 151–7.
18. Muller, O. J., Kaul, F., Weitzman, M. D., Pasqualini, R., Arap, W., Kleinschmidt, J. A., et al. (2003) Random peptide libraries displayed on adeno-associated virus to select for targeted gene therapy vectors. *Nat Biotechnol* **21,** 1040–6.
19. Maheshri, N., Koerber, J. T., Kaspar, B. K., and Schaffer, D. V. (2006) Directed evolution of adeno-associated virus yields enhanced gene delivery vectors. *Nat Biotechnol* **24,** 198–204.
20. Guntas, G., and Ostermeier, M. (2004) Creation of an allosteric enzyme by domain insertion. *J Mol Biol* **336,** 263–73.
21. Murakami, H., Hohsaka, T., and Sisido, M. (2002) Random insertion and deletion of arbitrary number of bases for codon-based random mutation of DNAs. *Nat Biotechnol* **20,** 76–81.
22. Kazazian, H. H., Jr. (2004) Mobile elements: drivers of genome evolution. *Science* **303,** 1626–32.
23. Brune, W., Menard, C., Hobom, U., Odenbreit, S., Messerle, M., and Koszinowski, U. H. (1999) Rapid identification of essential and nonessential herpesvirus genes by direct transposon mutagenesis. *Nat Biotechnol* **17,** 360–4.
24. Hobom, U., Brune, W., Messerle, M., Hahn, G., and Koszinowski, U. H. (2000) Fast screening procedures for random transposon libraries of cloned herpesvirus genomes: mutational analysis of human cytomegalovirus envelope glycoprotein genes. *J Virol* **74,** 7720–9.
25. Vilen, H., Aalto, J. M., Kassinen, A., Paulin, L., and Savilahti, H. (2003) A direct transposon insertion tool for modification and functional analysis of viral genomes. *J Virol* **77,** 123–34.
26. Yu, J. H., and Schaffer, D. V. (2006) Selection of novel vesicular stomatitis virus glycoprotein variants from a peptide insertion library for enhanced purification of retroviral and lentiviral vectors. *J Virol* **80,** 3285–92.
27. Batard, P., Jordan, M., and Wurm, F. (2001) Transfer of high copy number plasmid into mammalian cells by calcium phosphate transfection. *Gene* **270,** 61–8.
28. Rabinowitz, J. E., Bowles, D. E., Faust, S. M., Ledford, J. G., Cunningham, S. E., and Samulski, R. J. (2004) Cross-dressing the virion: the transcapsidation of adeno-associated virus serotypes functionally defines subgroups. *J Virol* **78,** 4421–32.
29. Gigout, L., Rebollo, P., Clement, N., Warrington, K. H., Jr., Muzyczka, N., Linden, R. M., et al. (2005) Altering AAV tropism with mosaic viral capsids. *Mol Ther* **11,** 856–65.

11

Photochemical Enhancement of DNA Delivery by EGF Receptor Targeted Polyplexes

Anette Bonsted, Ernst Wagner, Lina Prasmickaite, Anders Høgset, and Kristian Berg

Summary

Photochemical internalization (PCI) is a physico-chemical targeting method that enables light directed delivery of nucleic acids into cells. The technology is based on photosensitizers that localize in the membranes of endocytic vesicles. A light activation of the photosensitizers induces photochemical reactions that lead to rupture of the vesicular membranes. This results in the release of endocytosed compounds (e.g., nucleic acids) into the cell cytosol. Physico-chemical and biological targeting techniques can be combined to promote efficient and specific gene delivery to target cells. The present protocol describes PCI of epidermal growth factor receptor (EGFR)-targeted DNA polyplexes. The DNA polyplexes made are small (50-100 nm in diameter), and they contain polyethylenimine (PEI) conjugated with the EGF protein as a cell-binding ligand for EGFR-mediated endocytosis and polyethylene glycol (PEG) for masking the polyplex surface charge. PCI of such targeted PEG-PEI/DNA polyplexes enables high and EGFR-specific gene transfer activity in cells. Although describing in detail PCI of DNA polyplexes, the methodology presented in this protocol is also applicable for PCI of other gene therapy vectors (e.g. viral vectors), peptide nucleic acids (PNA), small interfering RNA (siRNA), and for vectors targeted to alternate cell surface receptors. Generally, PCI can be applied whenever 100% survival of the treated cell population is not required.

Key Words: Gene therapy; epidermal growth factor; polyethylenimine; PEGylation; photochemical internalization.

1. Introduction

A major goal in the development of gene therapy protocols is to achieve efficient, site-specific gene delivery. Poor escape of nucleic acids from

From: *Methods in Molecular Biology, vol. 434: Volume 2: Design and Characterization of Gene Transfer Vectors*
Edited by: J. M. Le Doux © Humana Press, Totowa, NJ

endosomes and their consequent degradation in lysosomes is a barrier for obtaining efficient gene delivery. Consequently, many of the delivery methods developed are based on improved endocytic uptake and release of the therapeutic molecule from the endocytic vesicles into the cytosol. Photochemical internalization is a method for light-inducible permeabilization of endocytic vesicles *(1)*. The technology is based on photosensitizers localizing in the endocytic membranes. Light activation of the photosensitizers initiates photochemical reactions, which cause rupture of the vesicles. This leads to the release of endocytosed compounds (e.g. nucleic acids) from the endocytic vesicles into the cytosol where they may act on their target or further translocate to the nucleus.

PCI derives from the field of photodynamic therapy *(2)*, taking advantage of the photochemical effects induced by a photosensitizer, light and oxygen. Photosensitizers are compounds that upon absorption of light at specific wavelengths induce chemical or physical alterations in other chemical entities. The most applied photosensitizers for PCI are aluminium phthalocyanine with two sulfonate groups on adjacent phenyl rings ($AlPcS_{2a}$) and meso-tetraphenylporphine with two sulfonate groups on adjacent phenyl rings ($TPPS_{2a}$) (*see* **Fig. 1**). These photosensitizers are amphiphilic compounds and localize in the membranes of endocytic vesicles. Photosensitizers that localize to other cellular structures are not efficient for inducing the PCI effect *(1,3)*. The light source may be any source emitting light of wavelengths absorbed by the photosensitizer. A red light source with peak wavelength at 670 nm may be used for the excitation of $AlPcS_{2a}$ and a blue light source with peak wavelength at 420 nm for $TPPS_{2a}$. The photochemical reactions induced following excitation of the photosensitizer proceed mainly via formation of singlet oxygen ($^1O_2$) *(2)*. Singlet oxygen is highly reactive and is generated after interaction between the

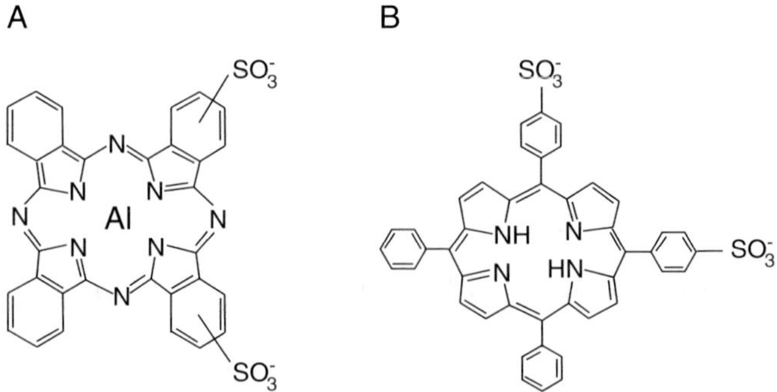

Fig. 1. Structure of the amphiphilic photosensitizers $AlPcS_{2a}$ (A) and $TPPS_{2a}$ (B).

excited photosensitizer in its triplet state and ground state molecular oxygen (O_2). Singlet oxygen has a short lifetime (0.04 μs) and a short range of action (10-20 nm) in cells ***(4)***. Accordingly, only structures very close to the photosensitizer will be affected following light exposure, whereas distant molecules will be left unaffected.

The use of the PCI technology has been documented for proteins ***(1)***, genes carried by non-viral and viral vectors ***(5–7)***, and peptide nucleic acids (PNA) ***(8,9)***. *In vivo*, PCI has been shown to enhance the therapeutic efficacy of the plant toxin gelonin ***(10)*** and the chemotherapeutic drug bleomycin ***(11)***. Recently, PCI-induced gene delivery with non-viral vectors was demonstrated *in vivo* ***(12,13)***. Biologically targeted gene vectors may also be applied in combination with PCI ***(14,15)***. The principle behind PCI of targeted DNA polyplexes is illustrated in **Fig. 2**. The biological targeting enables specific uptake of gene vectors in target cells, while the photochemical treatment permits light-directed escape from endocytic vesicles into the cytosol. Such dual targeting may reduce the risk of side effects due to inadvertent expression of transgenes in non-target cells. The present protocol describes (i) the formulation of small, neutrally charged epidermal growth factor receptor (EGFR)-targeted DNA complexes, and (ii) photochemical transfection.

Polycations are able to condense and protect DNA by electrostatic interactions. Such complexes of cationic polymers and DNA are termed polyplexes. Polyethylenimine (PEI) is a widely used gene carrier in polyplex formulations. PEI has an intrinsic ability to facilitate endosomal release, which is dependent on PEI concentration. Small PEI/DNA polyplexes with diameters of around 50-100 nm would be preferred for pharmacological reasons, particularly for *in vivo* gene therapy applications. However, PEI displays only moderate endosomolytic activity in such formulations ***(16)***, and therefore, the transfection may be enhanced by PCI ***(15)***.

PEI can be linked to targeting ligands to promote uptake through specific cell surface receptors. In addition, PEI can be coupled with polyethylene glycol (PEG) for masking the polyplex surface charge in order to reduce undesired unspecific interactions. Here, we describe epidermal growth factor receptor (EGFR)-targeted, PEG-shielded DNA polyplexes that contain polyethylenimine (PEI) for compacting the plasmid DNA, PEI-conjugated EGF as a cell-binding ligand for endocytosis, and PEI-conjugated polyethylene glycol (PEG). EGFR is a frequently used target due to its overexpression in many human tumors, and antibodies towards EGFR and Her2/neu are approved for clinical use.

PCI has been combined with several EGFR-targeted compounds, such as DNA polyplexes ***(15)***, adenovirus ***(17)*** and the protein toxin saporin ***(18)***. As PCI affects the endocytic vesicles, the improved gene transfer due to PCI retains EGF receptor specificity. Thus, dual targeting is achieved by combining

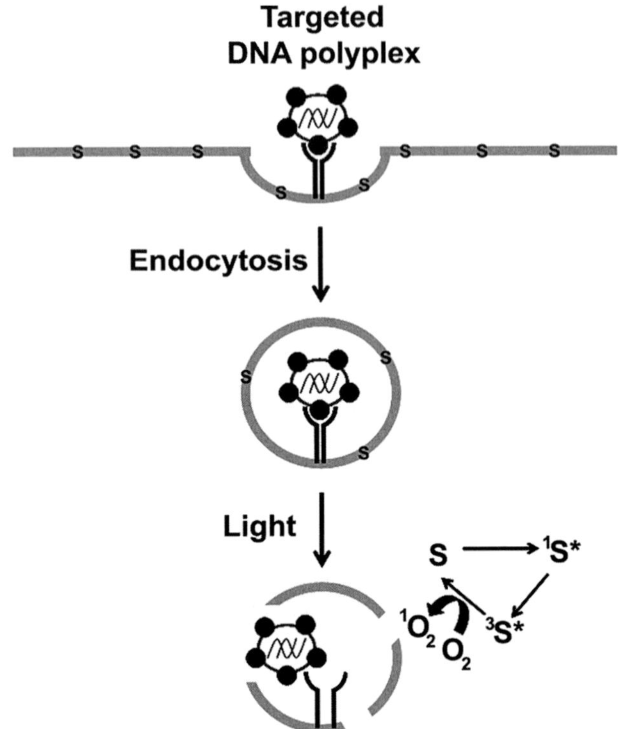

Fig. 2. Illustration of photochemical internalization (PCI) of DNA polyplexes targeted to a specific cell surface receptor. The DNA polyplexes bind the cell surface receptor and are endocytosed. Amphiphilic photosensitizers (S) localize to the membranes of the endocytic vesicles. Upon light exposure of the cells, the photosensitizer is converted to an excited singlet state ($^1S^*$), which may be converted to an excited triplet state ($^3S^*$). The triplet state excited photosensitizer transfers its energy to molecular oxygen ($^1O_2$), generating singlet oxygen ($^1O_2$). The oxidative damage of endocytic membranes by singlet oxygen promotes rupture of the vesicular membranes and the release of the DNA polyplex into the cytosol.

biological targeting via EGF-receptor interaction and physical targeting with light to direct a photochemical delivery of therapeutic genes to a desired location. The magnitude of enhancement due to photochemical treatment depends on the formulation of the targeted compound, the cell line, and the photochemical dose. In the case of photochemically enhanced gene delivery, the relative effect of PCI was most pronounced at suboptimal doses of polymer / DNA complexes (e.g. molar PEI nitrogen / DNA phosphate (N/P) ratios ≤ 6 and recombinant adenovirus vector doses at multiplicities of infections ≤ 20). The present protocol reviews the dissolving and stocking of the photosensitizer,

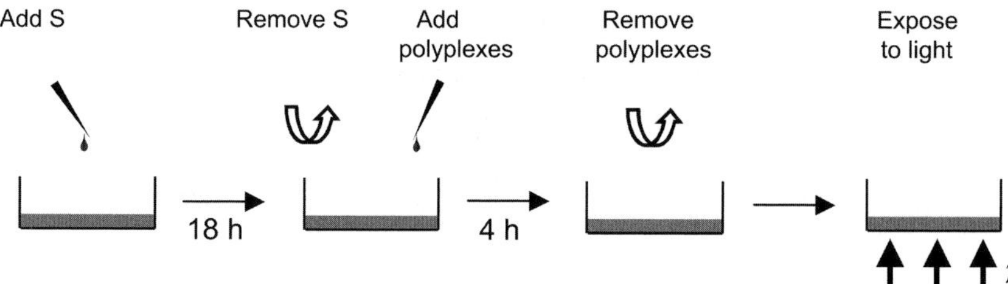

Fig. 3. Experimental scheme of PCI of DNA polyplexes on cultured cells. S: Photosensitizer.

the determination of the photochemical dose, the preparation of EGFR-targeted PEI-PEG/DNA polyplexes at N/P ratio 4 - 6, and the photochemical transfection (*see* **Fig. 3**).

2. Materials

2.1. Determination of the Light Dose

1. Stock solution of the photosensitizer $AlPcS_{2a}$ or $TPPS_{2a}$ (Frontier Scientific, Logan, UT). For $AlPcS_{2a}$: Dissolve 5 mg of $AlPcS_{2a}$ in 0.2 ml of 0.1 M NaOH. Dilute with PBS to a final volume of 1 ml. If complete solubilization of the photosensitizer is difficult, the photosensitizer solution may be sonicated for a few seconds. For $TPPS_{2a}$: Dissolve 1 mg of $TPPS_{2a}$ in 1 ml dimethyl sulfoxide. Sterilize the stock solution by filtration through a 0.2 μm filter. Store the stock solution in aliquots at –20 °C for up to 6 months (*see* **Note 1**).
2. Light source for excitation of $AlPcS_{2a}$ or $TPPS_{2a}$, e.g., LumiSource^{TM} (PCI Biotech AS, Oslo, Norway).

2.2. EGF-Receptor Targeted DNA Polyplexes

1. Linear polyethylenimine with an average molecular weight of 22 kDa (PEI22) ***(19)***. PEI22 can be synthesized by acid-catalysed deprotection of poly(2-ethyl-2-oxazoline) (50 kDa, Sigma Aldrich, St. Louis, MO, USA) ***(20)***. Complete deprotection of nitrogens is important and can be monitored by proton NMR analysis ***(21)***. For polyplex preparation, PEI22 was used at a 1 mg/ml working solution, neutralized with hydrochloric acid. Linear PEI is also available from Polyplus-transfection (Illkirch, France).
2. PEI22 modified with PEG of 20 kDa at equimolar ratio (PEI22-PEG20). PEI22-PEG20 can be synthesized as described in ref 22.
3. Conjugates of epidermal growth factor (EGF) attached to PEG and branched PEI with an average molecular weight of 25 kDa (EGF-PEG-PEI25). EGF-PEG-PEI25 can be synthesized as described in ref 23.

4. Plasmid DNA (*see* **Note 2**).
5. HEPES buffered glucose (HBG): 5 % (w/w) glucose, 20 mM HEPES, pH 7.1.

3. Methods

3.1. Determination of the Light Dose

The optimal photochemical dose has to be determined individually for every cell line, photosensitizer, and light source prior to photochemical transfection of the cells. For this purpose, survival of cells preloaded with the photosensitizer should be determined as a function of the light dose. Light doses generating 50-70% cell survival are recommended for photochemical transfection.

1. Seed the cells in cell culture dishes at a density of $2x10^4$ cells/cm^2 in the cell culture medium recommended for the cell line in use. Allow the cells to attach to the substratum for at least six hours at 37 °C in a CO_2 incubator.
2. In subdued light, add 5 µg/ml $AlPcS_{2a}$ or 0.2 µg/ml $TPPS_{2a}$ to the cells and incubate for 16-18 hours at 37 °C in a CO_2 incubator (*see* **Note 3**).
3. Wash the samples three times with photosensitizer-free cell culture medium. Chase for 4 hours at 37 °C in photosensitizer-free cell culture medium.
4. Expose the samples to different light doses with a suitable light source (*see* **Note 4**).
5. Measure cell survival 24–48 hours after illumination by one of the common cell survival tests such as the MTS/MTT-test, protein synthesis, clonogenic analysis, or another test established in individual laboratories (*see* **Note 5**).

3.2. Preparation of EGF-Receptor Targeted DNA Polyplexes

1. Mix the EGF-PEG-PEI25, PEG-PEI22, and free PEI22 polymers at a weight/weight/weight ratio of 10 % / 20 % / 70 % (based on PEI content of the conjugates).
2. Dilute the DNA and the PEI polymer conjugates separately in HBG in equal volumes to 40 µg/ml (DNA) and 31 µg/ml (PEI conjugates), respectively, in the case of a molar PEI nitrogen / DNA phosphate (N/P) ratio of 6. For other N/P ratios the concentration of conjugates has to be adjusted, for example 21 µg/ml PEI conjugates for N/P 4.
3. Add the DNA to an equal volume of PEI polymer buffer solution and. mix carefully with a pipette. This results in a 20 µg/ml DNA polyplex solution. More concentrated polyplexes (up to 200 µg/ml DNA) can be generated by applying more concentrated DNA and polymer solutions. Allow the polyplexes to form for 20 min at room temperature before use (*see* **Note 6**). Shortly before transfection, dilute the polyplexes in cell culture medium to the appropriate DNA concentrations (0.2 – 5 µg/ml).

3.3. PCI-Enhanced Transfection with EGF-Receptor Targeted DNA Polyplexes

1. Seed the cells in cell culture dishes at a density of $2x10^4$ cells/cm^2 in the cell culture medium recommended for the cell line in use. Allow the cells to attach to the substratum for at least six hours at 37 °C in a CO_2 incubator.
2. In subdued light, add 5 μg/ml $AlPcS_{2a}$ or 0.2 μg/ml $TPPS_{2a}$ to the cells and incubate for 16-18 hours at 37 °C in a CO_2 incubator (*see* **Note 3**).
3. Wash the cells three times with photosensitizer-free cell culture medium. Add the EGFR-targeted DNA polyplexes to the cells and incubate for 4 hours at 37 °C (*see* **Notes 7** and **8**).
4. Remove the polyplexes, wash the cells once with cell culture medium. Expose the cells to light from an appropriate light source; use the light dose(s) empirically determined as described in section 3.1. (*see* **Note 9**).
5. Grow the cells further in the dark for 24-48 hours and analyze for transgene expression or transgene effect. If a plasmid encoding luciferase is applied, transgene expression may be analyzed by a commercial luciferase assay (e.g., Promega). If a plasmid encoding eGFP is applied, transgene expression may be analyzed by fluorescence microscopy or flow cytometry.

4. Notes

1. Long-term storage or repeated freezing and thawing may cause aggregation and hence reduce the efficacy of the photosensitizer. The photosensitizer solution should also be protected from light to avoid photoinduced damage to the photosensitizer.
2. PCI has no restriction on the size of DNA to be delivered to the cell. Thus, any required plasmid may be applied. For the EGFR-targeted polyplexes, we used the plasmid pCMVLuc (Photinus pyralis luciferase gene under the control of the CMV enhancer/ promoter) ***(24)***, and the plasmid pEGFP-N1 (encoding enhanced green fluorescent protein (eGFP) under the control of CMV promoter), purchased from Clontech Laboratories, Inc. (Palo Alto, CA).
3. In order to avoid uncontrollable activation of the photosensitizer and to protect the cells from undesirable photochemical damage, all the procedures starting from point 2 in sections 3.1. and 3.3. should be carried out in subdued light. For *in vitro* studies it may be sufficient to turn off the light in the sterile bench. To test if the light in the laboratory is sufficiently subdued, the toxicity of the photosensitizer without irradiating the cells with light from the light source could be analyzed. With the photosensitizer concentrations recommended here we have not experienced any significant toxicity from the photosensitizer in the cell lines tested thus far.
4. In principle all light sources that emit light absorbed by the photosensitizer may be used for PCI. The LumiSource^{TM} red lamp consists of 4 x 18 W Philips Fluotone 18/950 light tubes and a PMMA PSC-S110 filter, and it delivers light with an irradiance of 1.5 mW/cm^2. For cells that have been incubated with 5 μg/ml of

$AlPcS_{2a}$ for 18h, a typical range of light doses would be 0.9–1.4 J/cm^2 (i.e. 10–16 min). The LumiSource^{TM} blue lamp consists of 4 x 18 W Osram L 18/67 standard light tubes, and delivers blue light with an irradiance of 13.5 mW/cm^2. For cells that have been incubated with 0.2 μg/ml of $TPPS_{2a}$ for 18h, a typical range of light doses would be 0.5–1.0 J/cm^2 (i.e. 40–75 sec). Both lamps are air-cooled during light exposure, which prevents cells from being exposed to hyperthermia and keeps the irradiance stable over time.

5. Light doses enabling 50-70% survival of photosensitizer-treated cells are good starting points for photochemical transfection. A decrease in light dose would reduce both the cytotoxicity and the effect of the photochemical treatment on transfection. The cytotoxicity of the photochemical treatment may be a drawback for some gene therapy applications, but has been successfully exploited in photodynamic therapy of cancer. Work is ongoing to develop photosensitizers with less toxic effects upon illumination (12).
6. Normally, the polyplexes are freshly made before the application to the cells. However, PEGylated polyplexes formed in HBG can be snap-frozen in liquid nitrogen and stored at –80 °C until use. Other polymer vectors and other targeting ligands than those described here, may be applied in photochemical transfection. Readers are encouraged to test their transfection vector of choice while applying the same main principles of photochemical transfection as described in the current protocol. The size and charge of the polyplexes should be verified prior to transfection, for example by using a Malvern Zetasizer instrument (Malvern Instruments Ltd., Worcestershire, UK). Charge shielded complexes with diameters $\leq$ 200 nm are preferred for *in vitro* and *in vivo* applications where target specificity is required.
7. In photochemical transfection the cells are first preloaded with the photosensitizer, then treated with the DNA polyplexes followed by exposure to light ***(5)***. This procedure is explained in detail in the text. Alternatively, the cells may be treated with the photosensitizer for 18 hours, washed three times, chased 4 hours in photosensitizer-free medium, and exposed to light before the polyplexes are added. In the latter case, the polyplexes should be added within six hours after the light exposure ***(25)***.
8. For drug delivery by PCI, the cells may be incubated with the photosensitizer and the macromolecule at the same time (e.g. for 18 hours prior to removal of the photosensitizer). In the case of delivery of proteins, PNA or siRNA to cells, molecules have been added to the cells at the same time as the photosensitizer and incubated for 18 hours prior to 4 hours chase and light exposure ***(1,8)***. However, for gene delivery, incubation with the photosensitizer prior to the gene is recommended to enhance the photochemical effect on transfection (i.e. this protocol would maximize the amount of photosensitizer in the endocytic vesicles and minimize the degradation of the transgenes). Shorter incubation times than 4 hours with the DNA complex (e.g. 0.5-1 hour pulse) are also possible. Then the cells should be chased in photosensitizer-free medium before incubation

with the DNA complex, so that the total incubation time in photosensitizer-free medium before irradiation is 4 hours. For gene transfer with adenovirus and adeno-associated virus vectors, we usually chase the cells for 3 hours in photosensitizer-free medium prior to the adding of the virus to the cells and light exposure ***(6)***.

9. It is recommended to incubate the cells in photosensitizer-free medium for some time before exposure to light in order to reduce damage to the plasma membrane. The time needed to remove the bulk of photosensitizer from the plasma membrane may vary between cell lines, but usually 4 hours incubation in photosensitizer-free medium has been found to induce efficient photochemical transfection. This chase period assures that the light exposure do not induce extensive photochemical damage to the plasma membrane, which is lethal for the cells, and that the main effect on transfection is due to induced rupture of the endocytic vesicles.

References

1. Berg, K., Selbo, P. K., Prasmickaite, L., Tjelle, T. E., Sandvig, K., Moan, J., et al. (1999) Photochemical internalization: A novel technology for delivery of macromolecules into cytosol. *Cancer Res.* **59**, 1180–1183.
2. Dolmans, D. E., Fukumura, D., and Jain, R. K. (2003) Photodynamic therapy for cancer. *Nat. Rev. Cancer* **3,** 380–387.
3. Prasmickaite, L., Hø, A., and Berg, K. (2001) Evaluation of different photosensitizers for use in photochemical gene transfection. *Photochem. Photobiol.* **73**, 388–395.
4. Moan, J., and Berg, K. (1991) The photodegradation of porphyrins in cells can be used to estimate the lifetime of singlet oxygen. *Photochem. Photobiol.* **53**, 549–553.
5. Høgset, A., Prasmickaite, L., Tjelle, T. E., and Berg, K. (2000) Photochemical transfection: A new technology for light-induced, site-directed gene delivery. *Hum. Gene Ther.* **11**, 869–880.
6. Høgset, A., Engesæter, B. Ø., Prasmickaite, L., Berg, K., Fodstad, Ø., and Mælandsmo, G. M. (2002) Light-induced adenovirus gene transfer, an efficient and specific gene delivery technology for cancer gene therapy. *Cancer Gene Ther.* **9**, 365–371.
7. Bonsted, A., Høgset, A., Hoover, F., and Berg, K. (2005) Photochemical enhancement of gene delivery to glioblastoma cells is dependent on the vector applied. *Anticancer Res.* **25**, 291–298.
8. Folini, M., Berg, K., Millo, E., Villa, R., Prasmickaite, L., Daidone, M. G., Benatti. U., and Zaffaroni, N. (2003) Photochemical internalization of a peptide nucleic acid targeting the catalytic subunit of human telomerase. *Cancer Res.* **63**, 3490–3494.
9. Shiraishi, T., and Nielsen, P. E. (2006) Photochemically enhanced cellular delivery of cell penetrating peptide-PNA conjugates. *FEBS Lett.* **580**, 1451–1456.

10. Selbo, P. K., Sivam, G., Fodstad, O., Sandvig, K., and Berg, K. (2001) In vivo documentation of photochemical internalization, a novel approach to site specific cancer therapy. *Int. J. Cancer* **92**, 761–766.
11. Berg, K., Dietze, A., Kaalhus, O., and Høgset, A. (2005) Site-specific drug delivery by photochemical internalization enhances the antitumor effect of bleomycin. *Clin. Cancer Res.* **11**, 8476–8485.
12. Nishiyama, N., Iriyama, A., Jang, W. D., Miyata, K., Itaka, K., Inoue, Y., et al. (2005) Light-induced gene transfer from packaged DNA enveloped in a dendrimeric photosensitizer. *Nat. Mater.* **4**, 934–941.
13. Ndoye, A., Dolivet, G., Hogset, A., Leroux, A., Fifre, A., Erbacher, P. et al. (2006) Eradication of p53-Mutated Head and Neck Squamous Cell Carcinoma Xenografts Using Nonviral p53 Gene Therapy and Photochemical Internalization. *Mol. Ther.* **13**, 1156–1162.
14. Prasmickaite, L., Hø, A., Tjelle, T. E., Olsen, V. M., and Berg, K. (2000) Role of endosomes in gene transfection mediated by photochemical internalisation (PCI). *J. Gene Med.* **2**, 477–488.
15. Kloeckner, J., Prasmickaite, L., Hø, A., Berg, K., and Wagner, E. (2004) Photochemically enhanced gene delivery of EGF receptor-targeted DNA polyplexes. *J. Drug Target.* **12**, 205–213.
16. Ogris, M., Steinlein, P., Kursa, M., Mechtler, K., Kircheis, R., and Wagner, E. (1998) The size of DNA/transferrin-PEI complexes is an important factor for gene expression in cultured cells. *Gene Ther.* **5**, 1425–1433.
17. Bonsted, A., Engesæter, B. Ø., Hø, A., Mælandsmo, G. M., Prasmickaite, L., D'Oliveira, C., et al. (2006) Photochemically enhanced transduction of polymer-complexed adenovirus targeted to the epidermal growth factor receptor. *J. Gene Med.* **8**, 286–297.
18. Weyergang, A., Selbo, P. K., and Berg, K. (2006) Photochemically stimulated drug delivery increases the cytotoxicity and specificity of EGF-saporin. *J. Control. Release* **111,** 165–173.
19. Zou, S. M., Erbacher, P., Remy, J. S., and Behr, J. P. (2000) Systemic linear polyethylenimine (L-PEI)-mediated gene delivery in the mouse. *J. Gene Med.* **2**, 128–134.
20. Brissault, B., Kichler, A., Guis, C., Leborgne, C., Danos, O., and Cheradame, H. (2003) Synthesis of linear polyethylenimine derivatives for DNA transfection. *Bioconjug. Chem.* **14**, 581–587.
21. Thomas, M., Lu, J. J., Ge, Q., Zhang, C., Chen, J., and Klibanov, A. M. (2005) Full deacylation of polyethylenimine dramatically boosts its gene delivery efficiency and specificity to mouse lung. *Proc. Natl. Acad. Sci. USA* **102**, 5679–5684.
22. Kursa, M., Walker, G. F., Roessler, V., Ogris, M., Roedl, W., Kircheis, R., et al. (2003) Novel Shielded Transferrin-Polyethylene Glycol-Polyethylenimine/DNA Complexes for Systemic Tumor-Targeted Gene Transfer. *Bioconjug. Chem.* **14**, 222–231.
23. Wolschek, M. F., Thallinger, C., Kursa, M., Rossler, V., Allen, M., Lichtenberger, C., et al. (2002) Specific systemic nonviral gene delivery to human hepatocellular carcinoma xenografts in SCID mice. *Hepatology* **36**, 1106–1114.

24. Plank, C., Zatloukal, K., Cotton, M., Mechtler, K., and Wagner, E. (1992) Gene transfer into hepatocytes using asialoglycoprotein receptor mediated endocytosis of DNA complexed with an artificial tetra-antennary galactose ligand. *Bioconjug. Chem.* **3,** 533–539.
25. Prasmickaite, L., Høgset, A., Selbo, P. K., Engesæter, B. Ø., Hellum, M., and Berg, K. (2002) Photochemical disruption of endocytic vesicles before delivery of drugs: a new strategy for cancer therapy. *British J. Cancer* **86**, 652–657.

12

Reducing the Genotoxic Potential of Retroviral Vectors

Ali Ramezani, Teresa S. Hawley, and Robert G. Hawley

Summary

The recent development of leukemia in gene therapy patients with X-linked severe combined immunodeficiency disease because of retroviral vector insertional mutagenesis has prompted reassessment of the genotoxic potential of integrating vector systems. In this chapter, various strategies are described to reduce the associated risks of retroviral genomic integration. These include deletion of strong transcriptional enhancer-promoter elements in the retroviral long terminal repeats, flanking the retroviral transcriptional unit with enhancer blocking sequences and designing vectors with improved RNA 3′ end processing. Protocols are provided to evaluate the relative biosafety of the modified vectors based on their ability to immortalize hematopoietic progenitor cells and propensity to trigger clonal hematopoiesis or leukemogenesis following hematopoietic stem cell transplantation.

Key Words: Gammaretroviral vectors; lentiviral vectors; insertional mutagenesis; genotoxicity; self-inactivating vectors; enhancer blocking sequences; post-transcriptional regulatory elements; RNA 3′ end processing.

1. Introduction

Until 2003, retroviral vectors had been considered a relatively safe means for gene delivery to human hematopoietic stem cells (HSCs), with no serious adverse events related to the vector systems reported in at least 40 gene marking and gene therapy trials involving more than 200 patients ***(1,2)***. A caveat of these observations is that most of the early clinical studies were characterized by very low efficiencies of gene transfer. This proviso notwithstanding, no progression to clonal hematopoiesis or leukemia had been observed to this point in preclinical investigations involving nonhuman primates or dogs followed for up to 7 years after receiving retrovirally transduced HSCs, including instances

From: *Methods in Molecular Biology, vol. 434: Volume 2: Design and Characterization of Gene Transfer Vectors*
Edited by: J. M. Le Doux © Humana Press, Totowa, NJ

where significant levels of gene transfer had been achieved *(3)*. However, the subsequent emergence of leukemia in three gene therapy patients with X-linked severe combined immunodeficiency disease (X-SCID) because of insertional activation of the *LMO2* proto-oncogene has prompted reevaluation of the safety profile of retroviral-mediated gene delivery *(4–7)*. Other recent preclinical and clinical follow-up studies have provided additional evidence of the potentially dangerous consequences of retroviral chromosomal insertion events *(8–16)*.

While retroviral DNA integration is not site-specific, genome-wide analyses of integration sites have revealed that murine leukemia virus (MLV)-based gammaretroviral vectors and human immunodeficiency virus type 1 (HIV-1)-based lentiviral vectors do not integrate at random throughout the genome. Rather, they preferentially insert their cargos into open chromatin regions in proximity to or within genes *(6,17,18)*. Whereas MLV exhibits a strong bias to integrate within a 5-kb window upstream or downstream of transcription initiation sites, HIV-1 tends to target the transcriptional unit itself. It is clear therefore that gene activation or disruption as a side effect of the retroviral integration process is greater than what would be predicted based on the assumption of a random distribution of sites throughout the genome. In this context, it is important to emphasize that MLV belongs to the slow transforming group of retroviridae (aka oncoretroviruses) and that its ability to induce hematopoietic tumor formation in mice is precisely because of its capacity to activate cellular proto-oncogenes—ergo the origin of the term "retroviral insertional mutagenesis" *(19–21)*. Nonetheless, although MLV is known to transform primate cells when allowed to replicate *(22)*, in view of a multistep mechanism of malignant transformation in humans *(23,24)*, the risk of leukemogenesis from replication-defective MLV-based gammaretroviral vectors *(16)* was believed to be low. Similarly, clonal dominance—the predominant occurrence of a few retroviral vector-marked HSC-derived hematopoietic clones following bone marrow transplantation—had been thought to be a natural property of hematopoiesis *(25–27)*. However, insertion site analysis in recent preclinical and clinical gene transfer studies has indicated that retroviral integration may influence this process by activating genes involved in growth control—such as the *Evi1* oncogene *(8,13,14,28)*—leading in some cases to selective nonmalignant clonal outgrowth *(9,15)*. Consequently, the combinatorial effects of multiple gammaretroviral vector insertions within a cell could lead to oncogenic conversion, especially if the transgene product itself confers a survival or proliferative advantage to the target cell population *(8,10,29,30)*, as may be the case with the common gamma chain γ cytokine receptor gene in X-SCID *(31,32)*. Although most safety concerns associated with insertional mutagenesis have to do with gammaretroviral vector-induced transcriptional activation of genes, it is not unreasonable to assume that haploinsufficiency of tumor suppressor genes because of lentiviral vector-mediated gene disruption could predispose toward malignancy ***(12,33–37)***. In any case, obtaining a low copy number of vector integrants per cell is clearly desirable ***(10,35)***.

Besides reducing the frequency of multiple integration events per cell, several safety modifications of the retroviral vector design can be envisioned (for reviews see **refs.** ***6,7,38,39***). MLV transcriptional activation of cellular gene expression results from the strong enhancer-promoter elements in the retroviral long terminal repeats (LTRs), with enhancer-mediated gene activation being the most frequent mechanism ***(19–21)***. Therefore, deletion of the LTR enhancer elements would be presumed to provide some modicum of improvement in terms of safety. Retroviral vectors having this design are referred to as "self-inactivating" (SIN) vectors ***(40–43)***. With the exception of our HHAM vector platform ***(26,27,41)***, the titers of the initial versions of most SIN gammaretroviral vectors were too low to be of utility for HSC gene transfer. However, recent advances have resulted in SIN gammaretroviral vector titers approaching those of conventional LTR vectors ***(44–46)***.

In the absence of the LTR enhancer-promoter elements, SIN retroviral vectors utilize an internal enhancer-promoter to drive transgene transcription. Therefore, although effectively reducing the copy number of enhancer elements per vector by a factor of 2, there is still a potential for the internal enhancer in these vectors to activate cellular gene expression. A strategy being implemented to prevent this involves flanking the vector sequences with insulators or enhancer blocking sequences ***(38,47,48)***. We and others have introduced a 1.2-kb fragment containing a monomer of the chicken β-globin 5′ hypersensitive site 4 (5′HS4) insulator, which functions as a chromatin domain boundary, into the U3 region of retroviral LTRs ***(49–54)***. However, the titers of these 5′HS4 insulator-containing "double-copy" vectors ***(55,56)*** are generally reduced. Moreover, superior protection in cell culture transfection and *Drosophila* transgenic experiments is obtained when the transgene is flanked by two tandem copies of the 1.2-kb 5′HS4 insulator fragment ***(57)***. In extensive work by the Felsenfeld group, a GC-rich core element of the 5′HS4 insulator has been mapped to a 5′ 250-bp fragment ***(58)***, two tandem copies of which were shown to provide complete insulator function ***(59)***. Therefore, more recent approaches have involved insertion of a dimer of the smaller 250-bp 5′HS4 insulator core element into the LTR U3 regions ***(45,60)***. Because almost all of the enhancer blocking activity of the 1.2-kb 5′HS4 insulator fragment can be conferred by a 42-bp sequence that is bound by CTCF (CCCTC-binding factor), a highly conserved and ubiquitous DNA-binding protein implicated in both transcriptional silencing and activation ***(59,61)***, it might be possible to further minimize the 5′HS4 insulator sequences without compromising any improvements gained in safety with the larger fragments.

It is worth noting that many of the existing SIN vectors (including SIN lentiviral vectors) are not necessarily completely transcriptionally disabled, either because the LTR is not fully inactivated ***(40,45,62,63)*** or because of residual transcriptional regulatory elements within the 5′ untranslated region of viral RNA that may contribute to the activation of cryptic promoters ***(64,65)***. Moreover, transcription termination is especially leaky in the SIN vector format ***(66)***. In addition to promoting mRNA nuclear export, the eukaryotic splicing process results in enhanced 3′ end formation and polyadenylation ***(67)***. Therefore, engineering introns within the vector may also reduce 3′ RNA readthrough ***(44,45,66,68)***. An alternative approach is the inclusion of *cis*-acting RNA transport elements such as the woodchuck hepatitis virus post-transcriptional regulatory element, which also augments RNA 3′ end processing and polyadenylation ***(68–70)***. In this regard, a safety-modified

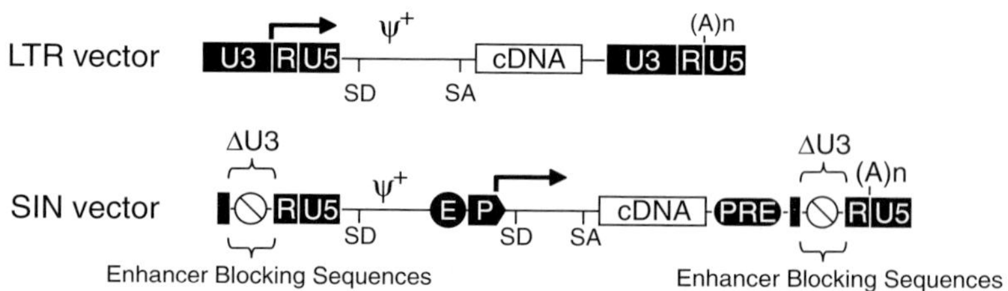

Fig. 1. Schematic representation of integrated forms of prototypical LTR and SIN retroviral vectors. For simplicity, gammaretroviral vectors are illustrated. Comparable lentiviral vectors contain additional components such as a central polypurine tract and the Rev response element involved in import of the preintegration complex into the nucleus and export of vector RNA into the cytoplasm, respectively ***(72,73)***. (Top) *LTR vector* The flanking intact LTRs comprise: U3, sequence unique to the 3′ vector RNA and repeated in the integrated vector DNA; R, short sequence repeated at both termini of the vector RNA; and U5, sequence unique to the 5′ vector RNA and repeated in the integrated vector DNA. Vector RNA transcripts initiate at the 5′ boundary of the R region in the 5′ LTR (arrow) and are polyadenylated [(A)n] at the 3′ boundary of the R region in the 3′ LTR. The packaging signal (Ψ^+) allows the full-length vector RNA to be efficiently encapsidated into budding vector particles. The protein coding region (cDNA) of the gene of interest is expressed as a spliced transcript (SD, splice donor; SA, splice acceptor). (Bottom) *SIN vector* The enhancer-promoter elements are deleted from the U3 regions of the LTRs (ΔU3) and an internal enhancer (E)-promoter (P) is used to drive transgene expression. The transgene pre-mRNA has been optimized for improved splicing and 3′ end formation by inclusion of an intron and a post transcriptional regulatory element (PRE). Enhancer-mediated interactions with cellular promoters are modulated by inclusion of enhancer blocking sequences ⃠ in the deleted U3 regions of the LTRs.

version of this post-transcriptional regulatory element devoid of potentially oncogenic promoter sequences and open reading frames that were present in the original version has recently been described ***(71)***. A prototypical SIN retroviral vector incorporating all of the features discussed above is illustrated in **Fig. 1**.

This chapter provides protocols to assess the relative genotoxicity of different retroviral vector designs and to systematically evaluate whether the proposed modifications enhance vector safety.

2. Materials

2.1. Vector Production

2.1.1. Transient Transfection

1. 293T/17 (293T) human embryonic kidney cell line (American Type Culture Collection, Manassas, VA, USA; cat. no. CRL-11268) (*see* **Note 1**).
2. 293T cell growth medium: Dulbecco's Modified Eagle's Medium (DMEM) supplemented with 4.5 g/l glucose, 2 mM L-glutamine, 50 IU/ml penicillin, 50 μg/ml streptomycin, and 10% heat-inactivated fetal bovine serum (FBS). Store at 4°C and warm up to 37°C before use. This growth medium is also used to culture NIH3T3 cells.
3. Gammaretroviral or lentiviral vector plasmid DNA containing the gene of interest (*see* **Note 2**).
4. Gammaretroviral (e.g., pEQPAM3-E; ***(74)***) or lentiviral (e.g., pCMV Δ R8.91; ***(75)***) packaging construct plasmid DNA.
5. Vesicular stomatitis virus G (VSV-G) glycoprotein (e.g., pMD.G; ***(76,77)***) or MLV ecotropic envelope plasmid DNA (e.g., pCAG4-Eco; ***(78,79)***).
6. 2.5 M $CaCl_2$: Dissolve 183.7 g $CaCl_2$ dihydrate (tissue culture grade) in deionized, distilled water. Bring the volume up to 500 ml and filter-sterilize using a 0.22-μm nitrocellulose filter. Stable at –20°C.
7. 2× *N*-(2-hydroxyethyl)piperazine-*N'*-(2-ethanesulfonic acid) (HEPES)-buffered saline (2× HBS): 50 mM HEPES (Sigma-Aldrich Corp., St. Louis, MO, USA; cat. no. H4034), 280 mM NaCl, 1.5 mM Na_2 PHO_4. Titrate to pH 7.05 with 5 N NaOH. Filter sterilize using a 0.22-μm nitrocellulose filter. Store as single use aliquots at –20°C.
8. 1 M HEPES in 0.85% NaCl (Cambrex Bio Science Walkersville, Inc., Walkerville, MD, USA; cat. no. 17-737F). Store at 4°C.

2.1.2. Collection and Concentration of Vector Particles

1. Low protein-binding Durapore (PVDF) 0.45-μm filter unit (Millipore Corp., Bedford, MA, USA).
2. Stericup 150 ml 0.45-μm filter unit (Millipore Corp.).

2.1.3. Titration of Vector Stocks

1. NIH3T3 murine fibroblasts (ATCC cat. no. CRL-1658).
2. Polybrene stock solution (Sigma-Aldrich Corp.; hexadimethrine bromide, cat no. H9268). Prepare stock of 6 mg/ml (1000×) in sterile deionized, distilled water; aliquot and store at –20°C.
3. Geneticin (Sigma-Aldrich Corp.; G418 disulfate salt, cat. no. A1720). Prepare stock of 40 mg/ml in sterile deionized, distilled water; aliquot and store at –20°C.

2.2. Insertional Mutagenesis Assays

2.2.1. Immortalization of Hematopoietic Progenitor Cells

1. Female C57BL/6 mice (6- to 8-week old; The Jackson Laboratory, Bar Harbor, ME, USA; cat. no. 000664) used as bone marrow donors. All procedures involving mice must follow the guidelines set forth in the National Institutes of Health Guide for the Care and Use of Laboratory Animals and be approved by an Institutional Animal Care and Use Committee.
2. 5-fluorouracil (5-FU; Sigma-Aldrich Corp., cat. no. F6627). Store the stock at room temperature, avoiding light. Prepare a fresh working solution of 15 mg/ml in phosphate-buffered saline (PBS) immediately before use.
3. Erythrocyte lysing solution: 154 mM NH_4Cl 10 mM $NaHCO_3$, and 0.082 mM sodium ethylenediaminetetraacetic acid (EDTA), pH 7.3. Commercial lysing solutions are also available. Store at room temperature.
4. Bone marrow progenitor cell growth medium: Iscove's modified Dulbecco's medium (IMDM) supplemented with 4.5 g/l glucose, 2 mM L-glutamine, 50 IU/ml penicillin, 50 μg/ml streptomycin, 15% heat-inactivated FBS, 100 ng/ml murine stem cell factor (SCF), 30 ng/ml murine interleukin-3 (IL-3), and 10 ng/ml murine IL-6. Store at 4°C and warm up to 37°C before use.
5. Recombinant fibronectin fragment (RetroNectin; Takara Mirus Bio, Madison, WI, USA).
6. Polybrene stock solution (1000×; 8 mg/ml) prepared as described in **Subheading 2.1.3**. Aliquot and store at –20°C.

2.2.2. Hematopoietic Clonal Evolution and Leukemogenesis Assays

1. Female C57BL/6 mice (6- to 8-week old; as listed in **Subheading 2.2.1.**) used as bone marrow donors and recipients.
2. Small animal gamma irradiator.
3. Hematology analyzer equipped with a veterinarian software package.
4. Flow cytometer.

3. Methods

3.1. Vector Production

293T [293 human embryonic kidney cells expressing simian virus 40 (SV40) large tumor (T) antigen] cells are highly transfectable such that transient co-transfection with current generation transfer vectors and packaging plasmids yields high-titer, replication-defective gammaretroviral and lentiviral vector particles ***(80)***. Retroviral vectors pseudotyped with the VSV-G glycoprotein have a broad host-cell range and can be utilized in the insertional mutagenesis assays described herein ***(76,77)***. However, VSV-G-pseudotyped virions are frequently associated with toxicity and transduction artifacts ***(81–83)***. Moreover, because the assays involve transduction of murine HSC and progenitor populations, the MLV ecotropic envelope can be conveniently used to pseudotype any gammaretroviral or lentiviral vectors to be evaluated ***(79,84,85)***.

3.1.1. Transient Transfection

1. Culture 293T cells in 293T/NIH3T3 cell growth medium at 37°C in a humidified atmosphere with 5% CO_2.
2. Passage cells every 3–4 days using a trypsin–EDTA solution to dissociate them. To trypsinize, remove the medium and rinse the cells with PBS (without Ca_2^+ and Mg_2^+). Remove the PBS and add enough trypsin–EDTA to cover the cells. Incubate the plate at room temperature until the cells round up and detach. To inactivate the trypsin, add an equal volume of DMEM containing 10% FBS. Collect the cells and centrifuge for 5 min at 375 × *g*. Resuspend cells in fresh culture medium and plate between 1:4 and 1:8.
3. Transfect 293T cells with plasmid DNAs using the calcium phosphate precipitation method ***(86)***. On the day before transfection, plate 293T cells in 7 ml 293T/NIH3T3 cell growth medium at a density of 5 × 10^6 cells per 100-mm plate.
4. Mix 15 μg of the transfer vector plasmid, 10 μg of the appropriate packaging plasmid and 5 μg of the MLV ecotropic envelope (or VSV-G glycoprotein) plasmid (*see* **Note 3**). Bring the volume up to 400 μl with sterile water. Add 100 μl of 2.5 M $CaCl_2$ and mix. Add the DNA/$CaCl_2$ solution dropwise to 500 μl of 2× HBS in a 15-ml conical tube. Use a second pipettor and a 2-ml pipet to bubble the 2× HBS as the DNA/ $CaCl_2$ solution is added. Vortex immediately for 5 s and incubate at room temperature for 20 min. Add the 1 ml DNA/calcium phosphate mixture directly to each 100-mm plate while swirling (*see* **Note 4**). Incubate the cells at 37°C overnight (16 h). Change the medium and culture for 24–48 h (this method usually results in transfection of 50–80% of the cells).

3.1.2. Collection and Concentration of Vector Particles

Collect the vector-conditioned medium 24–48 h after medium change, centrifuge at 2000× *g* for 10 min to remove cellular debris and filter through a 0.45-μm pore-size filter (use a 5-ml syringe or a 150-ml filter unit depending on the volume). Use directly for transductions or aliquot and store at –80°C (*see* **Note 5**).

Several procedures have been developed to concentrate retroviral vector particles *(73)*. However, concentration by centrifugation is the most commonly used method. While the stability of the VSV-G glycoprotein allows concentration of vector particles by ultracentrifugation, vector particles pseudotyped with the MLV ecotropic envelope should be concentrated by the following low-speed centrifugation protocol.

1. Centrifuge the vector supernatants overnight (16 h) at 13,000 × *g* and 4°C.
2. Remove supernatants and, using gentle pipetting, resuspend pellets in ~100 µl of medium appropriate for the downstream application. To allow complete resuspension, vortex gently overnight at 4°C.
3. To remove cellular debris, centrifuge concentrated supernatants at 10,000 × *g* and 4°C for 5 min in a microcentrifuge. Remove the supernatant, aliquot and freeze at –80°C (*see* **Note 6**).

3.1.3. Titration of Vector Stocks

The NIH3T3 murine fibroblast cell line can be used to determine the titer of ecotropic gammaretroviral and lentiviral vector preparations.

1. Culture NIH3T3 cells in 293T/NIH3T3 cell growth medium at 37°C in a humidified atmosphere with 5% CO_2.
2. Plate 2.5 × 10^5 NIH3T3 cells in each well of a 6-well plate 4–6 h before titrating the vector supernatants.
3. In a final volume of 1 ml, prepare serial dilutions for each vector preparation (e.g., 10^0, 10^{-1}, and 10^{-2} for unconcentrated vector supernatants; 10^{-2}, 10^{-3}, and10^{-4} for concentrated supernatants) using 293T/NIH3T3 cell growth medium. To each dilution, add 1 µl of 6 mg/ml (1000×) polybrene for a final concentration of 6 µg/ml. Remove the medium from the cells and add each of the 1-ml dilutions to the corresponding well. Incubate at 37°C for 4 h.
4. Remove the supernatants after the 4-h transduction and replace with 2 ml fresh 293T/NIH3T3 cell growth medium. Return to the CO_2 incubator and incubate at 37°C.

   a. After 48 h, determine the relative end-point vector titers [in transducing units per ml (TU/ml)] by flow cytometric analysis [i.e., if the vector contains a fluorescent reporter gene such as the green fluorescent protein (GFP)] *(87)*. To determine the vector titer, use the following equation:

   $$\text{Vector titer} = \text{number of NIH3T3 cells} \times \%\text{ of GFP}^{+}\text{cells} \times \text{dilution factor}$$

   (*see* **Note 7**).

   b. If the vector carries a drug resistance gene *(88)*, split cells 1:10 after 48 h and seed into 100-mm dishes in medium containing the recommended amount of the corresponding drug [e.g., 400 µg/ml geneticin if the neomycin phosphotransferase (*neo*) drug resistance gene is used]. Replace medium every 4–5 days for 2 weeks. Fix and stain cells with 0.3% crystal violet in 70% methanol and enumerate the drug-resistant colonies. To determine the vector titer, use the following equation:

$$\text{Vector titer} = \text{number of drug-resistant colonies} \times 10 \times \text{dilution factor.}$$

3.2. Insertional Mutagenesis Assays

3.2.1. Immortalization of Hematopoietic Progenitor Cells

In light of the leukemias that developed in 3 X-SCID patients as a result of insertional activation of the *LMO2* oncogene ***(4)***, it is imperative that all ensuing retroviral gene therapy protocols include stringent preclinical evaluations of vector genotoxicity. Thus, irrespective of therapeutic efficacy, any HSC-based gene transfer approach will only become clinically acceptable if there is a low risk-to-benefit ratio ***(5–7)***. Recently, a retroviral insertional mutagenesis assay involving immortalization of murine bone marrow progenitors was described ***(14)***. Interestingly, ~25% of the immortalized cell lines obtained contained vector insertions in the *Evi1* gene, where they activated expression of a truncated protein similar to what is observed in human leukemias ***(89)***. It is of relevance that retroviral vector insertions in *Evi1* have also been identified in at least 6 murine bone marrow transplant recipients ***(8,9)***, 9 rhesus macaque bone marrow transplant recipients ***(13)***, and in two patients in a chronic granulomatosis disease clinical trial ***(15)***.

Copeland and colleagues ***(14)*** established cell lines from 50% of murine bone marrow progenitor cell cultures when 1×10^6 cells were transduced with a gammaretroviral vector harboring intact LTRs that had a titer of 3×10^6 TU/ml. Approximately 30% of the cultures comprised two clones, and the average number of vector copies per cell was 5–6. However, when vector titers were reduced to 4×10^5 TU/ml, no immortalized cell lines were obtained ($n = 10$). In contrast, when bone marrow progenitor cells were transduced with a derivative of the vector carrying an Evi1 cDNA (encoding the short Evi1 isoform expressed in many of the cell lines as a result of insertional mutagenesis), all of the cultures gave rise to immortalized cell lines with a titer of 1×10^5 TU/ml. Subcloning of these cell lines indicated single copy integration, suggesting that inappropriate expression of *Evi1* alone is sufficient to promote the immortalization of murine hematopoietic progenitor cells under these conditions. We previously obtained similar results—single copy integration in immortalized hematopoietic progenitor cell lines—using a gammaretroviral vector carrying the *HOX11/TLX1* oncogene ***(90)***. Therefore, insertional mutagenesis experiments should be carried out in parallel with a vector expressing *Evi1* or *HOX11/TLX1* (1×10^5 TU/ml) as a positive control.

3.2.1.1. Isolation, Transduction, and Culture of Murine Bone Marrow Cells

1. Dissolve 5-FU in sterile PBS (15 mg/ml) immediately before use and intravenously inject into mice (150 mg/kg body weight) using a 27-G needle attached to a 1-ml syringe.
2. Harvest bone marrow cells 4 days after the 5-FU injection. Flush the hind limbs with PBS containing 2% FBS using a 21-G needle attached to a 5-ml syringe ***(91,92)*** (*see* **Note 8**).
3. Lyse erythrocytes by incubating total bone marrow cells in erythrocyte lysis solution for 10 min at room temperature followed by centrifugation at 375 × *g* for 5 min.
4. Coat 35-mm suspension culture plates with 2 μg/cm^2 recombinant fibronectin fragment ***(93)***. Transfer the nucleated cells to plates at a density of 5 × 10^5 cells/ml and culture for 48 h in bone marrow progenitor cell growth medium at 37°C in a humidified atmosphere containing 5% CO_2.
5. Transduce the bone marrow cells for 3 consecutive days (4 h each day) by incubation with gammaretroviral or lentiviral vector conditioned medium in the presence of 8 μg/ml polybrene supplemented with the same growth factors as used for prestimulation ***(94)***.
6. Culture the transduced cells in bone marrow progenitor cell growth medium minus murine IL-6. Passage cells every 3 days. Immortalized clones that arise will continue to propagate after 1 month of culturing ***(95)***.
7. Extract genomic DNA from the immortalized cells and subject to Southern blot analysis to determine vector copy number. Perform integration site analysis [e.g., by linear amplification-mediated polymerase chain reaction (LAM-PCR) ***(5)***].

3.2.2. Hematopoietic Clonal Evolution and Leukemogenesis Assays

The in vivo effects of insertional mutagenesis—evolution to hematopoietic clonal dominance or frank leukemia—can be examined in mice transplanted with retrovirally transduced HSCs ***(8–12)***. For example, the latency of leukemic conversion can be reduced by using a gammaretroviral vector carrying an oncogene ***(11)***. In one study, 60% of the mice transplanted with bone marrow transduced with a Sox4 gammaretroviral vector presented with myeloid leukemias between 4 and 7 months post-transplant, whereas under similar conditions, no malignancies developed in animals transplanted with bone marrow cells transduced with the vector backbone alone. Although the finding is not surprising, it further reinforces the point that caution is warranted when the transgene product in a gene therapy protocol confers a selective growth or survival advantage to the target cell population ***(8,10,29–32)***. Along these lines, we previously reported sporadic leukemia development in murine transplant recipients that received transduced HSCs carrying gammaretroviral vector-encoded cytokine genes: IL-11 (one instance of 10 primary and 18 secondary recipients) or FLT3 ligand (20 of 24 primary recipients) ***(29,30)***.

3.2.2.1. Transplantation of Retrovirally Transduced Bone Marrow Cells

1. Expose C57BL/6 recipient mice to an otherwise lethal dose of total body γ-irradiation using a gamma irradiator (1050 cGy; split dose with 3 h between doses).
2. Transplant the irradiated mice with 1–2 × 10^6 retrovirally transduced bone marrow cells (obtained as described in **Subheading 3.2.1.1.**) injected through the tail vein in 300 μl PBS plus 2% FBS using a 27-G needle attached to a 1-ml syringe ***(84)***. House the transplanted mice in sterile microisolator cages on laminar flow racks. Add an antibiotic to the drinking water (e.g., 2 cc/250 ml Baytril; Bayer Corp., Shawnee Mission, KS, USA) for 3–4 weeks as a prophylactic measure during hematopoietic recovery to prevent possible deaths because of adventitious infections.
3. At biweekly intervals after transplant, collect peripheral blood from the retro-orbital venous sinus and analyze hematological parameters—such as total leukocytes, total erythrocytes, hemoglobin, hematocrit, mean corpuscular volume, mean corpuscular hemoglobin, mean corpuscular hemoglobin concentration, red cell distribution width, total platelets, and mean platelet volume—using a hematology analyzer equipped with a veterinarian software package. Mouse bleeding is performed after inhalation isoflurane anesthesia and administration to the eye of one drop of a local anesthetic (e.g., 0.5% Tetracaine Ophthalmic Solution, Phoenix Pharmaceutical Inc., St. Joseph, MO, USA). Collect blood using micro-hematocrit capillary tubes and place ~100 μl into 1.3 ml micro-collection tubes containing potassium EDTA (1.6 mg/ml blood; Sarstedt Inc., Newton, NC, USA; cat. no. 41.1395.105). Cap the tube and gently mix the contents of the tube by flicking the side of the tube. Keep the blood sample at room temperature for at least 5 min before measurement. The sample may be analyzed up to several hours after collection.
4. If no hematopathologic changes are observed within 6 months, sacrifice the mice and isolate bone marrow cells to transplant into lethally irradiated secondary recipients (1–2 × 10^6 cells per mouse). Extract genomic DNA from bone marrow and/or spleen cells and perform integration site analysis (e.g., by Southern blot analysis and LAM-PCR).
5. Collect peripheral blood from the retro-orbital venous sinus of the secondary recipients at biweekly intervals after transplant and analyze hematological parameters.
6. If mice display hematopathologic changes (leukocytosis) or become moribund (showing substantial weight loss, ruffled fur, and a hunched posture) ***(29,30)***, collect peripheral blood from the retro-orbital venous sinus and place ~100 μl into micro-collection tubes. Centrifuge at 375 × *g* for 5 min and resuspend the pellet in 500 μl PBS containing 2% FBS plus 0.1% NaN_3. Sacrifice the mice and also collect cells from bone marrow, spleen, and thymus. Lyse the erythrocytes and prepare aliquots of ~5 × 10^4 cells/50 μl in PBS containing 2% FBS plus 0.1% NaN_3 for staining individually with fluorochrome-conjugated

lineage-specific (Gr1, Mac1, CD19, TER119, CD3, CD4, CD8) monoclonal antibodies. Incubate for 30 min at 4°C. Wash in 2 ml PBS containing 2% FBS plus 0.1% NaN_3. Centrifuge at 375 × *g* for 5 min. Decant supernatant and drain. Resuspend in 300 μl PBS containing 2% FBS plus 0.1% NaN_3 and analyze by flow cytometry (*see* **Note 9**). Transplant 1 × 10^6 bone marrow cells into nonirradiated recipients to determine whether any leukemias that arise are transplantable.

7. Sacrifice any mice when they display hematopathologic changes or become moribund and all asymptomatic secondary recipients 6 months after transplantation. Extract genomic DNA from bone marrow and/or spleen cells and perform integration site analysis (e.g., by Southern blot analysis and LAM-PCR).

4. Notes

1. The 293T/17 cell line is a derivative of the 293T (293tsA1609neo) cell line ***(96)***. 293T is a highly transfectable derivative of the 293 human embryonic kidney cell line into which the temperature sensitive gene for SV40 large T antigen was inserted. Human 293 cells express the adenovirus serotype 5 E1A 12S and 13S gene products, which strongly transactivate transcription from expression vectors containing the human cytomegalovirus immediate early region enhancer-promoter elements ***(97)***. In addition, the SV40 large T antigen may stimulate extrachromosomal replication of plasmids containing the SV40 origin of replication during transient transfection.
2. Besides academic sources, retroviral vector backbones and packaging systems are available commercially (e.g., Clontech Laboratories, Inc., Mountain View, CA, USA; Invitrogen Corp., Carlsbad, CA, USA; Stratagene Corp., La Jolla, CA, USA).
3. A so-called "third generation" HIV-1 lentiviral vector system has been developed in which the *rev* gene has been deleted from the *gag-pol* packaging construct ***(98)***. Since both the expression of the *gag* and *pol* genes and lentiviral vector transcripts are dependent on *trans*-complementation by a separate Rev expression construct, use of this system requires cotransfection of four plasmids. For further details on HIV-1 lentiviral vector designs and protocols, see **refs.** *72,73*.
4. The vector titer depends on several factors including the vector backbone design, the size, and nature of any inserted sequences as well as on the efficiency of transfection. Always use an exponentially growing culture of 293T cells for transfection (50–70% confluent) and make sure that the cells are trypsinized well during plating, so that they form a uniform monolayer. Another factor that affects the transfection efficiency is the quality of the plasmid DNA used. Use a commercially available plasmid DNA purification kit (e.g., from Qiagen, Valencia, CA, USA) to obtain highly purified endotoxin-free supercoiled plasmid DNA (traditionally obtained by purifying on two separate cesium chloride gradients). Sterilize the DNA by ethanol precipitation, resuspend the air-dried pellet in sterile deionized,

distilled water, and determine its concentration and quality by spectrophotometric analysis and gel electrophoresis. Also of importance with regard to the transfection efficiency is the pH of the 2× HBS solution, which should be between 7.05 and 7.12. Once the transfection mixture has been added to the cells, a fine precipitate should develop within a few minutes.

5. Vector titers are also influenced by factors that affect the stability of the vector particles, which include the pH, temperature, and freeze-and-thaw frequency ***(99)***. Avoid pH changes in vector supernatants as this could lead to significant loss of titer. pH changes can be prevented by adding HEPES buffer to the 293T/NIH3T3 cell growth medium at a final concentration of 10 mM. Once collected, vector supernatants should be kept on ice at all times and, if they are not being used immediately, stored in aliquots at –80°C. Avoid repeated freezing and thawing. The peak of vector particle production by transient transfection is on days 2 and 3 post-transfection. Therefore, the optimal time to collect the supernatant is 48 h after addition of fresh medium to the transfected cells.
6. Small (~2- to 5-mm diameter) pellets should be visible after concentration by centrifugation. Pellets may not resuspend completely. Vortex overnight at low speed and 4°C to facilitate resuspension. Expect a 50–75% recovery following vector concentration.
7. When GFP is used as the reporter gene ***(87)***, it is useful to wait ~5 days before analyzing the transduced cells by flow cytometry. This will minimize the contribution of false positive signals because of pseudotransduction, which is the direct transfer of reporter protein present in the vector supernatants or incorporated into the vector particles to the target cells; this is particularly problematic for VSV-G-pseudotyped vectors ***(82,83)***. Note that it has also been shown that transgenes can be efficiently transiently expressed from unintegrated lentiviral vectors during this timeframe ***(100)***.
8. It should be possible to obtain 3–4 × 10^6 bone marrow cells from each 5-FU-treated mouse. The bone marrow cells can be used directly or further enriched for HSC/progenitors using various magnetic- or fluorescence-activated cell sorting procedures ***(91,92)***. When culturing bone marrow cells, keep the cell density at 0.5 × 10^6 cells/ml.
9. Detailed protocols for staining of murine hematopoietic cells and detection of cell surface antigens by immunofluorescence flow cytometric analysis may be found on the websites of the monoclonal antibody manufacturers.

Acknowledgments

This work was supported in part by National Institutes of Health grants R01HL65519, R01HL66305, and R24RR16209, and by the King Fahd Endowment Fund (The George Washington University School of Medicine and Health Sciences). R.G.H. receives royalties derived from the licensing of MSCV retroviral gene transfer technology to Clontech Laboratories, Inc.

References

1. Stewart, A. K., Dubé, I. D., and Hawley, R. G. (1999) Gene marking and the biology of hematopoietic cell transfer in human clinical trials, in *Blood Cell Biochemistry 8, Hematopoiesis and Gene Therapy* (Fairbairn, L. J. and Testa, N., eds.), Kluwer Academic/Plenum Publishers, NY, pp. 243–268.
2. Kohn, D. B., Sadelain, M., Dunbar, C., Bodine, D., Kiem, H. P., Candotti, F. et al. (2003). American Society of Gene Therapy (ASGT) ad hoc subcommittee on retroviral-mediated gene transfer to hematopoietic stem cells. *Mol. Ther.* **8**, 180–187.
3. Kiem, H. P., Sellers, S., Thomasson, B., Morris, J. C., Tisdale, J. F., Horn, P. A. et al. (2004). Long-term clinical and molecular follow-up of large animals receiving retrovirally transduced stem and progenitor cells: no progression to clonal hematopoiesis or leukemia. *Mol. Ther.* **9**, 389–395.
4. Hacein-Bey-Abina, S., von Kalle, C., Schmidt, M., McCormack, M. P., Wulffraat, N., Leboulch, P. et al. (2003). LMO2-associated clonal T cell proliferation in two patients after gene therapy for SCID-X1. *Science* **302**, 415–419.
5. von Kalle, C., Fehse, B., Layh-Schmitt, G., Schmidt, M., Kelly, P., and Baum, C. (2004). Stem cell clonality and genotoxicity in hematopoietic cells: gene activation side effects should be avoidable. *Semin. Hematol.* **41**, 303–318.
6. Baum, C., Kustikova, O., Modlich, U., Li, Z., and Fehse, B. (2006). Mutagenesis and oncogenesis by chromosomal insertion of gene transfer vectors. *Hum. Gene Ther.* **17**, 253–263.
7. Nienhuis, A. W., Dunbar, C. E., and Sorrentino, B. P. (2006). Genotoxicity of retroviral integration in hematopoietic cells. *Mol. Ther.* **13**, 1031–1049.
8. Li, Z., Dullmann, J., Schiedlmeier, B., Schmidt, M., von Kalle, C., Meyer, J. et al. (2002). Murine leukemia induced by retroviral gene marking. *Science* **296**, 497.
9. Kustikova, O., Fehse, B., Modlich, U., Yang, M., Dullmann, J., Kamino, K. et al. (2005). Clonal dominance of hematopoietic stem cells triggered by retroviral gene marking. *Science* **308**, 1171–1174.
10. Modlich, U., Kustikova, O. S., Schmidt, M., Rudolph, C., Meyer, J., Li, Z. et al. (2005). Leukemias following retroviral transfer of multidrug resistance 1 (MDR1) are driven by combinatorial insertional mutagenesis. *Blood* **105**, 4235–4246.
11. Du, Y., Spence, S. E., Jenkins, N. A., and Copeland, N. G. (2005). Cooperating cancer-gene identification through oncogenic-retrovirus-induced insertional mutagenesis. *Blood* **106**, 2498–2505.
12. Montini, E., Cesana, D., Schmidt, M., Sanvito, F., Ponzoni, M., Bartholomae, C. et al. (2006). Hematopoietic stem cell gene transfer in a tumor-prone mouse model uncovers low genotoxicity of lentiviral vector integration. *Nat. Biotechnol.* **24**, 687–696.
13. Calmels, B., Ferguson, C., Laukkanen, M. O., Adler, R., Faulhaber, M., Kim, H. J. et al. (2005). Recurrent retroviral vector integration at the Mds1/Evi1 locus in nonhuman primate hematopoietic cells. *Blood* **106**, 2530–2533.

14. Du, Y., Jenkins, N. A., and Copeland, N. G. (2005). Insertional mutagenesis identifies genes that promote the immortalization of primary bone marrow progenitor cells. *Blood* **106**, 3932–3939.
15. Ott, M. G., Schmidt, M., Schwarzwaelder, K., Stein, S., Siler, U., Koehl, U. et al. (2006). Correction of X-linked chronic granulomatous disease by gene therapy, augmented by insertional activation of MDS1-EVI1, PRDM16 or SETBP1. *Nat. Med.* **12**, 401–409.
16. Seggewiss, R., Pittaluga, S., Adler, R. L., Guenaga, F. J., Ferguson, C., Pilz, I. H. et al. (2006). Acute myeloid leukemia is associated with retroviral gene transfer to hematopoietic progenitor cells in a rhesus macaque. *Blood* **107**, 3865–3867.
17. Wu, X., Li, Y., Crise, B., and Burgess, S. M. (2003). Transcription start regions in the human genome are favored targets for MLV integration. *Science* **300**, 1749–1751.
18. Bushman, F., Lewinski, M., Ciuffi, A., Barr, S., Leipzig, J., Hannenhalli, S. et al. (2005). Genome-wide analysis of retroviral DNA integration. *Nat. Rev. Microbiol.* **3**, 848–858.
19. van Lohuizen, M. and Berns, A. (1990). Tumorigenesis by slow-transforming retroviruses—-an update. *Biochim. Biophys. Acta* **1032**, 213–235.
20. Askew, D. S., Bartholomew, C., and Ihle, J. N. (1993). Insertional mutagenesis and the transformation of hematopoietic stem cells. *Hematol. Path.* **7**, 1–22.
21. Akagi, K., Suzuki, T., Stephens, R. M., Jenkins, N. A., and Copeland, N. G. (2004). RTCGD: retroviral tagged cancer gene database. *Nucleic Acids Res.* **32**, 523–527.
22. Donahue, R. E., Kessler, S. W., Bodine, D., McDonagh, K., Dunbar, C., Goodman, S. et al. (1992). Helper virus induced T cell lymphoma in nonhuman primates after retroviral mediated gene transfer. *J. Exp. Med.* **176**, 1125–1135.
23. Rangarajan, A., Hong, S. J., Gifford, A., and Weinberg, R. A. (2004). Species- and cell type-specific requirements for cellular transformation. *Cancer Cell* **6**, 171–183.
24. Armstrong, S. A. and Look, A. T. (2005). Molecular genetics of acute lymphoblastic leukemia. *J. Clin. Oncol.* **23**, 6306–6315.
25. Lemischka, I. R., Raulet, D. H., and Mulligan, R. C. (1986). Developmental potential and dynamic behavior of hematopoietic stem cells. *Cell* **45**, 917–27.
26. Capel, B., Hawley, R., Covarrubias, L., Hawley, T., and Mintz, B. (1989). Clonal contributions of small numbers of retrovirally marked hematopoietic stem cells engrafted in unirradiated neonatal W/W^{v} mice. *Proc. Natl. Acad. Sci. U. S. A.* **86**, 4564–4568.
27. Capel, B., Hawley, R. G., and Mintz, B. (1990). Long- and short-lived murine hematopoietic stem cell clones individually identified with retroviral integration markers. *Blood* **75**, 2267–2270.
28. Mucenski, M. L., Taylor, B. A., Ihle, J. N., Hartley, J. W., Morse III, H. C., Jenkins, N. A. et al. (1988). Identification of a common ecotropic viral integration site, Evi-1, in the DNA of AKXD murine myeloid tumors. *Mol. Cell. Biol.* **8**, 301–308.

29. Hawley, R. G., Fong, A. Z. C., Ngan, B. Y., de Lanux, V. M., Clark, S. C., and Hawley, T. S. (1993). Progenitor cell hyperplasia with rare development of myeloid leukemia in interleukin 11 bone marrow chimeras. *J. Exp. Med.* **178**, 1175–1188.
30. Hawley, T. S., Fong, A. Z. C., Griesser, H., Lyman, S. D., and Hawley, R. G. (1998). Leukemic predisposition of mice transplanted with gene-modified hematopoietic precursors expressing flt3 ligand. *Blood* **92**, 2003–2011.
31. Dave, U. P., Jenkins, N. A., and Copeland, N. G. (2004). Gene therapy insertional mutagenesis insights. *Science* **303**, 333.
32. Sternberg, D. W. and Gilliland, D. G. (2004). The role of signal transducer and activator of transcription factors in leukemogenesis. *J. Clin. Oncol.* **22**, 361–371.
33. Song, W. J., Sullivan, M. G., Legare, R. D., Hutchings, S., Tan, X., Kufrin, D. et al. (1999). Haploinsufficiency of CBFA2 causes familial thrombocytopenia with propensity to develop acute myelogenous leukaemia. *Nat. Genet.* **23**, 166–175.
34. Yan, H., Dobbie, Z., Gruber, S. B., Markowitz, S., Romans, K., Giardiello, F. M. et al. (2002). Small changes in expression affect predisposition to tumorigenesis. *Nat. Genet.* **30**, 25–26.
35. Woods, N. B., Muessig, A., Schmidt, M., Flygare, J., Olsson, K., Salmon, P. et al. (2003). Lentiviral vector transduction of NOD/SCID repopulating cells results in multiple vector integrations per transduced cell: risk of insertional mutagenesis. *Blood* **101**, 1284–1289.
36. Payne, S. R. and Kemp, C. J. (2005). Tumor suppressor genetics. *Carcinogenesis* **26**, 2031–2045.
37. Themis, M., Waddington, S. N., Schmidt, M., von, K. C., Wang, Y., Al-Allaf, F. et al. (2005). Oncogenesis following delivery of a nonprimate lentiviral gene therapy vector to fetal and neonatal mice. *Mol. Ther.* **12**, 763–771.
38. Hawley, R. G. (2001). Progress toward vector design for hematopoietic stem cell gene therapy. *Curr. Gene Ther.* **1**, 1–17.
39. Baum, C., Schambach, A., Bohne, J., and Galla, M. (2006). Retrovirus vectors: toward the plentivirus? *Mol. Ther.* **13**, 1050–1063.
40. Yu, S. F., von Rüden, T., Kantoff, P. W., Garber, C., Seiberg, M., Rüther, U. et al. (1986). Self-inactivating retroviral vectors designed for transfer of whole genes into mammalian cells. *Proc. Natl. Acad. Sci. U. S. A.* **83**, 3194–3198.
41. Hawley, R. G., Covarrubias, L., Hawley, T., and Mintz, B. (1987). Handicapped retroviral vectors efficiently transduce foreign genes into hematopoietic stem cells. *Proc. Natl. Acad. Sci. U. S. A.* **84**, 2406–2410.
42. Miyoshi, H., Blomer, U., Takahashi, M., Gage, F. H., and Verma, I. M. (1998). Development of a self-inactivating lentivirus vector. *J. Virol.* **72**, 8150–8157.
43. Zufferey, R., Dull, T., Mandel, R. J., Bukovsky, A., Quiroz, D., Naldini, L. et al. (1998). Self-inactivating lentivirus vector for safe and efficient in vivo gene delivery. *J. Virol.* **72**, 9873–9880.
44. Kraunus, J., Schaumann, D. H., Meyer, J., Modlich, U., Fehse, B., Brandenburg, G. et al. (2004). Self-inactivating retroviral vectors with improved RNA processing. *Gene Ther.* **11**, 1568–1578.

45. Ramezani, A., Hawley, T. S., and Hawley, R. G. (2006). Stable gammaretroviral vector expression during embryonic stem cell-derived in vitro hematopoietic development. *Mol. Ther.* **14**, 245–254.
46. Schambach, A., Mueller, D., Galla, M., Verstegen, M. M., Wagemaker, G., Loew, R. et al. (2006). Overcoming promoter competition in packaging cells improves production of self-inactivating retroviral vectors. *Gene Ther.* **13**, 1524–1533.
47. Felsenfeld, G., Burgess-Beusse, B., Farrell, C., Gaszner, M., Ghirlando, R., Huang, S. et al. (2004). Chromatin boundaries and chromatin domains. *Cold Spring Harb. Symp. Quant. Biol.* **69**, 245–250.
48. Ellis, J. (2005). Silencing and variegation of gammaretrovirus and lentivirus vectors. *Hum. Gene Ther.* **16**, 1246.
49. Emery, D. W., Yannaki, E., Tubb, J., and Stamatoyannopoulos, G. (2000). A chromatin insulator protects retrovirus vectors from chromosomal position effects. *Proc. Natl. Acad. Sci. U. S. A.* **97**, 9150–9155.
50. Rivella, S., Callegari, J. A., May, C., Tan, C. W., and Sadelain, M. (2000). The cHS4 insulator increases the probability of retroviral expression at random chromosomal integration sites. *J. Virol.* **74**, 4679–4687.
51. Ma, Y., Ramezani, A., Lewis, R., Hawley, R. G., and Thomson, J. A. (2003). High-level sustained transgene expression in human embryonic stem cells using lentiviral vectors. *Stem Cells* **21**, 111–117.
52. Ramezani, A., Hawley, T. S., and Hawley, R. G. (2003). Performance- and safety-enhanced lentiviral vectors containing the human interferon-β scaffold attachment region and the chicken β-globin insulator. *Blood* **101**, 4717–4724.
53. Li, C. and Emery, D. W. (2006). The cHS4 chromatin insulator blocks epigenetic interactions between integrated vectors and the surrounding genome. *Mol. Ther.* **13 (Suppl. 1)**, S405.
54. Evans-Galea, M. V., Wielgosz, M. M., Hanawa, H., and Nienhuis, A. W. (2006). Suppression of clonal dominance in cultured human lymphoid cells by addition of the 5′cHS4 insulator to a lentiviral vector. *Mol. Ther.* **13 (Suppl. 1)**, S405–S406.
55. Hantzopoulos, P. A., Sullenger, B. A., Ungers, G., and Gilboa, E. (1989). Improved gene expression upon transfer of the adenosine deaminase minigene outside the transcriptional unit of a retroviral vector. *Proc. Natl. Acad. Sci. U. S. A.* **86**, 3519–3523.
56. Wiznerowicz, M., Fong, A. Z. C., Mackiewicz, A., and Hawley, R. G. (1997). Double-copy bicistronic retroviral vector platform for gene therapy and tissue engineering: application to melanoma vaccine development. *Gene Ther.* **4**, 1061–1068.
57. Chung, J. H., Whiteley, M., and Felsenfeld, G. (1993). A 5′ element of the chicken β-globin domain serves as an insulator in human erythroid cells and protects against position effect in Drosophila. *Cell* **74**, 505–514.
58. Chung, J. H., Bell, A. C., and Felsenfeld, G. (1997). Characterization of the chicken β-globin insulator. *Proc. Natl. Acad. Sci. U. S. A.* **94**, 575–580.

59. Recillas-Targa, F., Pikaart, M. J., Burgess-Beusse, B., Bell, A. C., Litt, M. D., West, A. G. et al. (2002). Position-effect protection and enhancer blocking by the chicken β-globin insulator are separable activities. *Proc. Natl. Acad. Sci. U. S. A.* **99**, 6883–6888.
60. Yao, S., Osborne, C. S., Bharadwaj, R. R., Pasceri, P., Sukonnik, T., Pannell, D. et al. (2003). Retrovirus silencer blocking by the cHS4 insulator is CTCF independent. *Nucleic Acids Res.* **31**, 5317–5323.
61. Bell, A. C., West, A. G., and Felsenfeld, G. (1999). The protein CTCF is required for the enhancer blocking activity of vertebrate insulators. *Cell* **98**, 387–396.
62. Guild, B. C., Finer, M. H., Housman, D. E., and Mulligan, R. C. (1988). Development of retrovirus vectors useful for expressing genes in cultured murine embryonal cells and hematopoietic cells in vivo. *J. Virol.* **62**, 3795–3801.
63. De, P. M., Montini, E., de Sio, F. R., Benedicenti, F., Gentile, A., Medico, E. et al. (2005). Promoter trapping reveals significant differences in integration site selection between MLV and HIV vectors in primary hematopoietic cells. *Blood* **105**, 2307–2315.
64. Logan, A. C., Haas, D. L., Kafri, T., and Kohn, D. B. (2004). Integrated self-inactivating lentiviral vectors produce full-length genomic transcripts competent for encapsidation and integration. *J. Virol.* **78**, 8421–8436.
65. Hanawa, H., Persons, D. A., and Nienhuis, A. W. (2005). Mobilization and mechanism of transcription of integrated self-inactivating lentiviral vectors. *J. Virol.* **79**, 8410–8421.
66. Zaiss, A. K., Son, S., and Chang, L. J. (2002). RNA 3′ readthrough of oncoretrovirus and lentivirus: implications for vector safety and efficacy. *J. Virol.* **76**, 7209–7219.
67. Bentley, D. L. (2005). Rules of engagement: co-transcriptional recruitment of pre-mRNA processing factors. *Curr. Opin. Cell Biol.* **17**, 251–256.
68. Ramezani, A., Hawley, T. S., and Hawley, R. G. (2000). Lentiviral vectors for enhanced gene expression in human hematopoietic cells. *Mol. Ther.* **2**, 458–469.
69. Zufferey, R., Donello, J. E., Trono, D., and Hope, T. J. (1999). Woodchuck hepatitis virus post transcriptional regulatory element enhances expression of transgenes delivered by retroviral vectors. *J. Virol.* **73**, 2886–2892.
70. Loeb, J., Harris, M., and Hope, T. (2000). The woodchuck hepatitis virus post transcriptional regulatory element increases transgene expression by enhancing the 3′-end metabolism of mRNAs. *Mol. Ther.* **1 (Suppl. 1)**, S142.
71. Schambach, A., Bohne, J., Baum, C., Hermann, F. G., Egerer, L., von, L. D. et al. (2006). Woodchuck hepatitis virus post-transcriptional regulatory element deleted from X protein and promoter sequences enhances retroviral vector titer and expression. *Gene Ther.* **13**, 641–645.
72. Ramezani, A. and Hawley, R.G. (2002) Overview of the HIV-1 lentiviral vector system, in *Current Protocols in Molecular Biology* (Ausubel, F., Brent, R., Kingston, B., Moore, D., Seidman, J., Smith, J. A., and Struhl, K., eds.), John Wiley & Sons, Inc., Somerset, NJ, pp. 16.21.1–16.21.15.

73. Ramezani, A. and Hawley, R.G. (2002) Generation of HIV-1-based lentiviral vector particles, in *Current Protocols in Molecular Biology* (Ausubel, F., Brent, R., Kingston, B., Moore, D., Seidman, J., Smith, J. A., and Struhl, K., eds.), John Wiley & Sons, Inc., Somerset, NJ, pp. 16.22.1–16.22.15.
74. Kelly, P. F., Vandergriff, J., Nathwani, A., Nienhuis, A. W., and Vanin, E. F. (2000). Highly efficient gene transfer into cord blood nonobese diabetic/severe combined immunodeficiency repopulating cells by oncoretroviral vector particles pseudotyped with the feline endogenous retrovirus (RD114) envelope protein. *Blood* **96**, 1206–1214.
75. Zufferey, R., Nagy, D., Mandel, R. J., Naldini, L., and Trono, D. (1997). Multiply attenuated lentiviral vector achieves efficient gene delivery in vivo. *Nat. Biotechnol.* **15**, 871–875.
76. Burns, J. C., Friedmann, T., Driever, W., Burrascano, M., and Yee, J.-K. (1993). Vesicular stomatitis virus G glycoprotein pseudotyped retroviral vectors: concentration to very high titer and efficient gene transfer into mammalian and nonmammalian cells. *Proc. Natl. Acad. Sci. U. S. A.* **90**, 8033–8037.
77. Naldini, L., Blomer, U., Gallay, P., Ory, D., Mulligan, R., Gage, F. H. et al. (1996). In vivo gene delivery and stable transduction of nondividing cells by a lentiviral vector. *Science* **272**, 263–267.
78. Markowitz, D., Goff, S., and Bank, A. (1988). A safe packaging line for gene transfer: separating viral genes on two different plasmids. *J. Virol.* **62**, 1120–1124.
79. Hanawa, H., Kelly, P. F., Nathwani, A. C., Persons, D. A., Vandergriff, J. A., Hargrove, P. et al. (2002). Comparison of various envelope proteins for their ability to pseudotype lentiviral vectors and transduce primitive hematopoietic cells from human blood. *Mol. Ther.* **5**, 242–251.
80. Pear, W. S., Nolan, G. P., Scott, M. L., and Baltimore, D. (1993). Production of high-titer helper-free retroviruses by transient transfection. *Proc. Natl. Acad. Sci. U. S. A.* **90**, 8392–8396.
81. Chen, S.-T., Iida, A., Guo, L., Friedmann, T., and Yee, J.-K. (1996). Generation of packaging cell lines for pseudotyped retroviral vectors of the G protein of vesicular stomatitis virus by using a modified tetracycline inducible system. *Proc. Natl. Acad. Sci. U. S. A.* **93**, 10057–10062.
82. Liu, M. L., Winther, B. L., and Kay, M. A. (1996). Pseudotransduction of hepatocytes by using concentrated pseudotyped vesicular stomatitis virus G glycoprotein (VSV-G)-Moloney murine leukemia virus-derived retrovirus vectors: comparison of VSV-G and amphotropic vectors for hepatic gene transfer. *J. Virol.* **70**, 2497–2502.
83. Gallardo, H. F., Tan, C., Ory, D., and Sadelain, M. (1997). Recombinant retroviruses pseudotyped with the vesicular stomatitis virus G glycoprotein mediate both stable gene transfer and pseudotransduction in human peripheral blood lymphocytes. *Blood* **90**, 952–957.
84. Hawley, R. G., Hawley, T. S., Fong, A. Z., Quinto, C., Collins, M., Leonard, J. P. et al. (1996). Thrombopoietic potential and serial repopulating ability of murine

hematopoietic stem cells constitutively expressing interleukin 11. *Proc. Natl. Acad. Sci. U. S. A.* **93**, 10297–10302.

85. Schambach, A., Galla, M., Modlich, U., Will, E., Chandra, S., Reeves, L. et al. (2006). Lentiviral vectors pseudotyped with murine ecotropic envelope: increased biosafety and convenience in preclinical research. *Exp. Hematol.* **34**, 588–592.
86. Hawley, T. S., Sabourin, L. A., and Hawley, R. G. (1989). Comparative analysis of retroviral vector expression in mouse embryonal carcinoma cells. *Plasmid* **22**, 120–131.
87. Hawley, T. S., Herbert, D. J., Eaker, S. S., and Hawley, R. G. (2004). Multiparameter flow cytometry of fluorescent protein reporters. *Methods Mol. Biol.* **263**, 219–238.
88. Hawley, R. G., Lieu, F. H., Fong, A. Z., and Hawley, T. S. (1994). Versatile retroviral vectors for potential use in gene therapy. *Gene Ther.* **1**, 136–138.
89. Nucifora, G., Laricchia-Robbio, L., and Senyuk, V. (2006). EVI1 and hematopoietic disorders: history and perspectives. *Gene* **368**, 1–11.
90. Hawley, R. G., Fong, A. Z. C., Lu, M., and Hawley, T. S. (1994). The HOX11 homeobox-containing gene of human leukemia immortalizes murine hematopoietic precursors. *Oncogene* **9**, 1–12.
91. Eaker, S. S., Hawley, T. S., Ramezani, A., and Hawley, R. G. (2004). Detection and enrichment of hematopoietic stem cells by side population phenotype. *Methods Mol. Biol.* **263**, 161–180.
92. Hawley, R.G., Ramezani, A., and Hawley, T.S. (2006) Hematopoietic stem cells, in *Methods in Enzymology, Adult Stem Cells* **419** (Lanza, R. and Klimanskaya, I., eds.), Elsevier/Academic Press, San Diego, CA. pp. 149–179.
93. Donahue, R. E., Sorrentino, B. P., Hawley, R. G., An, D. S., Chen, I. S. Y., and Wersto, R. P. (2001). Fibronectin fragment CH-296 inhibits apoptosis and enhances ex vivo gene transfer by murine retrovirus and human lentivirus vectors independent of viral tropism. *Mol. Ther.* **3**, 359–367.
94. Moayeri, M., Ramezani, A., Morgan, R. A., Hawley, T. S., and Hawley, R. G. (2004). Sustained phenotypic correction of hemophilia A mice following oncoretroviral-mediated expression of a bioengineered human factor VIII gene in long-term hematopoietic repopulating cells. *Mol. Ther.* **10**, 892–902.
95. Hawley, T. S., McLeish, W. A., and Hawley, R. G. (1991). Establishment of a novel factor-dependent myeloid cell line from primary cultures of mouse bone marrow. *Cytokine* **3**, 60–71.
96. DuBridge, R. B., Tang, P., Hsia, H. C., Leong, P. M., Miller, J. H., and Calos, M. P. (1987). Analysis of mutation in human cells by using an Epstein-Barr virus shuttle system. *Mol. Cell. Biol.* **7**, 379–387.
97. Gorman, C. M., Gies, D., McCray, G., and Huang, M. (1989). The human cytomegalovirus major immediate early promoter can be trans-activated by adenovirus early proteins. *Virology* **171**, 377–385.
98. Dull, T., Zufferey, R., Kelly, M., Mandel, R. J., Nguyen, M., Trono, D. et al. (1998). A third-generation lentivirus vector with a conditional packaging system. *J. Virol.* **72**, 8463–8471.

99. Higashikawa, F. and Chang, L. (2001). Kinetic analyses of stability of simple and complex retroviral vectors. *Virology* **280**, 124–131.
100. Nightingale, S. J., Hollis, R. P., Pepper, K. A., Petersen, D., Yu, X. J., Yang, C. et al. (2006). Transient gene expression by nonintegrating lentiviral vectors. *Mol. Ther.* **13**, 1121–1132.

13

Evaluation of Promoters for Use in Tissue-Specific Gene Delivery

Changyu Zheng and Bruce J. Baum

Summary

Vectors used in gene therapy require an expression cassette. The expression cassette consists of three important components: promoter, therapeutic gene and polyadenylation signal. The promoter is essential to control expression of the therapeutic gene. A tissue-specific promoter is a promoter that has activity in only certain cell types. Use of a tissue-specific promoter in the expression cassette can restrict unwanted transgene expression as well as facilitate persistent transgene expression. Therefore, choosing the correct promoter, especially a tissue-specific promoter, is a major step toward achieving successful therapeutic transgene expression. Ideally, the elements of the natural promoter region, necessary for obtaining the required level of the gene expression while retaining tissue-specificity, should be known. Also, it is important to understand whether interactions occur between the promoter region and the rest of the vector genome that could affect promoter activity and specificity. To assess this, it is helpful to select a suitable vector system that will be used in further gene therapy studies. Second, have one or several candidate tissue-specific promoters available for use. Third, ideally have an in vitro cell model suitable to evaluate tissue-specificity. Fourth, have a convenient in vivo animal model to use. Fifth, select a good reporter gene system. Next, using conventional recombinant DNA techniques create different promoter constructs with the selected vector system. Lastly, have a suitable transfection method to test the plasmid constructs in both the in vitro and the in vivo models.

Key Words: Tissue-specific promoter; transcription; expression cassette; gene therapy vector.

From: *Methods in Molecular Biology, vol. 434: Volume 2: Design and Characterization of Gene Transfer Vectors*
Edited by: J. M. Le Doux © Humana Press, Totowa, NJ

1. Introduction

In mammalian cells, each gene has its own promoter, and some promoters can only be activated in a specific cell type *(1)*. The promoter is a specific genetic region involved in the binding of a RNA polymerase to initiate transcription and is located 5′ from the transcription start site *(2)*. Therefore, the location of a promoter determines the template strand for each gene transcription. In eukaryotic nuclei, there are three RNA polymerases, RNA polymerase I, II and III. RNA polymerase II is involved in transcribing the most cellular genes; however, in eukaryotic cells, it cannot initiate transcription on a DNA template. This requires many nuclear proteins, called general transcription factors, that are designated as transcription factor for polymerase II (TFII) including TFIIA, TFIIB, etc., to assemble at the promoter region with the RNA polymerase II and initiate transcription *(2)*. A DNA sequence called the TATA box exists in the promoter regions of most genes and typically is located approximately 25 nucleotides upstream from the transcription start site. The TATA box signals the start of transcription. TFIID recognizes and binds to the TATA box, and causes other general transcription factors to assemble at the promoter, helping to position RNA polymerase II correctly at the promoter *(2)*. Although transcription in undifferentiated mouse embryos (at the 2- to 8-cell stages) does not require a TATA box, it becomes critical for efficient transcription in differentiated cells *(3)*. This means that the TATA box itself also can directly be involved in the regulation of gene transcription. In eukaryotic nuclei, RNA polymerase II also requires activator, mediator and chromatin-modifying proteins to correctly transcribe DNA. The transcription activators also bind to specific sequences in DNA and help to attract RNA polymerase II to the start point of transcription. Mediators form a protein complex that allows the activator proteins to communicate properly with RNA polymerase II and with general transcription factors *(2)*. Transcription initiation in the cell often requires the local recruitment of chromatin-modifying enzymes, including chromatin remodeling complexes and histone acetylases, which allow greater accessibility of RNA polymerase II to the DNA present in chromatin *(2)*.

In aggregate, eukaryotic transcription initiation is a very complex process that uses many regulatory proteins. Furthermore, some regulatory proteins can bind to DNA thousands of nucleotides away from the promoter, which means that a single promoter can be controlled by an almost unlimited number of regulatory sequences scattered along the DNA. Also, each regulatory protein usually contributes to the control of many genes *(2)*. Although some gene regulatory proteins are fairly specific and only expressed in one or a few cell types, most are found in a variety of cell types, in many tissues, and at several times during development. This type of combinatorial gene control makes it possible to generate considerable biological complexity with a relatively defined number of regulatory molecules ***(2)***.

The promoter region in eukaryotic genes is commonly a relatively large DNA fragment. Most gene therapy vectors have a size limitation in the capacity of an expression cassette; thus, only the essential part of a promoter can be used. Therefore, any use of a tissue-specific promoter requires prior careful evaluation in specific and faithful expression systems. Doing so helps to minimize toxicity, maximize efficient gene expression in the desired target cell/tissue and optimize the gene therapy vector. Herein, we use our previous studies ***(4,5)*** evaluating salivary gland tissue-specific promoters in gene therapy vectors as an example of how to assess the tissue-specific promoters in vitro and in vivo.

In salivary glands, two general epithelial cell types are present—acinar cells and ductal cells in primarily a densely packed monolayer ***(6)***. Both are involved in the formation of saliva. Acinar cells are water permeable and salt secreting, whereas ductal cells are relatively water impermeable and salt absorbing. The epithelial cells in salivary glands are easy to access in vivo in animals, as well as humans. As noted above, having a convenient in vivo animal assay for studying promoters is essential. Vector delivery is performed through local cannulation of the main excretory ducts of the targeted glands. The orifices are accessible directly in the mouth, and in animals anesthesia is required only for restraint. For these studies, we have employed mammalian plasmid expression vectors and adenoviral vectors. Replication-deficient recombinant adenoviral vectors are useful for delivering exogenous genes into both salivary epithelial cell types ***(7,8)*** in vivo. Because acinar and ductal cells can be differentially affected by salivary disorders, it is useful to target each cell type specifically ***(9)***. In our studies, we have used many different promoters. Some promoters, such as human cytomegalovirus (CMV), Rous sarcoma virus (RSV), simian virus 40 (SV40) and mammalian elongation factor 1α (EF1α), are non-specific promoters and are commonly used in gene therapy vectors. Other promoters, such as cytokeratins 18 and 19, are epithelial cell-specific and should have activities in both acinar and ductal cells ***(10–13)***. The tissue kallikrein promoter is considered ductal cell specific in salivary glands ***(5)***, while the amylase 1C and aquaporin-5 (AQP5) promoters should be relatively acinar cell specific ***(4,14,15)***.

2. Materials

2.1. Serial Deletion Analysis and Plasmid Construction

1. Restriction endonuclease enzymes, chosen according to the promoter sequence and the plasmid vector sequence.

2. Taq and Pfu DNA polymerases (Stratagene, La Jolla, CA, USA) and EasyStart polymerase chain reaction (PCR) Mix-in-a Tube (Molecular BioProducts, San Diego, CA, USA).
3. Klenow Fill-In kit (Stratagene) and T4 DNA polymerase (Invitrogene, Carlsbad, CA, USA).
4. Bacterial alkaline phosphastase (Invitrogen).
5. T4 DNA ligase (Invitrogen).
6. DH5α^{TM} cells (Invitrogen).
7. Agarose (Invitrogen).
8. Ethidium bromide (Invitrogen).
9. SOC and LB media (Invitrogen).
10. Wizard Plus Minipreps DNA Purification System, Wizard Plus Midipreps DNA Purification System, and Maxipreps DNA purification system (Promega, Madison, WI, USA).

2.2. Recombinant Adenoviral Vector Production

1. 293 cell line (Microbix, Ontario, Canada).
2. pJM17 plasmid (Microbix).
3. Calcium Phosphate Transfection Kit (Invitrogen).
4. Cesium Chloride (CsCl) gradients (1.25, 1.33 and 1.4 mg/ml) for adenovirus purification. CsCl (Invitrogen) is dissolved in TD buffer (140 mM NaCl, 5 mM KCl, 25 mM Tris–HCl and 0.7 mM Na_2HPO_4; adjust pH to 7.4 with HCl).
5. Virus dialysis buffer: 100 mM Tris–HCl (pH 7.4), 10 mM $MgCl_2$ and 10% (v/v) glycerol.
6. Virus dilution buffer: 5 mM $MgCl_2$, 10 mM Tris–HCl and 20% (v/v) glycerol, pH 7.4.
7. SYBR Green PCR Master Mix (Applied Biosystems, Foster City, CA, USA).

2.3. Cell Culture

1. McCoy's 5A medium (Invitrogen) supplemented with 10% fetal bovine serum (HyClone, Logan, UT, USA), 100 u/ml penicillin G (Invitrogen) and 100 ug/ml streptomycin (invitrogen) for the A5 cell line, which is a rat submandibular ductal cell line ***(16)***.
2. IMEM medium (Invitrogen) supplemented with 10% fetal bovine serum (HyClone), 100 u/ml penicillin G (Invitrogen) and 100 ug/ml streptomycin (Invitrogen) for 293 cell line, which is a human embryonic kidney cell line ***(17)***.
3. Note, there is no salivary acinar cell line available.

2.4. In Vivo Animal Experiments

1. PE 10 tube for cannulation.
2. 1 cc insulin syringe U-100 28G1/2" (Becton Dickinson, Franklin Lakes, NJ, USA)
3. Dexamethasone (SIGMA, St. Louis, MO, USA)

4. Ketamine (100 mg/ml, Phoenix Scientific, St. Joseph, MO, USA) and xylazine (20 mg/ml, Phoenix Scientific).
5. Restraint cannulation board for proper positioning of animal's oral cavity.

2.5. Plasmid Delivery

1. Adcontrol, which is a replication-deficient adenovirus without any transgene in the E1 region. Ad5. Null is a similar vector to Adcontrol that can be obtained from Qbiogene, Irvine, CA, USA.
2. Polyethylenimine (PEI), high molecular weight, water-free (Cat. No. 40872-7, ALDRICH, Milwaukee, WI, USA)

2.6. Protein Assay

1. BCA protein Assay Reagents A and B (PIERCE, Rockfold, IL, USA)

2.7. Luciferase Assay

1. Luciferase cell Culture Lysis Reagent 5× (Promega).
2. Luciferase Assay Substrate and Luciferase Assay Buffer (Promega).

2.8. Immunohistochemistry

1. Hydrogen peroxide 30% (w/w) solution (SIGMA)
2. ImmunoCruz™ Staining System for rabbit primary antibody (Santa Cruz Biotechnology, Inc., Santa Cruz, CA, USA)

3. Methods

There are many ways to evaluate tissue-specific promoters (*see* **Note 1**). The following (**steps 1–5**) are the most common.

1. Serial DNA deletion (*see* **Fig. 1**) by either restriction endonuclease digestion or PCR.
2. Subclone different promoter fragments into the plasmid vector.
3. Check the different constructs in the cell and, ideally, animal models by transfection.
4. Detect reporter gene expression with an appropriate and convenient assay.
5. Locate site of protein expression in cells or tissue using immunohistochemistry.

Analysis of the rat AQP5 promoter *(5)* is given as an example. For this study, we selected two reporter genes. One was luciferase, which is a very sensitive reporter gene with a low background, but for which we have found no good antibodies available. The second reporter gene, encoding enhanced green fluorescence protein (EGFP), has good antibodies available. EGFP can be directly observed as well as examined by immunohistochemical staining.

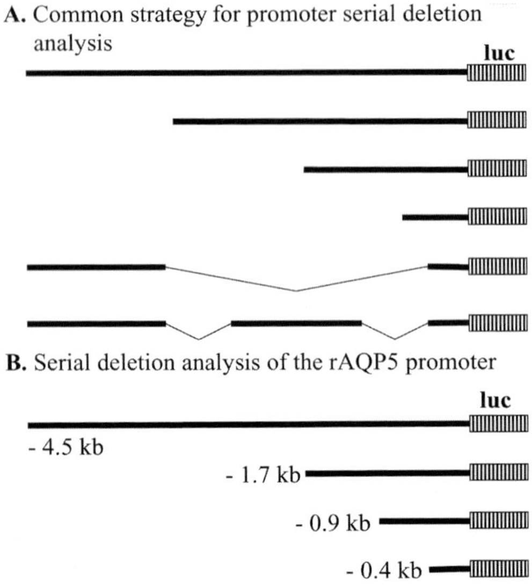

Fig. 1. Serial deletion studies for promoter analysis. (A) is a general diagram showing a common strategy for promoter serial deletion analysis. (B) is a specific example of such an analysis that we performed to evaluate the rat AQP5 (rAQP5) promoter in our vector system ***(5)***. Luc, luciferase, which is the reporter gene used.

3.1. Serial Deletion Analysis

Before performing a serial deletion analysis for a promoter, it is best to conduct some database searches on the sequence, e.g., PubMed, Gene Bank, and tissue-specific promoter database (TiProD) (http://tiprod.cbi.pku.edu.cn:8080/index.html) ***(18)*** to determine whether this promoter has previously been studied, and if so, what the results were and how it was performed. If not, it is best to try to learn as much about the promoter sequence as possible from gene bank and other databases, e.g., how many and what kind of transcription factor-binding sites are present, is there a TATA box, where is it, what is the guanine/cytosine content. The rAQP5 promoter had been described by Borok *et al.* ***(19)***. They made a series of progressive unidirectional 5′→ 3′ AQP5-luciferase deletion constructs beginning at –1716 with an identical 3′-end (starting at –6 bp) by using a combination of Exonuclease III, the Erase-a-Base System (Promega) and *Bal*31 nuclease digestion. The Erase-a-Base System is designed for the rapid construction of a plasmid containing progressive unidirectional deletions of any inserted DNA. This system ***(20,21)*** uses exonuclease III to specifically digest inserted DNA from a 5′ protruding or blunt end restriction site. The adjacent sequencing primer-binding site is protected from digestion by a four base 3′ overhang restriction site or by an α-phosphorothioate-filled end. This method makes it rapid to construct nested deletions from plasmids, and useful to evaluate a tissue-specific promoter, especially an unknown promoter.

On the basis of the studies of Borok *et al.* ***(19)***, we decided to do three serial deletions in the 4.5-kb rAQP5 promoter fragment (1.791, 0.9 and 0.391 kb) and evaluate those rAQP5 promoter fragments in our plasmid and adenoviral vector systems (*see* **Fig. 1B**). We received the plasmid pUC-AQP5-2 thanks to the generosity of Dr. David Ann, University of Southern California. After restriction endonuclease analysis of the 4.5-kb fragment, we recognized we could use *Eco*RI/*Hin*dIII to get the 4.5-kb rAQP5 promoter fragment, *Fok*I/*Hin*dIII for the 1.791-kb fragment, *Sau*3AI/*Hin*dIII for the 0.9-kb fragment, and *Del*I/*Hin*dIII for the 0.391-kb fragment. In brief, each fragment was filled in with Klenow Fill-In kit (Stratagene), cleaned of the Klenow enzyme, precipitated with 100% of ethanol, resuspended with ddH_2O, ligated with *Hin*dIII linker, cleaned again, then digested with *Hin*dIII. We used *Hin*dIII to open the plasmid vector, pAC-luc, cleaned this, precipitated with 100% of ethanol, resuspended with ddH_2O, dephosphorylated, then ligated it with the above individual rAQP5 promoter fragments. Ligated plasmids were transformed into DH5α competent cells, colonies screened by preparing plasmid DNA and restriction endonuclease digestions performed. These procedures resulted in construction of the following plasmids: pACrAQP5-4.5-luc, pACrAQP5-1.7-luc, pACrAQP5-0.9-luc, and pACrAQP5-0.4-luc.

3.2. Plasmid Constructions

To construct the rAQP5 tissue-specific promoter into the plasmid expression vector, the following common recombinant DNA techniques were used. Detailed protocols for each can be obtained from commercial sources, as well as from various standard molecular biology texts ***(22,23)***. Accordingly, these techniques will not be described further.

1. Plasmid DNA preparations using Wizard Plus Midipreps DNA Purification System and Maxipreps DNA purification system (Promega).
2. Restriction endonuclease digestion of plasmid DNA (*see* **Note 2**) followed by agarose gel electrophoresis.
3. PCR to amplify a special fragment of the tissue-specific promoter with Taq or Pfu DNA polymerases (Stratagene).
4. Fill-in or removal of both 5′ or 3′ overhangs to make blunt ends using Klenow Fill-in kit (Stratagene) or T4 DNA polymerase (Invitrogen).
5. Plasmid DNA precipitation with 2 vol of 100% cold ethanol, then wash twice with 70% ethanol.
6. Dephosphorylation of 5′-phosphorylated termini of vector DNA to prevent self-ligation using bacterial alkaline phosphatase (Invitrogen).

7. Plasmid DNA and promoter DNA fragment ligation using T4 DNA ligation (Invitrogen).
8. Transformation using MAX Efficiency DH5 α Chemically Competent Cells (Invitrogen), then culture in a shaker at 37°C for 1 h, streak out competent cells on an LB agar or agarose ampcillin plate and grow the bacteria overnight at 37°C to obtain colonies.
9. Colony screening. Pick up single colonies from the above plate and transfer to a bacterial culture tube with 3 ml LB medium, then incubate in a shaker at 37°C overnight; perform mini plasmid preparation using Wizard plus Minipreps, digest DNA samples using selected restriction endonuclease enzymes, run the digested samples in a 1% agarose gel with ethidium bromide (50 ng/100 ml), then select a correct clone based on the electrophoresis results for use in evaluating the promoter activity in vitro and in vivo.

3.3. Recombinant Adenoviral Vector Production

A common gene transfer vector system that we use is based on the serotype 5 adenovirus. We therefore also evaluated the rAQP5 promoter in the adenoviral vector system. Recombinant adenoviral vectors are approximately 36 kb, replication-deficient DNA virus. For these studies, the expression cassette, including the rAQP5 promoter, was inserted into the deleted E1 gene region of the adenovirus. Because remaining portions of the adenoviral genome can possibly affect and/or interfere with the rAQP5 promoter, i.e., *cis*-acting effects, it was important for us to evaluate the rAQP5 promoter in an adenoviral vector context. Our in vitro and in vivo plasmid expression results demonstrated that pACrAQP5-0.4-luc was of interest for further evaluation. Therefore, we used this plasmid to make an adenoviral vector. The following is the protocol we used to produce the adenoviral vector, AdrAQP5-0.4-luc.

1. Make a maxipreparation using Wizard Plus Maxipreps of pACrAQP5-0.4-luc constructed based on **Subheadings 3.1** and **3.2**.
2. Make a maxipreparation of the plasmid pJM17 using Wizard Plus Maxipreps (*see* **Note 3**). pJM17 consists of most of the adenovirus serotype 5 (Ad5) genome except for the E1 region. The pJM17 vector and the pAC shuttle vector contain overlap regions (0.0–1.3 and 9.3–17.0 map units) that permit homologous exchange when co-transfected into 293 cells ***(24)***.
3. Grow 293 cells in IMEM (*see* **Note 4**). The 293 cell line is a human embryonic kidney cell line that contains an integrated copy of the leftmost Ad5 genome, which complements the defect in E1-deficient vectors ***(17)***.
4. Cotransfection of pJM17 and pACrAQP5-0.4-luc into 293 cells by calcium phosphate co-precipitation using the Calcium Phosphate Transfection Kit (Invitrogen). Add 15 μg of pJM17, 5 μg of pACrAQP5-0.4-luc, 36 μl of 2 M $CaCl_2$ and ddH_2O to 300 μl in a 14-ml polypropylene tube, mix well. Next, quickly add 300 μl 2× HEPES-buffered saline and bubble with a pipet for 1 min, then

incubate at room temperature for 30 min to form a fine precipitate. Add the entire mixture to a 100-mm plate of 293 cells and gently rotate the plate to mix well. Put the plate back into the tissue culture incubator and replace the growth medium every 2 days. After approximately 10–12 days, plaques (holes in the 293 monolayer) are observed, which indicate the replication of the recombinant adenovirus, AdrAQP5-0.4-luc. Continually and carefully replace the medium until approximately 50% of 293 cells are lysed. Harvest the entire plate (293 cells and growth medium) by pipetting into a 50-ml conical test tube.

5. Completely lyse the 293 cells by freezing (dry ice) and thawing (37°C water bath) five times to release adenoviral vector from the cells. Centrifuge at approximately 2,000 × *g* for 5 min. Transfer supernatant [crude viral lysate (CVL)] to a new 50-ml tube. Use 50 µl of the CVL to infect one well of 293 cells (10^5 cells/well in 96-well plate). Measure the luciferase activity in the lysate of these cells after 24 h to confirm that the CVL contained functional virus.
6. Large scale preparation to propagate AdrAQP5-0.4-luc. Use the above CVL to infect three fresh 100-mm plates of 293 cells. Harvest at day 3. Freeze and thaw five times as above, combine and again centrifuge to collect supernatant. Use this CVL to infect twenty 150-mm plates of 293 cells. Harvest the cells at day 3, centrifuge the harvested cells and use 6 ml of the CVL to resuspend the cell pellet. Freeze and thaw five times, then centrifuge at 3500 × *g* for 10 min.
7. Transfer supernatant to the first CsCl gradient [place 2.5 ml of density 1.25 CsCl in sterile ultra-clear centrifuge tubes (Beckman Coulter, No. 344059, Fullerton, CA, USA), then slowly underlay 2.5 ml of density 1.40 CsCl]. Spin the tubes in a SW41 rotor at 210,000 × *g*, 22°C for 1 h (balance carefully). Clean the outside of the tube with 70% alcohol, then use a 21-G needle and syringe to poke through the side of the tube to collect the lower opalescent band, which contains the recombinant Ad5 vector. Transfer this vector band to a second CsCl gradient [place 8 ml of density 1.33 CsCl into sterile ultra-clear centrifuge tubes (Beckman Coulter, No. 344059)]. Spin in a SW41 rotor at 210,000 × *g*, 22°C for 18 h (balance carefully). As above, clean the outside of the tube with 70% alcohol and use a 21-G needle and syringe to collect the lower opalescent band.
8. Transfer the vector into a PIERCE Slide-A-Lyzer dialysis cassette (Product No. 66425) and dialyze in 500 ml dialysis buffer for 30 min, twice, and then in 1000 ml dialysis buffer for 1 h, thrice. Remove the vector suspension from the dialysis cassette and aliquot in sterile Eppendorf tubes at approximately 100 µl/tube (*see* **Note 5**).
9. Use real time PCR (QPCR) to titer the AdrAQP5-0.4-luc with primers from the E2 region of adenovius: E2q1 (5´-GCAGAACCACCAGCACAGTGT-3´) and E2q2 (5´-TCCACGCATTTCCTTCTAAGCTA-3´). Titers are expressed as particles per milliliter. The plasmid pACrAQP5-0.4-luc was used as a standard for QPCR, with 1 µg of the plasmid (10132 bp) being equivalent to 9.0 × 10^{10} molecules. Standard curves are established from 10^2 molecules to 10^9 pACrAQP5-0.4-luc molecules, and adenoviral vectors are tested at three dilutions over a 100-fold range. QPCR assays are typically carried out with the SYBR Green PCR Master

Mix from Applied Biosystems using an ABI prism 7700 Sequence Detector (Applied Biosystems), with the following conditions: stage 1, 95°C for 2 min; stage 2, 95°C for 10 min; stage 3, 95°C for 15 s, 60°C for 1 min, repeated 40 times.

3.4. Plasmid Delivery

It is difficult to deliver plasmid alone, or with typical in vitro transfection reagents, to rat submandibular glands in vivo because of minimal efficiency. Therefore, an adenovirus–PEI–plasmid complex is used to deliver plasmids containing the promoters being studied, both in vitro and in vivo (*see* **Note 6**). PEI is an organic macromolecule with the highest cationic-charge-density potential. PEI plays a bridge role in this complex to connect plasmid DNA and the adenovirus. Receptors on eukaryotic cell membranes recognize the adenovirus, facilitating cell entry and nuclear targeting. The adenovirus used by us is Adcontrol, as noted an E1-deleted replication-deficient vector without any transgene.

1. PEI stock solution (10 mM): mix 9 mg of PEI with 10 ml ddH_2O, adjust pH to 7.0 with HCl and filter to sterilize.
2. Add 2×10^{10} molecules (*see* **Note 7**) of plasmid DNA (e.g., pACrAQP5-0.4-luc) into 50 µl of 20 mM HEPES (pH 7.5), then add 50 µl of PEI solution (1 µl of PEI stock solution in 49 µl of 20 mM HEPES (pH 7.5). Mix well and incubate at room temperature for 20 min.
3. Add 3×10^{10} particles of Adcontrol, mix well and incubate at room temperature for 20 min.
4. Take 20 µl of the above two mixtures, add to A5 cells, plated 24 h previously (2 $\times 10^5$ cells/well in 96-well plate) and allow transfection for 1 h.
5. Add 180 µl of fresh growth medium, return to tissue culture incubator, and measure the luciferase activity after 24 h.
6. For in vivo rat submandibular gland delivery, use the following: 4.35×10^{12} molecules of plasmid DNA/gland, 0.5 mM PEI and 1×10^{11} particles of Adcontrol in a volume of 150 µl. The complexes are delivered into rat submandibular glands by local cannulation (*see* **Subheading 3.6**). After 3 days, the tissues are harvested to measure luciferase activity and to perform histological staining (e.g., hematoxylin and eosin) or immunohistochemistry staining (for EGFP).

3.5. In Vitro Viral Infection

1. A5 cells in suspension are infected at 100 particles/cell, 37°C for 1 h.
2. Plate A5 cells at 5×10^5 cells/200 µl/well in a 96-well plate and incubate for 24 h.
3. Measure the luciferase activity after 24 h as described below in **Subheading 3.7**.

3.6. In Vivo Viral Infection (Rat Submandibular Gland Cannulation)

1. Using dilution buffer dilute AdrAQP5-0.4-luc to an intended dose of 10^9 particles/150 μl/gland.
2. Rat is anesthetized with ketamine (60 mg/kg) + xylazine (8 mg/kg).
3. Place the rat on a special board to immobilize and access main excretory ducts of the submandibular glands. Insert a tapered PE 10 cannula into the main excretory duct under a microscope and fix the cannula into position with superglue.
4. Inject atropine (1 mg/kg) intramuscularly to inhibit saliva secretion.
5. 10 min after atropine injection, use an insulin syringe to connect the other end of the PE 10 tubing to inject 150 μl of AdrAQP5-0.4-luc (or Adcontrol-PEI-plasmid complex, see Subheading 3.4 Plasmid Delivery) into the rat submandibular gland.
6. After 10 min, remove the cannula, wait for the rat to awake from anesthesia, and return the rat back to its cage.
7. After 3 days, the rat submandibular glands are harvested to measure luciferase activity and to perform histological or immunohistochemistry staining.

3.7. Luciferase Assay

1. To prepare the Luciferase Assay Reagent, add 10 ml of Luciferase Assay Buffer to the vial containing the lyophilized Luciferase Assay Substrate. To prepare 1× Cell Culture Lysis Reagent, add 4 vol of ddH_2O to 1 vol of 5× Cell Culture Lysis Reagent.
2. For cell culture that includes non-transfected and transfected cells in 96-well plate, add 25 μl of 5× Cell Culture Lysis Reagent into each well and mix three times by pipetting, incubate at room temperature for 15 min. For animal tissue, add 500 μl of 1× Cell Culture Lysis Reagent to approximately 30 mg of tissue, homogenize for approximately 30 s with a polytron, incubate at room temperature for 15 min, and then centrifuge at 16,000 × *g* for 20 s.
3. Mix 50 μl of cell or tissue lysates with 100 μl Luciferase Assay Reagent in a 12 × 75-mm glass test tube, then place this tube in an OPTOCOMP I luminometer (GEM Instruments, inc., Hamden, CT, USA) to measure the light emitted for 10 s. The result is reported as relative light units.

3.8. Protein Assay

1. Working reagent. Add 1 part reagent B to 50 parts Reagent A. Mix well.
2. Pipet 0.1 ml of each standard or unknown protein sample into the appropriately labeled test tube.
3. Add 1 ml working reagent to each tube. Mix well.
4. Incubate tubes at 37°C for 30 min, cool tubes to room temperature.
5. Measure the absorbance at 562 nm versus water reference. Subtract the absorbance of the blank from the value found.

3.9. Immunohistochemistry

Immunohistochemistry is done with the ImmunoCruz™ Staining System for rabbit primary antibody.

1. Primary antibody, anti-GFP (Rabbit polyclonal), was from Abcam, Inc. (Cambridge, MA, USA)
2. There are two types of controls (*see* **Note 8**): one is a reagent control using normal rabbit IgG instead of the EGFP primary antibody, and the second is a negative tissue control (i.e., does not receive any adenoviral vector containing the rAQP5 promoter and EGFP reporter gene).
3. Paraffin-embedded tissue sections were deparaffinized with xylene. Slides containing tissue sections are put in xylene for 15 min, twice.
4. Deparaffinized slides are rehydrated in a graded series of ethanol, 100, 100, 95, 90, 80, 70% for 5 min each. Wash sections in PBS for 5 min.
5. Peroxidase quenching. Submerge the slides in 15% hydrogen peroxide (25 ml ddH_2O and 25 ml of 30% of hydrogen peroxide) for 30 min.
6. Rinse the slides with ddH_2O, then PBS three times. Drain all PBS from the slides. Draw a circle outside the tissue on the slide with a PAP Pen (Research Products International Corp.; Mount Prospect, IL, USA)
7. Add serum-blocking solution (PBS + 30% goat serum) into each circle and block for 20 min.
8. Drain the serum-blocking solution from the slide and add approximately 200 μl of primary antibody at a 1:200 dilution [antibody dilution buffer: 2% bovine albumin (SIGMA) in PBS] into the circles for 1 h.
9. Wash the slides with PBS for 2 min, three times.
10. Incubate the slides in biotinylated second antibody for 30 min.
11. Wash the slides with PBS for 2 min, three times.
12. Incubate the slides in HRP–streptavidin complex for 30 min.
13. Wash the slides with PBS for 2 min, three times.
14. Add 1–3 drops of HRP substrate to each circle on the slides for 20 min.
15. Rinse with ddH_2O and then wash the slides in ddH_2O for 2 min.
16. Hematoxylin counterstain and mount slides.

4. Notes

1. Serial deletion analysis is a very powerful method to evaluate tissue-specific promoters, but, it is not only way to do this. The gene trap method can be also used. This system is especially helpful to evaluate a tissue-specific promoter in vivo. Gene trap constructs utilize the *lacZ* gene, encoding beta-galactosidase, or GFP, both to disrupt a natural gene's function and to monitor gene expression simultaneously. Gene trapping is rapid, cost-effective, and produces a large variety of insertional mutations throughout genome ***(25)***. Transgenic mouse models can also be used to evaluate tissue-specific promoters in vivo. Note that promoter methylation often is involved in the regulation of tissue-specific promoters ***(26)***.

2. For restriction endonuclease digestion, keep the enzyme volume to $\leq$ 10% in the reaction mixture.
3. pJM17 is an approximately 40-kb plasmid. Maxipreparation of this plasmid is not easy. Make fresh ampcillin and add 100 ng/ml in the bacterial culture. During the plasmid DNA extraction, carefully resuspend DNA pellet with TE buffer (pH 8.0).
4. Use lower passage 293 cells (<50) for the cotransfection to make adenoviral vectors.
5. For aliquots of adenoviral vectors, minimize the number of freeze and thaw episodes to prevent loss of vector activity. Keep vector aliquots in a –80°C freezer.
6. Make fresh PEI stock solution each time to obtain reproducible results. PEI activity is decreased if kept as a stock solution at 4°C.
7. To calculate the number of plasmid molecules use the following: 1 μg of 1000-bp DNA = 9.1×10^{11} molecules.
8. For the immunohistochemistry, control groups are very important.

Acknowledgments

We thank Drs. Biman Paria and Gabor Racz for their careful reading of, and helpful comments on, an earlier version of this manuscript. Our research is supported by the Division of Intramural Research, National Institute of Dental and Craniofacial Research NIH.

References

1. Eloranta, J.J. and Goodbourn, S. (1996) Positive and negative regulation of RNA polymerase II transcription. In *Eukaryotic Gene Transcription* (Goodbourn, J.J., ed.), Oxford University Press, Walton Street, Oxford, pp. 1–33.
2. Alberts, B., Johnson, A., Lewis, J., Raff, M., Roberts, K. and Walter, P. (2002) *Molecular Biology of the Cell*. Garland Science, New York, NY.
3. Majumder, S. and DePamphilis, M.L. (1994) TATA-dependent enhancer stimulation of promoter activity in mice is developmentally acquired. *Mol. Cell. Biol.* **14,** 4258–4268.
4. Zheng, C., Hoque, A.T., Braddon, V.R., Baum, B.J., and O'Connell, B.C. (2001) Evaluation of salivary gland acinar and ductal cell-specific promoters in vivo with recombinant adenoviral vectors. *Hum. Gene. Ther.* **12,** 2215–2223.
5. Zheng, C. and Baum, B.J. (2005) Evaluation of viral and mammalian promoters for use in gene delivery to salivary glands. *Mol. Ther.* **12,** 528–536.
6. Baum, B.J., Wellner, R.B., and Zheng, C. (2002) Gene transfer to salivary gland. *Int. Rev. Cytol.* **213,** 93–146.
7. Mastrangeli, A., O'Connell, B., Aladib, W., Fox, P.C., Baum, B.J., and Crystal, R.G. (1994) Direct in vivo adenovirus-mediated gene transfer to salivary glands. *Am. J. Physiol.* **266,** G1146–G1155.

8. Delporte, C., Redman, R.S., and Baum, B.J. (1997) Relationship between the cellular distribution of the alpha(v)beta3/5 integrins and adenoviral infection in salivary glands. *Lab. Invest.* **77,** 167–173.
9. Baum, B.J., Wang, S., Cukierman, E., Delporte, C., Kagami, H., Marmary, Y., Fox, P.C., Mooney, D.J., and Yamada, K.M. (1999) Re-engineering the functions of a terminally differentiated epithelial cell in vivo. *Ann. N.Y. Acad. Sci.* **875,** 294–300.
10. Moll, R., Franke, W.W., Schiller, D.L., Geiger, B., and Krepler, R. (1982) The catalog of human cytokeratins: patterns of expression in normal epithelia, tumors and cultured cells. *Cell.* **31,** 11–24.
11. Chow, Y.H., O'Brodovich, H., Plumb, J., Wen, Y., Sohn, K.J., Lu, Z., Zhang, F., Lukacs, G.L., Tanswell, A.K., Hui, C.C., Buchwald, M., and Hu, J. (1997) Development of an epithelium-specific expression cassette with human DNA regulatory elements for transgene expression in lung airways. *Proc. Natl. Acad. Sci. U.S.A.* **94,** 14695–14700.
12. Brembeck, F.H. and Rustgi, A.K. (2000) The tissue-dependent keratin 19 gene transcription is regulated by GKLF/KLF4 and Sp1. *J. Biol. Chem.* **275,** 28230–28239.
13. Kagaya, M., Kaneko, S., Ohno, H., Inamura, K., and Kobayashi, K. (2001) Cloning and characterization of the 5'-flanking region of human cytokeratin 19 gene in human cholangiocarcinoma cell line. *J. Hepatol.* **35,** 504–511.
14. Ting, C.N., Rosenberg, M.P., Snow, C.M., Samuelson, L.C., and Meisler, M.H. (1992) Endogenous retroviral sequences are required for tissue-specific expression of a human salivary amylase gene. *Genes Dev.* **6,** 1457–1465.
15. He, X., Tse, C.M., Donowitz, M., Alper, S.L., Gabriel, S.E., and Baum, B.J. (1997) Polarized distribution of key membrane transport proteins in the rat submandibular gland. *Pflugers Arch.* **433,** 260–268.
16. Brown, A.M., Rusnock, E.J., Sciubba, J.J., and Baum, B.J. (1989) Establishment and characterization of an epithelial cell line from the rat submandibular gland. *J. Oral. Pathol. Med.* **18,** 206–213.
17. Graham, F.L., Smiley, J., Russell, W.C., and Nairn, R. (1977) Characteristics of a human cell line transformed by DNA from human adenovirus type 5. *J. Gen. Virol.* **36,** 59–74.
18. Chen, X., Wu, J.M., Hornischer, K., Kel, A., and Wingender, E. (2006) TiProD: the tissue-specific promoter database. *Nucleic Acids Res.* **34,** D104–D107.
19. Borok, Z., Li, X., Fernandes, V.F., Zhou, B., Ann, D.K., and Crandall, E.D. (2000) Differential regulation of rat aquaporin-5 promoter/enhancer activities in lung and salivary epithelial cells. *J. Biol. Chem.* **275,** 26507–26514.
20. Henikoff, S. (1984) Unidirectional digestion with exonuclease III creates targeted breakpoints for DNA sequencing. *Gene.* **28,** 351–359.
21. Putney, S.D., Benkovic, S.J., and Schimmel, P.R. (1981) A DNA fragment with an alpha-phosphorothioate nucleotide at one end is asymmetrically blocked from digestion by exonuclease III and can be replicated in vivo. *Proc. Natl. Acad. Sci. U.S.A.* **78,** 7350–7354.

22. Sambrook, J. and Russell, D.W. (eds.) (2001) *Molecular Cloning A Laboratory Manual*. Cold Spring Harbor Laboratory, New York.
23. Ausubel, F.M., Brent, R., Kingston, R.E., Moore, D.D., Scidman, J.G., Smith, J.A., and Struhl, K. (eds.) (2006) *Current Protocols in Molecular Biology*. Harvard Medical School. John Wiley & Sons, Inc.
24. McGrory, W.J., Bautista, D.S., and Graham, F.L. (1988) A simple technique for the rescue of early region I mutations into infectious human adenovirus type 5. *Virology* **163,** 614–617.
25. Bundschu, K., Gattenlohner, S., Knobeloch, K.P., Walter, U., and Schuh, K. (2006) Tissue-specific Spred-2 promoter activity characterized by a gene trap approach. *Gene Expr. Patterns* **6,** 247–255.
26. Bohwan, J., Seong, J.K., and Ryu, D.Y. (2005) Tissue-specific and *De Novo* promoter methylation of the mouse glucose transporter 2. *Biol. Pham. Bull.* **28,** 2054–2057.

14

Adenovirus-Mediated Transduction of Auto- and Dual-Regulated Transgene Expression in Mammalian Cells

Valeria Gonzalez-Nicolini and Martin Fussenegger

Summary

Transduction of therapeutic transgenes using multiply attenuated viral vectors is considered an essential technology for gene therapy scenarios. While first-generation viral transduction systems were engineered for constitutive expression of a single therapeutic transgene, most advanced viral gene-transfer technologies enable regulated expression of several transgenes. Efficient transfer of numerous transgenes enables co-expression of therapeutic transgenes along with marker or selection determinants, production of multi-subunit protein complexes, or combinatorial expression of a particular set of genes to treat multigenic disorders. Likewise, adjustable transcription control is fundamental to adapt therapeutic protein production to the changing daily dosing regimes of a patient, to titrate expression of protein pharmaceuticals into the therapeutic window, and to reverse dosing upon completion of the therapy. Also, conditional transcription dosing has been successfully used for production of difficult-to-express protein therapeutics in biopharmaceutical manufacturing and for sophisticated gene-function analysis in basic research programs. By way of example, we provide detailed design (auto-regulated and binary dual-regulated expression configurations), production (generation, purification, and quality control of transgenic adenovirus particles), and handling (transduction) protocols for adenovirus vectors that enable transduction of mammalian cells for regulated expression of several transgenes.

Key Words: Adenovirus; PIP system; pristinamycin; SAMY; SEAP; streptogramin; TET system; tetracycline.

From: *Methods in Molecular Biology, vol. 434: Volume 2: Design and Characterization of Gene Transfer Vectors*
Edited by: J. M. Le Doux © Humana Press, Totowa, NJ

1. Introduction

The combination of an efficient viral transduction technology with regulated transcription control modalities, which can independently fine-tune expression of several therapeutic transgenes, provides a powerful tool for future gene therapy scenarios.

A variety of different viral transduction systems are pre-clinically licensed for gene therapy studies ***(1–3)***. Recombinant adenoviruses have shown several advantages over other viral transduction systems, because they support high-level transduction of a wide variety of different cell types, production of high-titer particle batches, multi-transgene expression because of extended packaging size, and display a reduced risk for insertional mutagenesis as they remain in an episomal state ***(4–8)***. Adenoviruses are medium-sized (90–100 nm), non-enveloped icosohedral viruses containing a linear (~36 kb) double-stranded genome, which contains five early (E1A, E1B, E2, E3, E4) and five late (L1-5) transcription regions flanked by inverted terminal repeats (ITRs). Replication-incompetent adenovirus vectors are typically devoid of E1 or E3 or E1 and E3 regions, which thus increase their transgene cargo capacity to about 5–8 kb ***(9–14)***.

Gene-expression systems, regulated by clinically licensed antibiotics, are widely used, because they allow the selective and stringent regulation of transgene expression, not only in vitro, but also in vivo following oral administration ***(15–17)***. Among others, the streptogramin- and tetracycline-responsive expression systems (PIP and TET systems) have been studied extensively ***(18–20)***. Both these systems consist of antibiotic-dependent transactivators (PIT, tTA) derived from bacterial response regulators (PIP, TetR) fused to a eukaryotic transactivation domain (VP16; PIT, PIP-VP16; tTA, TetR-VP16), which bind and activate their respective target promoters (P_{PIR}, $P_{hCMV^*_{-1}}$) in an antibiotic-dependent manner. The antibiotic-responsive promoters (P_{PIR}, $P_{hCMV^*_{-1}}$) were constructed by cloning PIT-/tTA-specific tandem operators (PTR, $tetO_7$) adjacent to a minimal eukaryotic promoter (e.g., $P_{hCMVmin}$; P_{PIR}, PTR-$P_{hCMVmin}$; $P_{hCMV^*_{-1}}$, $tetO_7$-$P_{hCMVmin}$). The PIP and the TET systems have been shown to be fully compatible and could be combined for independent regulation of different transgenes ***(21–26)***.

We have successfully designed transgenic adenovirus particles engineered for autoregulated streptogramin-responsive one-vector and dual-regulated streptogramin- and tetracycline-adjustable two-vector configurations. Adenovirus vectors harboring autoregulated expression cassettes contain a central bidirectional streptogramin-responsive promoter driving divergent transcription of PIT and the gene of interest. Low leaky transcription inherent to the central promoter will result in production of a few PIT proteins that bind and auto-activate the central promoter resulting in high-level co-expression of PIT and the gene of interest in the absence of antibiotics. Addition of streptogramin antibiotics prevents PIT binding to the divergent promoter, which shuts down bidirectional gene expression ***(27)***. In this configuration, a single adenovirus can transduce streptogramin-responsive transgene expression. Adenovirus-based engineering of mammalian cells for dual-regulated transgene expression requires co-transduction of two types of adenovirus particles, one encoding constitutive dicistronic expression of PIT and tTA (P_{SV40}-PIT-IRES-tTA-pA) and the other harboring divergently oriented P_{PIR} and $P_{hCMV^{*}-1}$ promoters separated by an orientation-dependent stuffer that prevents transcription interferences and enables independent control of two gene of interest ***(23,25,26)***.

Here, we provide design considerations and detailed protocols for production of engineered adenovirus particles and their use as transduction vectors that promote auto- or dual-regulated expression of different transgenes.

2. Materials

2.1. Molecular Biology

1. Molecular biological standard consumables were used to construct the desired plasmids: restriction enzymes, competent cells (DH5α), plasmid DNA purification (Wizard miniprep kit, Promega, Madison, WI, USA; Jet-Star 2.0 maxiprep, Genomed AG, Göttingen, Germany).

2.2. Adenovirus Shuttle Vectors

1. Adenovirus shuttle vectors pdE1sp1A and pdE1sp1B (Microbix, Toronto, Canada) were used.

2.3. Production of Adenovirus Particles

1. Adenovectors encoding desired transgenes (e.g., pVN24, pVN28, pVN10, pVN43). A detailed description of all vectors is provided in **Table** 1.
2. Adenovirus type 5 genomic vector pJM17 (Microbix).
3. Human embryonic kidney cells (HEK-293; Microbix), transgenic for the Ad5-derived E1 region, were used at the lowest passage possible (P30-P33).
4. Dulbecco's modified Eagles medium (DMEM; Invitrogen, Carlsbad, CA, USA) containing 10 or 2% heat-inactivated fetal calf serum (FCS; PAN Biotech GmbH, Germany) and 1% penicillin-streptomycin solution (SIGMA, St. Louis, MO, USA).
5. Trypsin solution: Trypsin–ethylenediaminetetraacetic acid (EDTA) solution in PBS (PAN Biotech GmbH, Aidenbach, Germany).
6. Sterile $CaCl_2$ (2 M) solution should be prepared fresh before use.

Table 1
Plasmids Used and Designed in this Study

Plasmid	Description and Cloning Strategy	Reference or Source
pΔE1sp1A	Adenoviral transfer vector	Microbix
pΔE1sp1B	Adenoviral transfer vector	Microbix
pJM17	Adenoviral type 5 genomic vector	Microbix
pBHG10	Adenoviral type 5 genomic vector	Microbix
pCF18	Dual-regulated expression vector. $pA_I - CFP \leftarrow P_{hCMV^*_{-1}}$-stuffer-$P_{PIR} \rightarrow YFP$-$pA_{II}$ (pDuoRex8)	*(33)*
pCF86	Dual-regulated expression vector. pA_I-SAMY$\leftarrow P_{hCMV^*_{-1}}$-stuffer-$P_{PIR} \rightarrow$SEAP-$pA_{II}$	*(26)*
pCF89	Mammalian expression vector. $P_{hCMVmin}$-MCS-pA_I-MCS-pA_{II}	*(26)*
pCF130	Bidirectional expression vector. pA_I-YFP$\leftarrow P_{hCMVmin}$-PTR-$P_{hsp70min} \rightarrow$PIT-pA_{II} (pBiRex4)	*(27)*
pCF138	Bidirectional expression vector. pA_I-SEAP$\leftarrow P_{hCMVmin}$-PTR-$P_{hsp70min} \rightarrow$PIT-pA_{II} (pBiRex5)	*(27)*
pMF171	Dicistronic expression vector. $P_{SV40} \leftarrow$ PIT-IRES-tTA-pA_{SV40} (pTWIN1)	*(20)*
pVN10	Adenovector encoding a constitutive dicistronic PIT- and tTA-encoding expression unit ($P_{SV40}\Delta$PIT-IRES-tTA-pA_{SV40}). $P_{SV40} \leftarrow$PIT-IRES-tTA-pA_{SV40} was excised from pMF171 using *Cla*I/*Mfe*I and cloned in adenoviral forward orientation into the compatible sites (*Eco*RI/*Cla*I) of pΔ E1sp1A.	*(26)*
pVN24	Adenovector encoding a streptogramin-responsive autoregulated SEAP expression unit (pA_I-SEAP$\leftarrow P_{hCMVmin}$-PTR-$P_{hsp70min} \rightarrow$ PIT-pA_{II}). pA_I-SEAP$\leftarrow P_{hCMVmin}$-PTR-$P_{hsp70min} \rightarrow$PIT-pA_{II} was excised from pCF138 using *Eco*RV/*Acl*I and cloned into the compatible sites (*Eco*RV/*Cla*I) of pΔE1sp1A.	*(26)*
VN25	Dual-regulated intermediate vector encoding divergent $P_{hCMV*-1}$ – driven SAMY and P_{PIR} – *driven* SEAP expression units (*SAMY* $\leftarrow P_{hCMV*-1}$ – stuffer – $P_{PIR} \rightarrow$ SEAP).SAMY $\leftarrow P_{hCMV*-1}$ – stuffer – $P_{PIR} \rightarrow$ SEAP was excised from pCF86 using *Bgl*II and cloned in sense orientation into the compatible site (*Bam*HI) of pΔE1sp1A.	*(26)*

pVN27	Dual-regulated intermediate vector encoding divergent $P_{hCMV^{*}-1}$-driven SAMY and P_{PIR}-driven SEAP expression units (SAMY ← $P_{hCMV^{*}-1}$-stuffer-P_{PIR} →SEAP-pA_{II}). pA_{II} was excised from pCF150 using *Xho*I and cloned into the compatible sites (*Sal*I/*Xho*I) of pVN25.	**(26)**
pVN28	Dual-regulated adenovector encoding divergent $P_{hCMV^{*}-1}$-driven SAMY and P_{PIR}-driven SEAP expression units (pA_{I}-SAMY ← $P_{hCMV^{*}-1}$-stuffer-P_{PIR} →SEAP-pA_{II}). pA_{I} was excised from pCF89 using *Bcl*I/*Bgl*II and cloned into the compatible site (*Bcl*I) of pVN27.	**(26)**

IRES, internal ribosome entry site of polioviral origin; pA_{I}, pA_{II}, minimal synthetic polyadenylation sites; pA_{SV40}, SV40-derived polyadenylation site; $P_{hCMVmin}$, minimal version of the human cytomegalovirus immediate early promoter; $P_{hCMV^{*}-1}$, tetracycline-responsive promoter; $P_{hsp70min}$, minimal version of the *Drosophila melanogaster* heat-shock protein 70 promoter; PIT, streptogramin-dependent transactivator; P_{PIR}, streptogramin-responsive promoter; PTR, PIT-specific operator; P_{SV40}, promoter of the simian virus 40; SAMY, *Bacillus stearothermophilus*-derived secreted α-amylase; SEAP, human placental secreted alkaline phosphatase; tTA, tetracycline-dependent transactivator.

7. Sterile 2× HBS solution: 0.28 M NaCl, 0.05 M HEPES, 1.5 mM Na_2HPO_4, pH 7.2. If stored for long periods of time, the pH should be checked and adjusted before use.
8. PBS (SIGMA).
9. Low-TE Buffer: 1 M Tris–HCl, pH 7.5, 0.5 M EDTA, pH 8; filter sterile.

2.4. Analysis of the Endonuclease Restriction Pattern of Viral DNA

1. 1× Tris–EDTA, pH 8: 10 mM Tris–HCl, 1 mM EDTA.
2. SDS 10% (in dH_2O).
3. 0.5 M EDTA (Sigma).
4. Proteinase K (20mg/ml) (Qiagen GmbH, Hilden, Germany).
5. 5 M NaCl.
6. Phenol : chloroform : isoamyl alcohol (25:24:1) (Sigma).
7. Isopropanol (pure).
8. 0.1mg/ml RNAse A (diluted in dH_2O) (Qiagen GmbH).

2.5. Caesium Chloride Purification

1. Caesium Chloride (CsCl) buffer: 5mM Tris–HCl (pH 8), 1 mM EDTA.
2. CsCl 1.33 g/ml density, add 8.4 g CsCl to 16 ml CsCl buffer.
3. CsCl 1.45 g/ml density, add 8.4 g CsCl to 11.5 ml CsCl buffer.

4. Buffer TE: 10 mM Tris–HCl, 1 mM EDTA, adjust to pH 8.0 with NaOH, and sterilize by autoclaving.
5. 16-G needles (Becton Dickinson, NJ, USA).
6. Buffer A: 10 mM Tris–HCl, 1 mM $MgCl_2$, 135 mM NaCl, pH 7.5.
7. Buffer B: 10mM Tris–HCl, pH 7.5 + 10% glycerol.
8. Dialysis tubing (M.W. 12,400).
9. Dialysis closures (Sigma).

2.6. Quantification of Adenovirus Vector Concentration

1.Virus lysis solution: 0.1% SDS; 10 mM Tris–HCl, pH 7.4, 1 mM EDTA.

2.7. RCA Assay

1. Human adenocarcinoma cells (HeLa, ATCC:CCL2) were cultivated in DMEM containing 10% heat-inactivated FCS and 1% penicillin-streptomycin solution.

2.8. Evaluation of Regulated Transgene Expression

1. Primary human aortic fibroblasts (HAFs) (described in **ref.** ***30***) were cultured in DMEM supplemented with 10% FCS.
2. Pristinamycin I (PI; Pyostacin®, Sanofi-Aventis Inc., Frankfurt, Germany)
3. 2× secreted alkaline phosphatase (SEAP) buffer: 20 mM homoarginine, 1 mM $MgCl_2$, 21% (w/v) diethanolamine-HCl, pH 9.8; store in the dark at 4°C for up to 6 months.
4. pNPP solution: 120 mM *p*-nitrophenylphosphate (Sigma 104 Phosphatase substrate, Sigma Cat. No. 104-0), in 2× SEAP buffer. Store in aliquots for single use at –20°C.
5. Wild-type Chinese hamster ovary cells (CHO-K1, ATCC:CCL61) were cultivated in FMX-8 medium (Cell Culture Technologies, Zurich, Switzerland) supplemented with 10% FCS (PAA Laboratories GmbH, Linz, Austria).
6. Tetracycline (TET; Sigma Chemicals, St. Louis, MO, USA).
7. Blue starch Phadebas® assay (Pharmacia Upjohn, Peapack, NJ, USA).
8. 0.5 M NaOH.

3. Methods

3.1. Construction of a Streptogramin-Responsive Auto-Regulated Adenovector

1. To construct the desired adenovectors, we used standard molecular biology consumables (restriction enzymes, competent cells, DNA purification kits). Details on vector construction are outlined in **Table** 1. We inserted a central

pristinamycin-dependent transactivator (PIT)-specific operator (PTR) between two divergently oriented minimal versions of the human cytomegalovirus immediate early promoter and the *Drosophila melanogaster* heat-shock protein 70 promoter ($P_{hCMVmin}$; $P_{hsp70min}$). SEAP cDNA was inserted 3´ of $P_{hCMVmin}$ and PIT 3´ of the $P_{hsp70min}$ promoter; a synthetic polyadenylation site was cloned at the end of each transcription unit [pVN24, pA_I-SEA 0P$\leftarrow P_{hCMVmin}$-PTR-$P_{hsp70min}\rightarrow$PIT-pA_{II}; (26)]. pVN24 as well as the commercial plasmid pJM17 was isolated using silica-based anion-exchange DNA purification kits according to the manufacturer's protocol (Genomed Jetstar maxiprep kit); the DNA concentration was determined by means of a spectrophotometer (*Eppendorf, Vaudax, CH*).

3.2. Construction of a Binary Dual-Regulated Vector

1. We designed a binary adenovector set for independent streptogramin- and tetracycline-responsive expression of two different transgenes. **Table** 1 outlines the cloning procedure for each plasmid; standard molecular biology consumables were used as in **Subheading 3.1**. pVN10 contains a dicistronic expression cassette for constitutive expression of PIT and tTA and was constructed by inserting PIT right after the simian virus 40 promoter (P_{SV40}), followed by an internal ribosome entry site of polioviral origin (IRES), tTA and a synthetic polyadenylation site. This expression cassette (P_{SV40}->PIT-IRES-tTA-pA) must be inserted in forward orientation into the adenovector for optimal regulation and to minimize leakiness.

The dual-regulated adenovector, pVN28, harbors the *Bacillus stearothermophilus*-derived α-amylase driven by the tetracycline-responsive promoter (P_{hCMV^*-1}) and SEAP controlled by the pristinamycin-responsive promoter P_{PIR} in divergent orientation separated by a central stuffer element to minimize transcriptional interference [pVN28, pA_I-SAMY<–P_{hCMV^*-1}-stuffer-P_{PIR}->SEAP-pA_{II}; (26)]. pVN10 and pVN28 as well as the commercial plasmid pJM17 were isolated using silica-based anion-exchange DNA purification kits according to the manufacturer's protocol (Genomed Jetstar maxiprep kit), and DNA concentration was determined using a Biophotometer (Eppendorf, Vaudax, CH).

3.3. Production of Adenoviruses (see Note 1)

1. Seed HEK-293 cells into 25 ml T-flasks containing DMEM-supplemented 10% FCS to achieve about 50–60% confluence after 24 h. The lower the confluence and the passage number of the HEK-293 culture are, the greater the chances are that the viral cytophatic effect (CPE) will occur within 10 days (*see* **Note 2**).

2. Combine low-TE buffer, 5 μg adenovector DNA (pVNs), 5 μg genomic vector (pJM17 for inserts up to 5000 kb or pBHG10 for inserts up to 8000 kb) and 30μl 2 M $CaCl_2$ to a final volume of 240 μl (*see* **Note 3**).
3. Add the mixture drop-wise to 240 μl of 2× HBS and bubble the mixture in the tube using a sterile Pasteur pipette attached to a standard pipettor until a fine precipitate forms.
4. Let the precipitate stand at room temperature for 20 min.
5. Replace the medium of the HEK-293 cultures with 2 ml DMEM containing 2% FCS.
6. Add the precipitate drop-wise to the culture and incubate at 37°C in a CO_2 incubator overnight (no longer than 18 h).
7. On the next morning remove the medium and replace it with 5 ml DMEM containing 10% FCS. Replace the medium thereafter every 3–4 days. CPE should appear after 7 days and is complete after another 3–5 days (*see* **Note 4**).
8. Collect the adenovirus after complete CPE by transferring the cells already detached (by tapping the flask) and the supernatant to a 10-ml Falcon tube. Centrifuge at 1700 × *g* for 5 min, discard the supernatant, and resuspend the cells in 500 μl PBS. Freeze and thaw the tube in liquid nitrogen three times, centrifuge again at 1700 × *g* for 10 min; transfer the supernatant containing the adenovirus particles to an Eppendorf tube. Use 20 μl of this viral stock to reinfect new HEK-293 monolayer cultures and produce adenoparticles for genomic restriction analysis (*see* **Subheading 3.4**), use 10 μl of viral stock for the plaque purification assay (*see* **Subheading 3.5**) and store the rest at –80°C. (*see* **Note 5**)

3.4. Adenovirus Screening

1. Infect HEK-293 cells cultivated in a T25 flask with 20 μl adenovirus stock solution (*see* **Subheading 3.3, step 8**) and wait for CPE to occur (no longer than 48 h or viral DNA will start to degrade).
2. After CPE, collect the cells carefully by tapping the flask and centrifuge at them at 200 × *g* for 5 min. Rinse the pellet with PBS and centrifuge again for 5 min at 200 × *g*.
3. Resuspend the pellet in 360 μl of 1× Tris–EDTA (pH 8) and transfer the cells to an Eppendorf tube. Add 25 μl 10% SDS, 8 μl 0.5 M EDTA, and 4 μl of proteinase K (20 mg/ml). Incubate for 2 h at 37°C.
4. Add 100 μl 5 M NaCl and mix gently by inverting the tube twice. Keep the sample on ice for 3 h.
5. Centrifuge at 15,000 × *g* for 1 h at 4°C, harvest the supernatant, and transfer it to an Eppendorf tube.
6. Add an equal volume of phenol : chloroform : isoamyl alcohol (25:24:1) to the supernatant and mix gently. Centrifuge for 15 min at 15,000 × *g*.
7. Transfer the aqueous phase (upper phase) to an Eppendorf tube. Precipitate the DNA by adding 1 vol of isopropanol and centrifuge for 20 min at 15,000 × *g*.
8. Discard the supernatant and wash the pellet with 70% ethanol. Centrifuge for 5 min at 15,000 × *g*.

9. Dry the pellet in air and resuspend it in 30 μl of dH_2O containing 0.1 mg/ml RNAse.
10. Digest with the appropriate enzymes (usually *Hin*dIII specific for the adenovirus genome and a second enzyme that is specific for the transgenic insert) (*see* **Note 6**).

3.5. Purification of Adenovirus

1. Three rounds of plaque purification are recommended after the initial viral particles are obtained. Seed 6000 HEK-293 per well of 96-well plate and incubate them at 37°C overnight in a CO_2 incubator.
2. Prepare serial dilutions of an aliquot of the adenovirus stock obtained in **step 8** of **Subheading 3.3** in DMEM supplement with 10% FCS and adjust the volume to 300 μl. Dilutions ranging from 10^{-2} to 10^{-12} are recommended.
3. Aspirate the media from the HEK-293 cultures (*see* **Subheading 3.5.**, **step 1**) and transduce the cells by adding 100 μl of each dilution. Perform all transductions in triplicate.
4. After 24 h, add 100 μl of pre-warmed DMEM containing 10% FCS to each well.
5. After 7 days, identify the set of triplicate cultures, in which complete CPE occurred. Collect the cells and supernatant from each of the three cultures and freeze/thaw three times in liquid nitrogen. Store two tubes at –80°C. Take one of the tubes and repeat **steps 1–5** of **Subheading 3.5**. Repeat the overall procedure for a total of three times.

3.6. Production of Bulk Adenovirus Stocks (see Note 7)

1. Seed HEK-293 cells into 150 ml T-flasks so they reach 80% confluence within 24 h.
2. Infect the cells at a multiplicity of infection (MOI) of two using adenovirus produced in **step 5** of **Subheading 3.5** and examine daily for signs of CPE.
3. When CPE is almost complete, transfer the cells and supernatant of two 150-ml T-flasks (approximately 25 ml) into a 50-ml falcon tube, centrifuge at 1000 × *g* for 10 min, and discard the supernatant. Resuspend the pellet in 500 μl of TE buffer and collect the entire re-suspended sample in a 10-ml Falcon tube.
4. Freeze/thaw the sample three times in liquid nitrogen, centrifuge again at 1000 × *g* for 15 min, transfer the supernatant to a new tube, and store at –80°C. This crude adenovirus stock can be used for in vitro analysis after the determination of the titer and replication-competent adenovirus (RCA) assay (RCAs) (**step 3.9**) or it can be purified further by means of a CsCl gradient for larger in vitro studies or in vivo use.

3.7. CsCl Purification of Adenovirus Stocks

1. Prepare a CsCl gradient by slowly adding 2.5 ml CsCl (1.33 g/ml) to a 15-ml Beckman tube; top it up with 1.5 ml CsCl (1.45 g/ml).

2. Layer 4 to 6 ml viral stock (**step 4**, **Subheading 3.6.**) onto the CsCl gradient.
3. Centrifuge at 76,000 × *g* for 2 h in an ultracentrifuge with a swing-bucket rotor (SW40 rotor).
4. By this step three bands are visible. Recover the lowest adenovirus-containing band with a wide bore syringe needle (16 G) and transfer it to a 10-ml Falcon tube (~2 ml); dilute the adenovirus by adding half a volume of TE buffer (pH 7.8; ~1 ml).
5. Prepare a second CsCl gradient by slowly adding 1.5 ml CsCl (1.33 g/ml) to a 15-ml Beckman tube; top up with 1 ml CsCl (1.45 g/ml).
6. Layer the adenovirus from **Subheading 3.7** (**step 4**) (~3 ml) on top of the CsCl gradient and centrifuge at 76,000 × *g* overnight in an ultracentrifuge with a swing-bucket rotor.
7. Recover the lowest adenovirus-containing band (~1 ml) with a wide bore syringe, fit with a needle, and transfer it to a dialysis tube; close the tube with dialysis closures.
8. Dialyze the adenovirus for 1 h by submerging the closed tubing in a 2-l beaker filled with 1 l Buffer A. Then transfer the sample to a 2-l beaker with 1 l Buffer B and dialyse for 2 h.
9. Store aliquots of 10–20 µl at –80°C.

3.8. Quantification of Adenovirus Concentration

3.8.1. By Titration (see Note 8) (28)

1. Inoculate 69 wells of a 96-well plate with 8 × 10^4 HEK-293 cells/well and incubate at 37°C overnight.
2. Prepare serial dilutions of the adenovirus stock (*see* **Subheading 3.7.**, **step 9**) from 10^{-2} to 10^{-14} in DMEM supplemented with 10% FCS and adjust the volume to 300 µl.
3. Aspirate the media from the HEK-293 cultures, transduce the cells by adding 100 µl of each dilution. Perform all transductions in triplicate.
4. After 24 h, add 100 µl of pre-warmed DMEM containing 10% FCS to each well.
5. After 10 days, identify the set of triplicate cultures, in which complete CPE occurred. Collect the cells and supernatant from each of the three cultures and freeze/thaw three times in liquid nitrogen. Determine the adenovirus concentration related to the type of dilution performed, i.e. (where PFU = plaque-forming units):

Well	1	2	3	4
Dilution	10^{-2}	10^{-3}	10^{-4}	10^{-5}
PFU/200 µl	10^2	10^3	10^4	10^5
PFU/1 ml	5 × 10^2	5 × 10^3	5 × 10^4	5 × 10^5

3.8.2. By Optical Absorbance

1. Thaw an aliquot produced in **step 9** of **Subheading 3.7** and dilute the adenovirus in 19 vol of virion lysis solution.
2. Incubate the sample at 56°C for 10 min while shaking.
3. Transfer the sample to a quartz cuvette and determine the OD_{260}/OD_{280} ratio using a UV spectrophotometer.
4. Determine the adenovirus particle concentration as $OD_{260} \times 20/9.09 \times 10^{-13}$, where $9.09 \times 10^{-13}$is the extinction coefficient of the wild-type adenovirus. Wild-type adenovirus particles have an $OD_{260} = 1$ and $OD_{260}/OD_{280} = 1.3$ at an average concentration of 1×10^{12}.

3.9. RCA Determination of the Adenovirus Stock (29) (see Note 9*)*

1. Seed 9000 HeLa cells per well of a 96-well plate and incubate them overnight at 37°C in a CO_2 incubator.
2. Prepare serial dilutions of an aliquot of adenovirus obtained in **step 9** of **Subheading 3.7** using DMEM supplemented with 10% FCS and adjust the volume to 300 µl. Dilutions from 10^{-2} to 10^{-12} are recommended.
3. Aspirate the media from the HeLa cultures and transduce the cells by adding 100 µl of each dilution. Perform all transductions in triplicate.
4. After 24 h, add 100 µl of pre-warmed DMEM containing 10% FCS to each well.
5. Monitor for CPE each day. After 7–9 days, there should be no sign of CPE if the adenovirus vector stock was properly purified and does not contain any RCA.

3.10. Evaluation of Reversible Transcription Control Using an Auto-Regulated Adenovirus-Based Expression Configuration

1. Seed 20,000 HAFs, resuspended in 500 µl DMEM supplemented with 10% FCS and 1% penicillin–streptomycin solution, into each well of a 24-well plate 3–5 days before transduction. For reversibility studies, where the cell culture must be kept for more than 3 days, a primary culture of non-dividing cells or cells that divide very slowly is recommended. A detailed protocol for establishing HAFs cells in culture has been described in **ref. *30*** (*see* **Note 10**).
2. Add pVN24-derived adenovirus particles at an MOI of 20 and supplement the culture with 2 µg of PI to achieve the OFF status and no PI to trigger transgene expression. Incubate for 5 h at 37°C in a CO_2 incubator. Perform all experiments in triplicate and use mock-transduced cultures as control.
3. Aspirate the medium, wash once with PBS, and apply 500 µl DMEM supplemented with 10% FCS, 1% penicillin–streptomycin solution and 2 µg PI to

maintain the OFF state (see **Subheading 3.10., step 2**) and no PI to keep the ON state. Incubate for 48 h at 37°C in a CO_2 incubator.

4. Collect the culture supernatant and quantify SEAP activity or freeze the supernatants at –20ºC for later quantification.
5. Wash the cells twice for 5 min in PBS, then add 500 µl DMEM supplemented with 10% FCS and 1% penicillin–streptomycin solution. Reverse the PI status of the culture. Add no PI to the culture that received 2 µg/ml PI in **step 3** in this section and add 2µg/ml to the culture that received no PI in **step 3**. Incubate for 30 min (*see* **Note 11**).
6. Wash the cells again for 5 min with 500 µl PBS of the appropriate PI status and add 500 µl DMEM supplemented with 10% FCS, 1% penicillin–streptomycin solution and maintain the PI status by adding no or 2 µg PI as indicated in **step 5**. Incubate for 48 h at 37ºC in a CO_2 incubator.
7. Collect the supernatant and quantify SEAP activity or freeze the supernatants at –20ºC for later quantification.
8. Repeat **steps 5–6** once.
9. Collect the supernatant and quantify SEAP activity (**Fig. 1**).

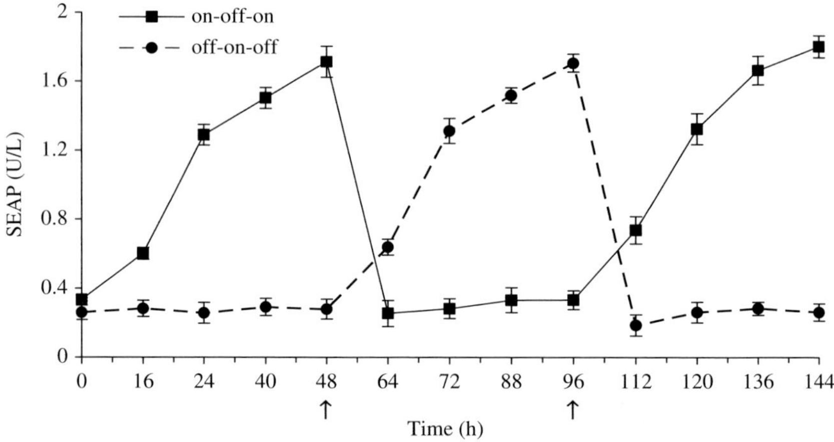

Fig. 1. pVN24-derived adenovirus enabling autoregulated streptogramin-responsive expression of the human placental secreted alkaline phosphatase (SEAP) (pA_{II}-PIT←$P_{hsp70min}$-PTR-$P_{hCMVmin}$→SEAP-pA_I inserted between 0–1 and 9–100 adenoviral map units) (*see* **Table** 1). Reversibility of streptogramin-inducible SEAP expression in human aortic fibroblast (HAF) monolayer cultures. pVN24-derived adenoviral particles were used to infect HAFs cells, which were kept for 48 h in the presence (2 µg/ml) or absence of PI. SEAP expression was reversed (ON to OFF, –PI to +PI; OFF to ON, +PI to –PI) by alternating addition or withdrawal of PI every 48 h (arrows).

3.10.1. Quantification of SEAP Activity

1. Add 100 µl 2× SEAP buffer to a flat-bottomed 96-well plate and keep it in the dark.
2. Incubate the supernatant (**steps 4**, **7**, and **10** of **Subheading 3.10**) at 65°C for 10 min to inactivate endogenous phosphatases and centrifuge at 15,000 × *g* for 2 min to remove cell debris.
3. Add 80 µl of supernatant from each sample to each well containing 2× SEAP buffer and then add 20 µl pNPP solution to each well using a multi-pipette and immediately determine the time course of light absorbance (Abs) at 405 nm (1 h reading is usually sufficient).
4. Determine the increase in optical density per minute within the linear part of the Abs curve and calculate SEAP activity (A) as (***31***) A(U/L) = D if Abs/min × v/(E × D) × 10^6, where v = dilution factor (200 µl total volume/80 µl sample volume), E = absorption factor = 18600 $M^{-1}cm^{-1}$ and D = light path = 0.5 cm in a 96-well plate.

3.11. Evaluation of Dose–Response Transcription Characteristics Using a Dual-Regulated Adenovirus-based Expression Configuration

1. Seed 20,000 CHO-K1 cells, re-suspended in 500 µl FMX-8 supplemented with 10% FCS and 1% penicillin–streptomycin solution, into each well of 24-well plate 1 day before transduction.
2. Co-transduce pVN28- and pVN10-derived adenovirus particles at an MOI of 50:75 and add increasing amounts of PI to individual wells before incubating for 5 h at 37°C in a CO_2 incubator. Perform all experiments in triplicate and use mock-transduce cultures as control.
3. Aspirate the medium, wash once with PBS and apply 500 µl FMX-8 supplemented with 10% FCS, 1% penicillin–streptomycin solution and maintain the previous PI concentration by addition of the appropriate amount. Incubate for 48 h at 37°C in a CO_2 incubator.
4. Collect the supernatant and quantify SEAP and SAMY activity or freeze the samples at –20°C for later quantification.
5. Follow steps of subheadings 3.10.1 to determine SEAP activity (Fig. 2).
6. To evaluate SAMY activity (32), prepare the substrate solution by diluting 1 blue starch tablet in 4 ml dH_2O.
7. Centrifuge the supernatant from **step 4** at 17,000 × *g* for 2 min to remove cell debris.
8. Add 25 µl supernatant to 500 µl substrate solution and incubate the samples at 70°C for 15 min while shaking.
9. Stop the reaction by adding 125 µl 0.5 M NaOH.
10. Add 200 µl of each sample to a well of a 96-well plate, read the Abs at 620 nm and determine the SAMY concentration as outlined in **ref. 32** (Fig. 2).

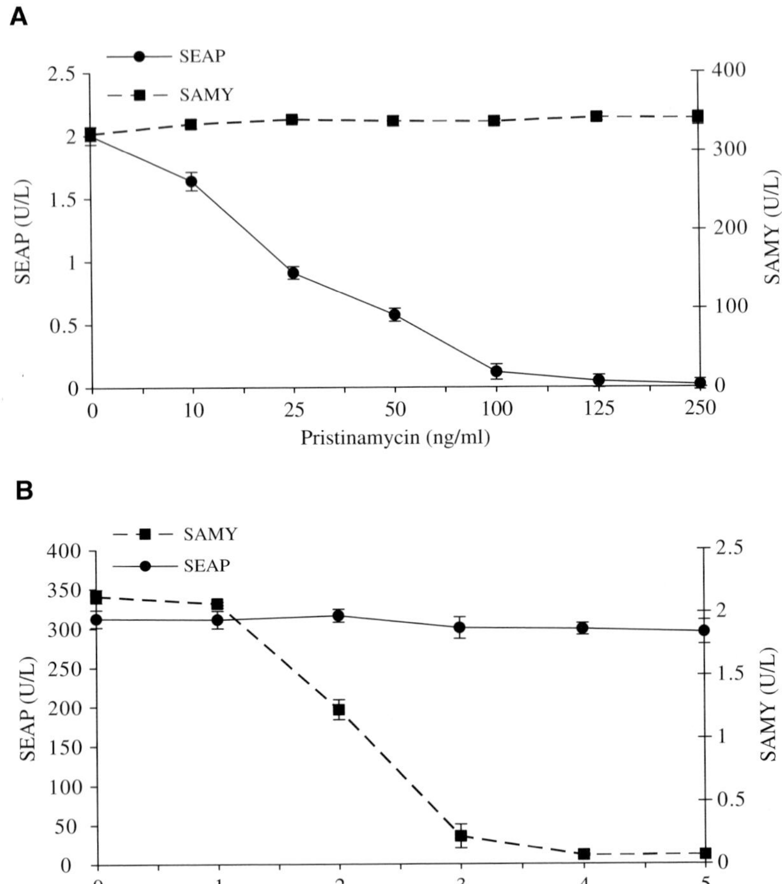

Fig. 2. Divergent P_{PIR} (pristinamycin-responsive promoter)-driven human placental secreted alkaline phosphatase (SEAP) and $P_{hCMV^{*}_{-1}}$ (tetracycline-responsive promoter)-driven SAMY expression units terminated by synthetic minimal polyadenylation sites (pA_{I}, pA_{II}) are separated by a stuffer fragment to minimize transcriptional crosstalk (pVN28-derived adenovirus). The dicistronic PIT-/tTA-encoding expression unit is driven by P_{SV40} (simian virus 40 promoter). Whereas PIT is transcribed in a classic cap-dependent manner, tTA production relies on translation-initiation by an internal ribosome entry site of polioviral origin (IRES) (*see* **Table** 1). The adjustment of pVN28-derived dual-regulated adenovirus enables the independent streptogramin-responsive control of SEAP)and the tetracycline-responsive control of the *Bacillus stearothermophilus*-derived secreted a-amylase (SAMY). Quantitative assessment of reporter protein profiles of CHO-K1 cotransduced with pVN28- and pVN10-derived adenoviruses and cultivated for 48 h in the presence and absence of increasing concentrations of regulating pristinamycin I (PI) (A) and tetracycline (TET) (B) antibiotics.

4. Notes

1. Adenovirus particles are classified at biosafety level 2, according to the National Institutes of Health; appropriate safety equipment must be used and governmental regulations met.
2. It is important that the HEK-293 cells do not reach confluence, seed them at low densities. HEK-293 should be used at the lowest possible number (typically n < P40).
3. Shuttle vectors and genomic plasmid preparations with an OD_{260}/OD_{280} coefficient of 1.8 were used for co-transfections.
4. CPE may appear after 15–18 days; the longer it takes the higher the chances are that the adenovirus has incorrectly recombined. Only a detailed restriction analysis of the viral DNA will reveal a correct recombination.
5. All the steps outlined in the extraction protocol for adenovirus DNA should be performed with great care to avoid breakdown of DNA.
6. When the adenovirus encodes a marker gene, a test transduction may reveal its correct expression. Also, if the restriction pattern remains inconclusive because bands are too close to each other, then it is advisable to perform a Southern blot to detect the gene of interest.
7. Always prepare any adenovirus stocks from plaque-purified adenovirus samples to minimize emergence of RCA variants.
8. Titration is the most reliable way to determine the adenovirus concentration. The procedure is cumbersome and will take an average of 15 days. Therefore, some people may prefer to determine the virus titer by an OD measurement, which is less reliable, but follow-up experiments could be started immediately.
9. In the RCA assay, the HeLa cells incubated with excessive adenovirus concentrations may become detached, which can be confused with CPE. Therefore, start the assay with dilutions of 10^{-4} to avoid false positives.
10. Upon arrival, it is important to defrost the FBS at 37°C and then to heat-inactivate it for 30 min at 56°C; store the aliquots at –20°C to achieve optimal transgene regulation.
11. It is extremely important to wash thoroughly several times to rid the supernatant of remnant marker molecules before the expression status is reversed.

References

1. Edelstein, M.L. Abedi, M.R. Wixon, J., and Edelstein, R.M. (2004) Gene therapy clinical trials worldwide 1989–2004-an overview. *J Gene Me.* **6**, 597–602.
2. Tomanin, R. and Scarpa, M. (2004) Why do we need new gene therapy viral vectors? Characteristics, limitations and future perspectives of viral vector transduction. *Curr Gene Ther;.* **4**, 357–372.
3. Vandendriessche, T.C. (2002) Clinical gene therapy - first international conference. *IDrugs.* **5**, 209–212.
4. Everts, M. and Curiel, D.T. (2004) Transductional targeting of adenoviral cancer gene therapy. *Curr Gene Ther.* **4**, 337–346.

5. Glasgow, J.N. Bauerschmitz, G.J. Curiel, D.T., and Hemminki, A. (2004) Transductional and transcriptional targeting of adenovirus for clinical applications. *Curr Gene Ther.* **4,** 1–14.
6. Gomez-Manzano, C. Yung, W.K. Alemany, R., and Fueyo, J. (2004) Genetically modified adenoviruses against gliomas: from bench to bedside. *Neurology.* **63,** 418–426.
7. Liu, M. Acres, B. Balloul, J.M. Bizouarne, N. Paul, S. Slos, P., and Squiban, P. (2004) Gene-based vaccines and immunotherapeutics. *Proc Natl Acad Sci USA.* **101**, 14567–14571.
8. Kamen, A. and Henry, O. (2004) Development and optimization of an adenovirus production process. *J Gene Med.* **6,** 184–192.
9. McConnell, M.J. and Imperiale, M.J. (2004) Biology of adenovirus and its use as a vector for gene therapy. *Hum Gene Ther.* **15,** 1022–1033.
10. Trapnell, B.C. (1993) Quantitative evaluation of gene expression in freshly isolated human respiratory epithelial cells. *Am J Physiol.* **264**, 199–212.
11. Volpers, C. and Kochanek, S. (2004) Adenoviral vectors for gene transfer and therapy. *J Gene Med.* **6**, 164–171.
12. Danthinne, X. and Imperiale, M.J. (2000) Production of first generation adenovirus vectors: a review. *Gene Ther.* **7**, 1707–1714.
13. Russell, W.C. (2000) Update on adenovirus and its vectors. *J Gen Virol.* **81**, 2573–2604.
14. Lusky, M. Christ, M. Rittner, K. Dieterle, A. Dreyer, D. Mourot, B. Schultz, H. Stoeckel, F. Pavirani, A., and Mehtali, M. (1998) In vitro and in vivo biology of recombinant adenovirus vectors with E1, E1/E2A, or E1/E4 deleted. *J Virol.* **72**, 2022–2032.
15. Weber, W. and Fussenegger, M. (2004) Approaches for trigger-inducible viral transgene regulation in gene-based tissue engineering. *Curr Opin Biotechnol.* **15**, 383–391.
16. Toniatti, C. Bujard, H. Cortese, R., and Ciliberto, G. (2004) Gene therapy progress and prospects: transcription regulatory systems. *Gene Ther.* **11,** 649–657.
17. Weber, W. Marty, R.R. Keller, B. Rimann, M. Kramer, B.P., and Fussenegger, M. (2002) Versatile macrolide-responsive mammalian expression vectors for multiregulated multigene metabolic engineering. *Biotechnol Bioeng.* **80,** 691–705.
18. Gossen, M. and Bujard, H. (1992) Tight control of gene expression in mammalian cells by tetracycline-responsive promoters. *Proc Natl Acad Sci USA.* **89,** 5547–5551.
19. Gossen, M. Freundlieb, S. Bender, G. Muller, G. Hillen, W. Bujard, H. (1995) Transcriptional activation by tetracyclines in mammalian cells. *Science.* **268,** 1766–1769.
20. Fussenegger, M. Morris, R.P. Fux, C. Rimann, M. von Stockar, B. Thompson, C.J., and Bailey, J.E. (2000) Streptogramin-based gene regulation systems for mammalian cells. *Nat Biotechnol.* **18,** 1203–1208.
21. Weber, W. and Fussenegger, M. (2004) Inducible gene expression in mammalian cells and mice. *Methods Mol Biol.* **267**, 451–66.

22. Fussenegger, M. Moser, S. Mazur, X., and Bailey, J.E. (1997) Autoregulated multicistronic expression vectors provide one-step cloning of regulated product gene expression in mammalian cells. *Biotechnol Prog*. **13**, 733–740.
23. Fux, C. Moser, S. Schlatter, S. Rimann, M. Bailey, J.E., and Fussencgger, M. (2001) Streptogramin- and tetracycline-responsive dual regulated expression of p27(Kip1) sense and antisense enables positive and negative growth control of Chinese hamster ovary cells. *Nucleic Acids Res*. **29**, E19.
24. Kramer, B.P. Viretta, A.U. Daoud-El-Baba, M. Aubel, D. Weber, W., and Fussenegger, M. (2004) An engineered epigenetic transgene switch in mammalian cells. *Nat Biotechnol*. **22,** 867–870.
25. Moser, S. Rimann, M. Fux, C. Schlatter, S. Bailey, J.E., and Fussenegger, M. (2001) Dual-regulated expression technology: a new era in the adjustment of heterologous gene expression in mammalian cells. *J Gene Med*. **3**, 529–49.
26. Gonzalez-Nicolini, V. and Fussenegger, M. (2005) A novel binary adenovirus-based dual-regulated expression system for independent transcription control of two different transgenes. *J Gene Med*. **7**, 1573–85.
27. Fux, C. and Fussenegger, M. (2003) Bidirectional expression units enable streptogramin-adjustable gene expression in mammalian cells. *Biotechnol Bioeng*. 83, 618–625.
28. Nyberg-Hoffman, C. Shabram, P. Li, W. Giroux, D., and Aguilar-Cordova, E. (1997) Sensitivity and reproducibility in adenoviral infectious titer determination. *Nat Med*. **3,** 808–811.
29. Dion, L.D. Fang, J., and Garver, R.I. Jr. (1996) Supernatant rescue assay vs. polymerase chain reaction for detection of wild type adenovirus-contaminating recombinant adenovirus stocks. *J Virol Methods*. **56**, 99–107.
30. Kelm, J.M. Djonov, V. Hoerstrup, S.P. Guenter, C.I. Ittner, L.M. Greve, F. Hierlemann, A. Sanchez-Bustamante, C.D. Perriard, J.C. Ehler, E., and Fussenegger, M. (2006) Tissue-transplant fusion and vascularization of myocardial microtissues and macrotissues implanted into chicken embryos and rats. *Tissue Eng*. **12**, 2541–53.
31. Berger, J. Hauber, J. Hauber, R. Geiger, R., and Cullen, B.R. (1988) Secreted placental alkaline phosphatase: a powerful new quantitative indicator of gene expression in eukaryotic cells. *Gene*. **66**, 1–10.
32. Schlatter, S. Rimann, M. Kelm, J., and Fussenegger M. (2002) SAMY, a novel mammalian reporter gene derived from Bacillus stearothermophilus alpha-amylase. *Gene*. **282**, 19–31.

15

Regulated Expression of Adenoviral Vectors-Based Gene Therapies

Therapeutic Expression of Toxins and Immune-Modulators

James F. Curtin, Marianela Candolfi, Mariana Puntel, Weidong Xiong, A. K. M. Muhammad, Kurt Kroeger, Sonali Mondkar, Chunyan Liu, Niyati Bondale, Pedro R. Lowenstein, and Maria G. Castro

Summary

Regulatable promoter systems allow gene expression to be tightly controlled *in vivo*. This is highly desirable for the development of safe, efficacious adenoviral vectors that can be used to treat human diseases in the clinic. Ideally, regulatable cassettes should have minimal gene expression in the "OFF" state, and expression should quickly reach therapeutic levels in the "ON" state. In addition, the components of regulatable cassettes should be non-toxic at physiological concentrations and should not be immunogenic, especially when treating chronic illness that requires long-lasting gene expression. In this chapter, we will describe in detail protocols to develop and validate first generation (Ad) and high-capacity adenoviral (HC-Ad) vectors that express therapeutic genes under the control of the TetON regulatable system. Our laboratory has successfully used these protocols to regulate the expression of marker genes, immune stimulatory genes, and toxins for cancer gene therapeutics, i.e., glioma that is a deadly form of brain cancer. We have shown that this third generation TetON regulatable system, incorporating a doxycycline (DOX)-sensitive $rtTA_2{}^S$-M2 inducer and tTS^{Kid} silencer, is non-toxic, relatively non-immunogenic, and can tightly regulate reporter transgene expression downstream of a TRE promoter from adenoviral vectors *in vitro* and also *in vivo*.

Key Words: Adenoviral vectors; tetracycline; TetON; inducible; regulatable; gene therapy; doxycycline; glioma; *Pseudomonas* exotoxin A (PE).

From: *Methods in Molecular Biology, vol. 434: Volume 2: Design and Characterization of Gene Transfer Vectors*
Edited by: J. M. Le Doux

1. Introduction

Gene expression can be regulated using a promoter that is sensitive to environmental (e.g., heat), physiological (e.g., steroids), or chemical (e.g., tetracycline) changes in the body ***(1–3)***. In general, chemical switches are preferable to either physiological or environmental switches because pharmacokinetic/pharmacodynamic properties can be determined, reducing the fluctuations that may otherwise interfere with environmental or physiological regulators. An ideal regulatable system for gene therapy requires that the following conditions are met: (1) regulation of gene expression *in vivo* should be achievable using a compound that is nontoxic; (2) the compound must be able to penetrate into the desired target tissue or organ at effective concentrations; (3) it should have a half-life of a few hours (as opposed to minutes or days), so that when withdrawn or added (depending on the regulatable system used), gene expression can be turned on or off quickly and effectively; (4) the genetic switches employed should ideally be non-immunogenic in the host; (5) expression in the "OFF" state should be minimal, and expression in the "ON" state should be sufficiently high for therapeutic efficacy of the transgene ***(4)***. We have developed a regulatable cassette based on the TetON system (*see* **Note 1**). In this cassette, therapeutic or reporter transgenes are under the control of the TRE promoter ***(5)***. The Tet-sensitive transactivator, $rtTA_2{}^S$-M2 (*see* **Note 2**) ***(6,7)***, and the repressor element tTS^{kid} ***(8)*** are both constitutively expressed by a murine CMV promoter. In the "OFF" state, tTS^{kid} suppresses unwanted activation of TRE and lowers basal therapeutic gene expression ***(9)***. In the "ON" state doxycycline (DOX) (a water soluble Tet analog that can be administered in food and water) binds with the $rtTA_2{}^S$-M2 transactivator, inducing a conformational change in $rtTA_2{}^S$-M2 that confers binding affinity for TetO repeat elements present in the TRE promoter, thus inducing expression of genes downstream from the TRE promoter (*see* **Fig. 1**). Tet regulatable vectors have been used in our laboratory for inducible expression of therapeutic transgenes *in vitro* ***(10)*** and *in vivo* ***(11)***. These Tet-dependent regulatable expression systems constitute an ideal platform to develop vectors that will allow regulated transgene expression for both basic and translational research applications and also to generate vectors expressing genes that could be toxic to the producer cell line, i.e., FasL, TNFα, *Pseudomona exotoxin A*, *Diphteria toxin*. In this chapter, we will describe in detail the protocols our laboratory has successfully used for (1) cloning plasmids containing transgenes under the control of the TetON regulatable cassette; (2) cloning mammalian expression plasmids where transgene expression is tightly regulated by the TetON promoter; (3) validation of cloning by Southern blot; (4) scale-up of HC-Ad vectors; (5) confirming DOX-inducible gene expression *in vitro*; and (6) *in vivo*.

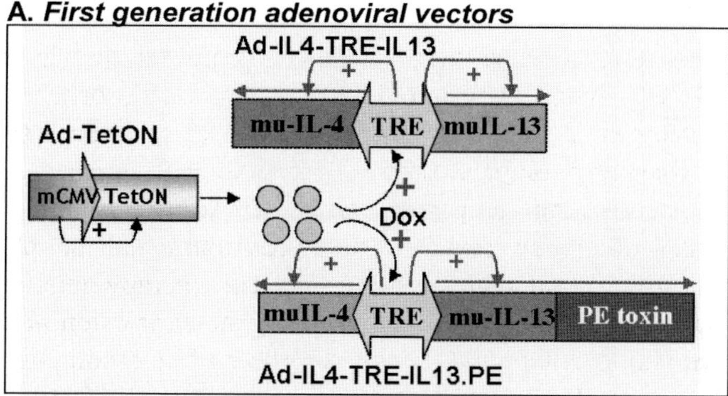

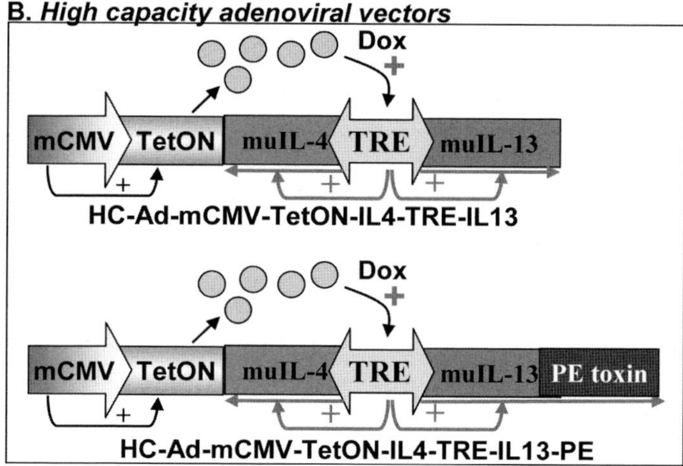

Fig. 1. Structure of regulatable first generation (**A**) and high capacity (**B**) adenoviral vectors. These novel adenoviral vectors express mulL4 and mlL13 or the cytotoxin mulL-13-PE under the control of the bidirectional TRE promoter. The TRE promoter is activated by the transactivator (TetON) in the presence of the inducer Dox. The expression of the TetON is driven by the strong murine CMV promoter from an additional first generation adenovirus that is administered with the therapeutic virus, or is included in the high capacity adenovirus, since it allows the expression of larger transgenes.

2. Materials

2.1. Cloning of Tet-ON Regulated Plasmids

1. Plasmids (see **Table 1** for description and sources).

    a. pΔE1sp1A.
    b. pAL120.
    c. pIRES-tTS^{kid}-pA.

Table 1
Plasmid Sources and Function

Name	Function	Reference
pΔE1sp1A	A first generation shuttle plasmid containing Xba1, Hind3 and Sal1 (among others) in the multiple cloning site.	Microbix Biosystems
pAL120	A first generation shuttle vector containing the mCMV promoter and a pA signal sequence. Sal1 is used for cloning transgenes downstream of the mCMV promoter and upstream of the pA.	Generated in our laboratory *(24)*
pIRES-tTS^{Kid}-pA	An intermediate plasmid in the cloning of the TetON cassette, it is a modified pIRES plasmid containing the tTSkid-pA sequence downstream of the IRES.	Generated in our laboratory *(2)*
pUHS 6-1	This plasmid contains the tTSkid repressor and polyadenylation (pA) signal flanked by Xba1 and Hind3.	H. Bujard at ZMBH, Germany *(8)*
pIRES	Cloning plasmid containing the Internal Ribosome Entry Site (IRES) from the encephalomyocarditis virus.	Clontech
pUHrT 62-1	This plasmid carries the $rtTA_2{}^S$-M2 Tet-sensitive transactivator gene flanked by EcoR1 and BamH1.	H. Bujard at ZMBH, Germany *(8)*
p$rtTA_2{}^S$M2-IRES-tTS^{kid}-pA	8200 bp plasmid based on pIRES containing $rtTA_2{}^S$M2 upstream and tTS^{kid} immediately downstream of the IRES sequence (i.e., the TetON cassette) flanked by Xho1 and Sal1.	Generated in our laboratory *(2)*
pSP72-Bgl2	Modified cloning plasmid (original from Clontech) with a multiple cloning site that contains Bgl2, Hind3 and EcoR1 among others.	Generated in our laboratory *(2)*.

pSP72-mCMV	pSP72-Bgl2 containing the mCMV promoter and pA sequence (from pAL120).	Generated in our laboratory *(2)*.
pSP72[mCMV-rtTA$_2$^{S}M2-IRES-tTS^{kid}-pA]	Intermediate plasmid containing the TetON cassette regulated by the mCMV promoter. The entire cassette is flanked on both ends by Bgl2.	Generated in our laboratory *(2)*
pBS2SK+ [Tre-MCS-pA]-Kana	Intermediate vector based on the bluescript 2 SK+ plasmid.	Generated in our laboratory (unpublished)
pmCMV-TetON	pBlueScript based vector containing the TetON cassette and also containing the Tet-sensitive promoter TRE upstream of a multiple cloning site (Age1, Afl2, Swa1, Bcl1) and pA.	Generated in our laboratory (unpublished)
pSP72Bgl2-MCS	Adaptor plasmid generated in our laboratory containing two copies each of the Age1, Afl2, Swa1, and Bcl1 restriction endonuclease sites. By cloning transgene through this vector, these rare restriction sites can be used to clone directly into pmCMV-TetON.	Generated in our laboratory (unpublished)
pSTK120	Plasmid used to generate high capacity adenoviral vectors. Plasmid only shares long terminal repeat (LTR) and packaging (Ψ) sequences of wild type Adenovirus. Several variants have been created, e.g., pSTK120.1, pSTK120.2, and pSTK120.3.	Generated by S. Kochanek *(25)*

d. pUHS 6-1.
e. pIRES.
f. pUHrT 62-1.
g. prtTA$_2$^{S}M2-IRES-tTS^{kid}-pA.
h. pSP72Bgl2.
i. pSP72-mCMV.
j. pSP72[mCMV-rtTA$_2$^{S}M2-IRES-tTS^{kid}-pA].
k. pBS2 SK(+) [Tre-MCS-pA]-Kana.

2. Restriction endonucleases (REs) and REs buffer (NEB, Ipswich, MA, USA).
3. 1% Agarose gel: 1% w/v agarose in 0.5× TBE buffer, 300 ng/ml ethidium bromide.
4. Gel Purification Kit (available from Qiagen, Valencia, CA, USA).
5. T4 DNA ligase and ligase buffer.
6. Maxi Prep kit (Qiagen).

2.2. Generation of pmCMV-TetON Plasmid Driving Inducible Expression of Gene of Interest

1. Plasmids (see **Table 1** for description and sources).

    a. pmCMV-TetON.
    b. pSP72Bgl2-MCS.

2. REs and buffer; Swa1, Bcl1, Afl2 or Age1.
3. Calf intestinal phosphatase (CIP) (NEB)
4. 1% Agarose gel: 1% w/v agarose in 0.5× TBE buffer, 300 ng/ml ethidium bromide.
5. Gel documentation system.
6. UV light box.
7. Gel Purification Kit (available from Qiagen).
8. Quick Ligase and Ligase buffer (NEB).
9. Maxi Prep kit (Qiagen).
10. Luria–Bertani media (LB): LB + 50 μg/ml Ampicillin (LB-AMP), LB-AMP agar plates *(12)*.

2.3. Confirming Insertion of Transgene by Southern Blot

1. 0.25 M HCl (500 ml).
2. 0.4 M NaOH (1 l).
3. Tris–HCl pH 9.00 (500 ml).
4. Tris–HCl pH 8.00 (500 ml).
5. 20× SSC: 3 M NaCl, 0.3 M Na-Citrate, pH 7.0.
6. Pre-hybridization solution (5× SSC): 1% blocking reagent, 0.1% N-lauroylsarcosine, 0.02% sodium dodecyl sulfate (SDS), heat to 60°C for 30 min or until fully dissolved. Store at –20°C.
7. Buffer 1: 100 mM Tris–HCl, 150 mM NaCl, pH 7.5.
8. Buffer 2: 100 mM Tris–HCl, 150 mM NaCl, 1% Blocking reagent (Roche, Indianapolis, IN, USA), pH 7.5.

9. Buffer 3: 100 mM Tris–HCl, 150 mM NaCl, 50 mM $MgCl_2$, pH 9.5.
10. Buffer 4: 10 mM Tris–HCl, 1 mM EDTA, pH 8.0.
11. Color substrate: 200 μl NBT/BCIP in 10 ml Buffer 3 (*see* **step 9**).

2.4. Scale-up of HC-Ad Vector Expressing Transgene Regulated by Tet-ON

1. 293 FLPe cells (293 cells stably transfected with Flippase expressing (FLPe) recombinase were generated by our laboratory ***(13)***).
2. 293 FLPe medium: 10% fetal bovine serum (FBS) (Gibco, Carlsbad, CA, USA), 1% non-essential amino acids (Gibco), 1% L-Gln (Gibco), 10 IU/ml Penicillin–Streptomycin (Gibco), 1.5 μg/ml Puromycin (Sigma, St. Louis, MO, USA).
3. First-generation FLPe-sensitive helper adenovirus (FL helper) (generated by our laboratory ***(13)***).
4. pSTK120 plasmid with TetON regulatable cassette driving transgene expression (see **Table 1** for description and sources).
5. Ultracentrifuge (Beckman Coulter, Fullerton, CA, USA).
6. 5% sodium deoxycholate in H_2O.
7. 2 M $MgCl_2$.
8. 3-ml sterile syringe (Becton Dickenson, Franklin lakes, NJ, USA).
9. 5-ml sterile syringe (Becton Dickenson).
10. 18-G, 3.5-inch spinal needle (Becton Dickenson).
11. 21-G, 1'-long needle (Becton Dickenson).
12. 1.33 g/ml = 8.349 g CsCl dissolved in 16 ml CsCl buffer.
13. 1.45 g/ml = 8.349 g CsCl dissolved in 11.4 ml CsCl buffer.
14. CsCl buffer: 5 mM Tris–HCl, pH 7.5, 1 mM EDTA, pH to 7.5 with 1 M sodium hydroxide, and then filter sterilize using a 0.45-μm filter.
15. TE buffer: 10 mM Tris–HCl, 1 mM EDTA, pH to 7.8 with 1 M sodium hydroxide, and then sterilize by autoclaving.
16. Mineral oil.
17. Buffer A: 7.092 g Tris–HCl, 0.4284 g $MgCl_2$, 35.5 g NaCl, in 4.5 l total volume with H_2O and autoclaved.
18. Buffer B: Buffer A with 10% glycerol and sterilized by autoclaving.

2.5. Confirming Inducible Gene Expression in Vitro

1. Cell lines.

   a. Cos7 (ATCC) African Green Monkey SV40-transfected kidney fibroblast cell line.
   b. GL26 (ATCC) C57BL/6 mouse glioma cell line.
   c. CNS1 (ATCC) Lewis rat glioma cell line.
   d. U87 MG (ATCC) human, caucasian female 44 years old; glioblastoma-astrocytoma.
   e. U251 (ATCC) human glioma cell line.

   f. IN859 (human biopsy primary culture obtained by our laboratory ***(14)***).
   g. IN2045 (human biopsy primary culture obtained by our laboratory ***(14)***).

2. DOX (Sigma).
3. Culture media; MEM media (Gibco) containing 10% FBS (Gibco), 1 mM L-Gln (Gibco), 1mM non essential amino acids (Gibco), 10 IU/ml Penicillin/Streptomycin (Gibco).
4. ELISA specific for transgene (Flt3L ELISA from R&D Systems, Minneapolis, MN, USA).
5. Immunoglobulins specific for Flt3L (generated in our laboratory ***(15)***), PE toxin (generated in our laboratory), IL4 (R&D), IL13 (R&D), and IL13-alphaR2 (R&D).
6. Immunoglobulins specific for β-galactosidase (generated in our laboratory ***(15)***).
7. Secondary anti-Ig-specific immunoglobulins-HRP conjugated (Dako Cytomation, Glostrup, Denmark).
8. SDS–PAGE equipment (BioRAD).
9. RIPA Lysis buffer: 50 mM Tris–HCl, pH 7.4 (50 μl of 1 M Stock), 150 mM NaCl (30 μl of 5 M Stock), 1 mM NaF (20 μl of 50 mM Stock), 1 mM $NaVO_4$ (8 μl of 125 mM Stock), 1 mM EGTA, pH 8.0 (20 μl of 50 mM Stock), 1% NP40 (10 μl), 0.25% sodium deoxycholate (25 μl), 1× PI (1 μl of 1000× Stock), 1× PMSF (1 μl of 1000× Stock), made up to 1 ml with 831 μl dH_2O.
10. Mini Trans-blot cell for Western Blotting (BioRad, Hercules, CA, USA).
11. PowerPac 300 Power supply (BioRad).
12. Film and automated film developing machine (Kodak, Rochester, NY, USA).

2.6. Confirming Inducible Gene Expression in Vivo

1. Male Fisher rats, 200–250 g (Harlan Sprague Dawley).
2. DOX (Sigma).
3. Stereotactic apparatus (Stoelting).
4. Anesthetics and analgesics.

   a. 100 mg/ml Ketamine HCl (Phoenix Pharmaceuticals Inc., St. Joseph, MO, USA).
   b. 1 mg/ml Medatomidine HCl (Pfizer, Exton, PA, USA).
   c. 0.3 mg/ml Buprinorphine (Reckitt Benckiser, Richmond, VA, USA).
   d. 5 mg/ml Atipamazole HCl (Pfizer).

5. First generation Ad and HC-Ad vectors.
6. Surgical equipment.

   a. Scalpel and sterile blades.
   b. Surgical Scissors, 14-cm long, straight, Sharp (World Precision Instruments, Sarasota, FL, USA).
   c. Iris Forceps, 10-cm long, serrated straight 0.8-mm tips (World Precision Instruments).

d. Wire Retractors, 5-cm long, 15-mm wire blades, maximum spread 25 mm (World Precision Instruments).
e. Nylon Ethicon Sutures 3-0 (Cardinal Health, Dublin, OH, USA).
f. Dremel Stylus cordless drill and drill bits (Dremel, Racine, WI, USA).

7. 26-G, 10-μl Hamilton needle and syringe.
8. Tyrode's solution: 0.14 M NaCl, 1.8 mM $CaCl_2$, 2.7 mM KCl, 0.32 mM NaH_2PO_4, 5.6 mM glucose, and 11.6 mM $NaHCO_3$.
9. 4% Paraformaldehyde (w/v) in PBS.
10. PBS + 0.1% sodium azide.
11. Electronic VT1000S vibrating blade vibratome (Leica, Wetzler, Germany).
12. Vectastain ABC Elite kit (Vector Laboratories, Burlingame, CA, USA).

3. Methods

3.1. Cloning of Tet-ON Regulated Plasmids

1. We have generated both Ad (TetOFF or TetON) and HC-Ad (TetON) vectors expressing inducible transgenes using Tet-regulatable system ***(5,16,17)***. Because of genomic size constraints (up to 7 kb can be inserted into first generation, E1a/E4 deleted viruses), first-generation Ad vectors require the Tet-regulatable cassette to be expressed by one adenoviral vector, and the transgenes under the control of the TRE promoter to be expressed by another vector. Thus, inducible gene expression in cells requires co-infection with Ad vectors expressing Tet-regulatable elements and Ad vectors expressing TRE-driven transgene elements. High multiplicity of infection (MOI) can be used in vitro, and first-generation Ad vectors are relatively easy to scale up, thus Ad vectors are useful for in vitro or in vivo studies of transgene biological activity or preclinical efficacy. The latest generation, high capacity, gutless Ad vectors (HC-Ad), with a theoretical capacity of ~35 kb, can encode for large transcriptional units, and transgene expression form of these vectors is stable even in the presence of a peripheral anti-Ad immune response as could be encountered in patients undergoing clinical trials. They are therefore the preferred choice for pre-clinical in vivo studies ***(13)***. Here we will describe in detail the generation of HC-Ad vectors; first-generation Ad vectors can also be generated by excising either TetON or Tre-transgene-pA elements and inserting them into first-generation Ad shuttle plasmids (e.g., pΔE1sp1A or pAL120).
2. To generate the TetON cassette, the transcriptional silencer carrying plasmid p[IRES-tTSKid-pA] was generated by digesting pUHS 6-1 with Xba1 and Hind3 to liberate tTS^{kid}, ligating tTS^{kid} into the shuttle plasmid pΔE1sp1A and digesting with Xba1 and Sal1 before ligation into pIRES digested with Xba1 and Sal1. This generated a plasmid containing IRES followed by tTSkid and a polyAdenylation (pA) signal.
3. The $rtTA_2{}^SM2$ transactivator was excised using EcoRI and BamHI from plasmid pUHrT 62-1; the BamHI site was adapted using short oligonucleotide dimers

to add an Mlu1 restriction site. The adapted rtTA$_2$^{S}M2 insert was directionally cloned into a previously generated plasmid p[IRES-tTSKid-pA], digested with EcoR1 and Mlu1, resulting in intermediate plasmid, p[rtTA2SM2-pIRES-tTSKid-pA]. This generated a plasmid where the rtTA$_2$^{S}M2 and tTS^{kid} coding sequences were separated by an IRES *(5)*.

4. Next, we placed a constitutive mCMV promoter upstream of rtTA$_2$^{S}M2 to drive expression of rtTA$_2$^{S}M2 and tTS^{kid}. We excised the mCMV promoter from pAL120 with EcoRI and HindIII and ligated into a pSP72-Bgl2 shuttle vector (Clontech, Mountain View, CA, USA), producing intermediary plasmid pSP72[mCMV]. Plasmid p[rtTA2SM2-pIREStTSKid-pA] was excised with XhoI and SalI to release cassette [rtTA2SM2-pIRES-tTSKid-pA] and subsequently cloned into its corresponding sites into plasmid pSP72[mCMV], generating the intermediate plasmid pSP72[mCMV-rtTA2SM2-pIRES-tTSKid-pA].
5. The [mCMV-rtTA2SM2-pIRES-tTSKid-pA] regulatable Tet-ON cassette was then excised with BglII and cloned into the BglII site of intermediate plasmid pBlueScript II SK(+)[TRE-MCS-polyA]-Kanamycin, thus generating the final intermediate plasmid, pBlueScript II SK(+)[TRE-MCS-polyA]-[mCMV-rtTA2SM2-IRES-tTSKid-pA]-Kanamycin.
6. The MCS contains REsites for Bcl1, Afl2, Age1, and Swa1 and can be used to directly insert transgenes downstream of TRE. A shuttle vector, pSP72-Bgl2-MCS, also exists that facilitates the cloning of transgenes into pmCMV-TetON.

3.2. Generation of pmCMV-TetON Plasmid Driving Inducible Expression of the Gene of Interest.

1. Digest 1 μg pmCMV-TetON (*see* **Note 3**) and 1 μg pSP72-Bgl2-MCS containing the transgene with compatible restriction enzyme (Swa1, Bcl1, Afl2, or Age1). The reaction mixture should contain 10 μl 10× RE buffer (*see* **Note 4**), 1 μg DNA, 5 μl RE, and H_2O bringing entire volume to 100 μl and needs to be digested at 37°C (55°C for Bcl1) for at least 2 h. This will linearize the pmCMV-TetON plasmid and drop out the transgene with compatible ends, allowing the ligation of the therapeutic gene downstream of the TRE promoter and upstream of pA sequence.
2. Digestion must be to completion, confirm that no undigested DNA remains by running a 1% agarose gel with 5 μl reaction mixture (and undigested DNA as a control in another gel lane) before continuing.
3. Run the entire sample on a 1% agarose gel and use a new razor blade to cut out the band of interest from the gel, visualizing the DNA using a UV illumination source (*see* **Note 5**).
4. Purify the DNA from Agarose gel using DNA purification columns from Qiagen.
5. Elute the DNA in 30 μl TE buffer, and run 2 μl on an agarose gel to quantify DNA concentration.
6. Dephosphorylate the ends of the vector using CIP. This prevents religation of vector.

7. Set up the ligation reaction with (10 min, room temperature) Quick Ligase (NEB). We generally ligate 200 ng vector with a threefold molar excess of insert.
8. Transform competent DH5α cells with ligated product (use 5 μl ligation reaction) and culture overnight at 37°C on Agar plates with LB + Ampicillin (50 μg/ml).
9. Pick colonies and grow for 8 h at 37°C in 2 ml LB + Ampicillin (50 μg/ml).
10. Transfer 1–200 ml LB + Ampicillin (50 μg/ml) and grow overnight at 37°C.
11. MAXI prep DNA from bacteria and use REs to confirm that the transgene has been inserted in the correct orientation. Expression of the transgene can be verified here by transient transfection before proceeding to insert the cassette into pSTK120 vector.

3.3. Confirming Insertion of Transgene by Southern Blot

1. *Probe preparation:* Digest 5–10 μg of plasmid DNA containing the sequence of preference (e.g., IL13) to liberate the probe. (10 μl DNA, 10 μl Buffer, 5 μl RE, 75 μl H_2O). Digest for 2 h at 37°C and run 1 μl to check completion of the digestion, together with undigested DNA and DNA ladder (Bioline Hyperladder I). Gel extract the desired fragment using Qiagen Gel Extraction kit in 30 μl total of milliQ water.
2. *Sample preparation:* Nucleic acids absorb strongly at 260 nm. The optical density at 260 nm (OD_{260}) of DNA samples (RNA free) can be measured using spectrophotometry to determine the concentration of DNA in a sample. Other major components of cells, i.e., proteins, lipids, etc., do not absorb strongly at 260 nm. Dilute 1 μl plasmid DNA with 100 μl H_2O, transfer to a quartz cuvette, and determine the OD_{260} after first blanking the machine with H_2O only. The concentration of DNA can be calculated by multiplying the OD_{260} of each sample by 5000 (Concentration will be μg/ml DNA). Digest 1 μg plasmid DNA with a restriction enzyme that drops out a band containing the DNA sequence (1 μg plasmid DNA; 3 μl HindIII; 2 μl Buffer 2; H_2O up to 20 μl). Incubate at 37°C for 2 h and run 1 μl to check that the digestion has proceeded to completion.
3. *Gel Electrophoresis:* Run DNA ladder, and Plasmid DNA in a 0.8% agarose gel at 95 V (Biorad, Minisub-cell GT; Biorad, PowerPac 300), for 1 h. Take a photograph of the gel using a gel documentation system (Alpha Innotech Corporation).
4. *Transfer:* Use a positively charged membrane to bind negatively charged DNA (Roche). Rinse the membrane with MilliQ water and soak in 0.25 M HCl, 15–39 min. Rinse the membrane twice with MilliQ water then soak in 0.4 M NaOH. Rinse with MilliQ water and set up the transfer apparatus. Trim off the edges of the gel that do not contain DNA, including the wells. Cut the top right corner of the gel with a blade (this will allow the orientation of the gel to be determined later). Cut thick Whatman 3 M filter paper with the exact measures of the gel. Cut the membrane to the same size as the gel and filter paper. Soak filters and membrane in 0.4 M NaOH and cut a piece of filter paper (Whatman, 3 M) long enough to form the "bridge" over the tray (Life Technologies, Blot transfer

system 20–25). Place the Whatman filter paper bridge on the Blot transfer system with the ends immersed in 0.4 M NaOH. Place the gel on this bridge and place the membrane above it taking care not to leave any bubbles between the gel and the membrane (*see* **Note 6**). Put the thick filter paper and paper towels on top of the membrane and place the weight on top of the paper towels. Transfer overnight.

5. *Labeling the probe:* [Note: 10 ng to 3 μg of DNA can be labeled (Roche, DNA labeling and detection kit).] Add 1 μg the DNA and MilliQ water to a final volume of 16 μl. Take OD260 of the purified probe and calculate the volume needed for 1 μg of DNA. Denature the diluted probe DNA in a boiling water bath for 10 min and quickly transfer to ice. Add the rest of the reagent needed for the labeling reaction (2 μl Hexanucleotide Mix 10×; 2 μl dNTPs labeling mix; 1 μl Klenow enzyme). Vortex, spin down and incubate overnight at 37°C a longer incubation will increase the yield of the labeled DNA. Stop the reaction by adding 2 μl of 0.2 M EDTA; pH 8.00 and precipitate labeled DNA with 2.5 μl of 4 M LiCl and 75 μl 100%EtOH. Mix well and leave 2 h at –20°C, centrifuge 15 min 15,000 g rinse the pellet with 50 μl of 70% ice cold EtOH, dry and resuspend the labeled pellet in 50 μl of MilliQ water. Store at –20°C until use.
6. *Prehybridization:* Briefly soak the membrane in 5× SSC buffer and transfer the membrane to a hybridization tube (TECHNE, FHB11). Prepare the prehybridization solution (10 ml/membrane) (15 ml 5× SSC; 150 mg blocking reagent (Roche DNA labeling and detection kit); 15 mg 0.1% N-lauril sarcosine and 3 μl 10% SDS). Incubate for 30 min at 65°C, until dissolved, add 10 ml of prehybridization solution taking care of avoiding air bubbles, again with the help of a pipette and incubate for 2 h, at least, at 68°C.
7. *Hybridization:* Denature the probe by heating for 10 min at 95°C and transfer to ice immediately. Remove the membrane from the incubator and add the 50 μl of labeled probe to the prehybridization solution. Incubate at 68°C overnight. Remove the membrane from the cylinder and wash in 2× SSC/0.1% SDS at room temp, 15 min on a shaker. Change washing solution to 0.2× SSC/0.1% SDS, incubate at room temperature, 15 min on a shaker and wash once with 0.1× SSC/0.1% SDS, at 65°C, 30 min.
8. *Immunodetection:* [Note: all the washes are done while shaking.] Wash the membrane in buffer 1, for 1 min and incubate the membrane for 30 min in buffer 2. Wash the membrane with buffer 1 for 5 min, then dilute 4 μl of anti-digoxigenin/Alkaline Phosphatase conjugate (vial 8, provided in the kit) in 10 ml of buffer 1 add to the membrane and incubate at room temperature, 30 min while shaking. Wash with buffer 1, for 30 min and incubate in buffer 3 for 2 min. Transfer the membrane to the color substrate solution (10 ml buffer 3 + 200 μl NBT/BCIP). Incubate in the dark until the bands are clearly visible and then stop the reaction using buffer 4. This incubation can be prolonged until bands are evident. Archive the image using a scanner or a gel documentation system. The image can be compared with the original agarose gel to determine sizes of the DNA (see **Fig. 2** as an example).

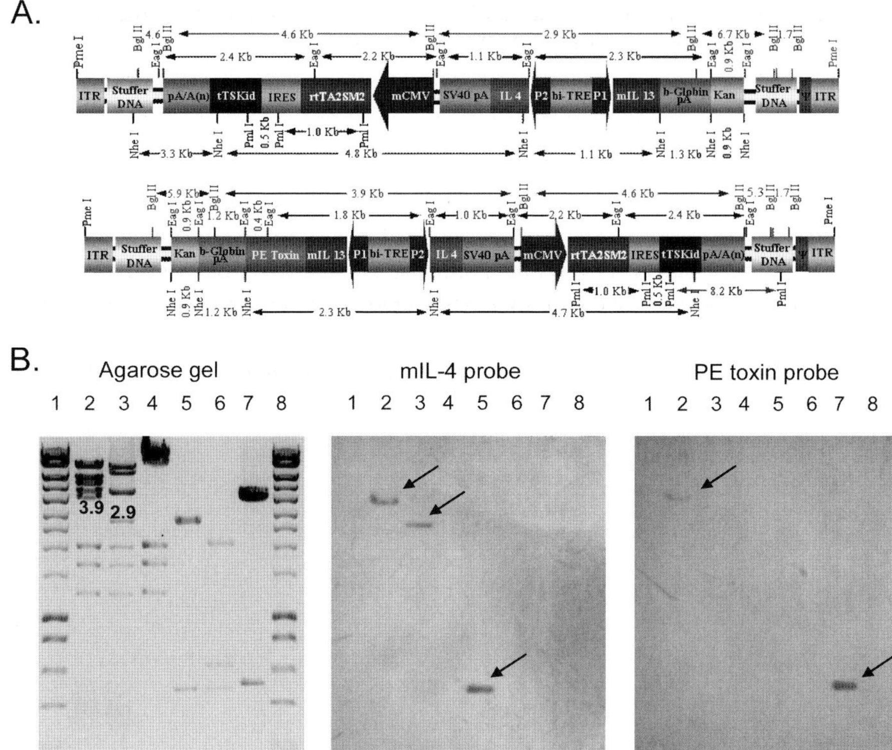

Fig. 2. **A)** Linear depiction of the HC-Ad vector encoding the mIL-4 and mIL-13 without or with the PE toxin transgene and mCMV driven regulatable TetON switch cassette. The constructs indicate the individual components and orientation of the cassettes and their promoters. Some restriction enzyme sites are shown with the appropriate size fragments which correspond to the sizes indicated in part **A. B)** Southern blot analysis of IL4 and PE toxin enconded by the therapeutic vector. HC-Ad plasmids were digested and lanes are as follows: lanes 1 and 8 Hypperladder; lane 2, pSTK120mA.mlL-4.mlL-13PE *Bgl ll* digest; lane 3, pSTK120mA.mlL-4.mIL-13 *Bgl ll* digest; lane 4, pSTK120mA Bgl 11 digest; lane 5, pLS.IL-4 *Sal l/Eagl* digest; lane 6, pLS.IL-13 *EcoRV/Hindlll* digest; lane 7, pRB39.1 PE toxin *EcoRI/Hindlll* digest. Southern blot hybridization was performed with specific probes to IL4 and PE toxin. Both of the recombinant HC-Ad vectors,and the control plasmid were positive for mlL4 (lanes 2, 3 and 5); the hybridization for PE toxin was positive only in the therapeutic HC-Ad vector and in the control plasmid (lanes 2 and 7).

3.4. Scale up of HC-Ad Vectors Expressing Transgenes Regulated by TetON

1. The generation of first-generation Ad vectors has been described in great detail before *(18)*. Here we will focus on the production and generation of HC-Ad vectors containing the TetON regulatable cassette.
2. The entire cassette encoding the transgene regulated by TRE promoter and the TetON cassette expressed by mCMV must be transferred into a plasmid that will allow production of adenoviral particles. We use pSTK120 to generate high-capacity Ad vectors. pSTK120 is a large plasmid that does not contain any known genes or regulatory sequences and is relatively stable; recombination events are low compared with other large plasmids. pSTK120 contains Amp resistance and also contains Ad-5 LTR sequences and the packaging sequence (Ψ), allowing production of HC-Ad vectors. pSTK120 is ~28 kDa but can be reduced in size if necessary by digestion with RE (e.g., Nhe1 and Nar1 can each remove 3 Kb from the final plasmid). Inserts can be cut with Eag1 compatible enzymes (e.g., Not1), or Nhe1 compatible enzymes (e.g., Avr2) (see **Table 2** for the complete list of pSTK120 plasmids and cloning capacities).
3. The insert [TRE-hsFlt3LpolyA]-[mCMV-rtTA2SM2-pIRES-tTSKid-pA]-Kana was excised from the shuttle plasmid using NotI and cloned into the compatible EagI site of HC-Ad plasmid pSTK120.2, generating pSTK120.2-[TRE-hsFlt3L-polyA]-[mCMV-rtTA2SM2-pIRES-tTSKid-pA]-Kana (HC-Ad-mCMV-TetON-Flt3L). The fragments were ligated overnight, transformed into DH5α cells, and plated on LB plates with 25 μg/ml kanamycin *(12)*. Plasmids were isolated from overnight cultures using miniprep and screened by HindIII digestion. Maxiprep purifications were performed on the correct clones (Qiagen). Restriction enzymes were used to confirm the integrity of the plasmids (*see* **Note 7**).
4. The major steps in HC-Ad vector production are rescue, amplification, scale up, and purification of HC-Ad vectors as previously described *(19)*. For a summary of this protocol (*see* **Fig. 3**).
5. *Rescue:* Seed 2 × 10^6 293 FLPe cells (*see* **Notes 8** and **9**) into 60-mm dish the day before transfection. (Note: 293 cells FLPe cells grow in MEM with

Table 2
HC-Ad Plasmids and Predicted Cloning Capacities for Transgene Cassettes

Plasmid	Cloning capacity (Kb)	Restriction sites for cloning in transgene cassette
pSTK120	0–7	Eag1
pSTK120.1	2–10	Nhe1, Eag1
pSTK120.2	4–13	Nhe1, Eag1, Nar1
pSTK120.3	8–18	Nhe1, Eag1, Nar1, Nco1

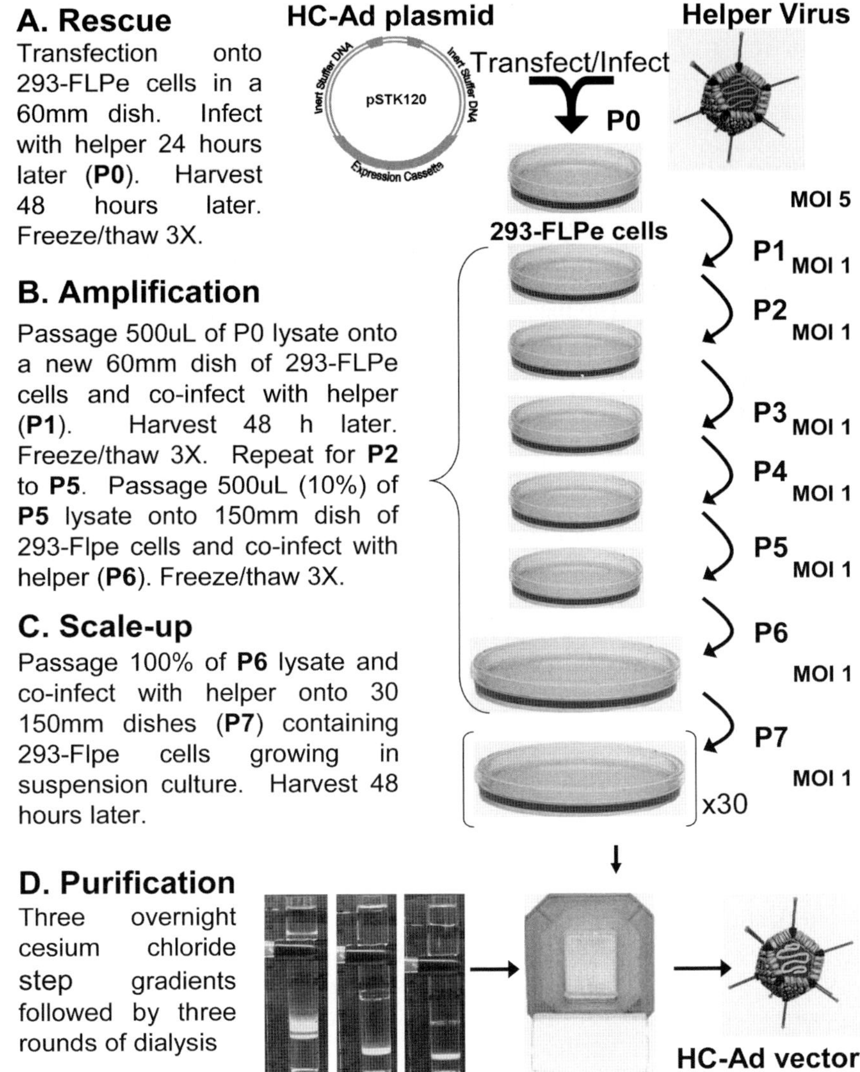

Fig. 3. Construction of High-Capacity adenoviral vectors (HC-Ad). **A**. HC-Ad vectors are rescued by transfecting 293 cells expressing FLPe recombinase with linearized pSTK120 plasmids containing two adenoviral ITR's, a packaging signal, an expression cassetle (Tet ON regulated), and inert stuffer DNA. The cells are infected with FL helper virus containing two loxP sites flanking its packaging signal 24 hours later (MOI 5 pfer per cell). The cells are harvested and lysed 48 hours later. **B**. The HC-Ad vectors are amplified on 293-FLPe cells by 5–7 more serial passages of 10% of the cell lysate from the previous passage, the last of which onto one 150 mm dish with the addition of fresh FL helper virus (MOI 1). The tites of the HC-Ad vectors

puromycin 1.5 μg/ml). Transfect the 293 FLPE cells (80% confluent cells) with 10 μg of gutless plasmid DNA linearized with PmeI, using the calcium phosphate precipitation technique (*see* **Note 10**). Incubate the cells at 37°C, 5% CO_2. (Include one flask transfected with pMV12, a plasmid expressing β-galactosidase as a transfection positive control, and another flask transfected with no plasmid DNA as a transfection negative control.) Sixteen hours later, wash the 293 FLPe cells first with PBS and second with 293FLPe culture media. Infect the cells with helper virus at an MOI 5 diluting the virus in 5 ml of medium. Incubate at 37°C, 5% CO_2. Harvest the cells when full CPE appears (usually 2–3 days). Tap the side of the dish to dislodge the cells (or use scraper) and lyse the cells by three freezing and thawing cycles.

6. *Amplification:* For each amplification passage (P1 to P6), seed 293 FLPe cells into a 60-mm dish the day before infection to obtain approximately 80% confluence of cells the next day. Cells are grown and incubated at 37°C, 5% CO_2. Infect using 10% of the volume collected from the previous passage lysate and coinfect with FL helper virus at MOI = 1. Harvest the cells when full CPE appears. Tap the side of the dish to dislodge the cells (or use scraper) and lyse the cells by three freezing and thawing cycles.
7. *Scale-up:* For P7, seed a 150-mm dish with 1×10^7 cells and infect using 100% of the lysate from P6 and co-infect with FL helper virus at MOI = 1. Keep 10% of the lysate (usually 50–100μl), for analysis, BFU, IU, molecular titration by duplex RT–PCR ***(18)***. Harvest the cells when full CPE appears and pellet down the cells by centrifugation, 10 min at 1500 g, 4°C; and resuspend in 500 μl of PBS. Seed 30 dishes (150 mm) with 1×10^7 cells the day before infection. Prepare the medium by adding the viral vector suspension and the FL helper virus at MOI 1, and infect the cells. Harvest the cells when full CPE appears (2–3 days) pellet down the cells by centrifugation 10 min at 1500 g, 4°C; and resuspend in 8 ml of 100 mM Tris–HCl, pH 8.0. Store at –80°C until CsCl purification is performed.
8. We have also used a similar method for rescuing and scaling up of HC-Ad vectors developed by Philip Ng and colleagues that utilizes a Cre recombinase sensitive helper adenovirus to package HC-Ad genomes in trans ***(19,20)***. We have found that this method, allows larger number of cells to be infected in the final scale-up step; thus, we routinely achieve higher titers (~one log higher

Fig. 3. *(Continued)* increases with each serial passage, eventually surpassing the titer of the FL helper which is not efficiently packaged due to the excision of its packaging signal. **C**. During scale up, 100% of P6 lysate is co-infected with fresh FL helper virus into 30 150 mm dishes containing 293-FLPe cells. Cells are harvested 48 hours later. **D**. Cells are lysed with 5% deoxycholate and DNaseI and HC-Ad vectors are purified over three CsCl step gradients. The vectors are dialyzed against three changes of dialysis buffer supplemented with 10% glycerol, aliquoted, and stored at −80°C.

titer) of HC-Ad vectors using the Cre method. This method has been described in detail by Ng et al. ***(19)***.

9. *Purification:* Cesium chloride (CsCl) purification has been described before for the purification of first generation adenoviral vectors ***(13,19)***. Briefly, 4 ml of 5% sodium deoxycholate is added to the cell pellets and incubated at room temperature for 30 min to lyse the cells. Add 400 µl of 2 M $MgCl_2$ and 100 µl of DNaseI. Incubate at 37°C for 1 h to remove genomic and non-packaged Ad DNA. Centrifuge at 20,000 × *g* for 15 min. Pipette 2.5 ml of CsCl at a density of 1.33g/ml in a Beckman 14-ml centrifuge tube (Beckman, 331374) (to fit a SW40 swing bucket rotor). Then, using a 5-ml syringe equipped with a wide bore 18-G needle, place the needle at the bottom of the centrifuge and very slowly inject 1.5 ml of CsCl at a density of 1.45 g/ml, so the less dense layer floats on top of the denser layer. (Densities were calculated by the determination of the refractive index.) CsCl was dissolved in 5 mM Trizma HCl, 1 mM EDTA pH 7.8. Overlay 7 mL of the virus solution on the CsCl gradient. Overlay mineral oil on the viral layer until the meniscus reaches 2 mm from the top of the tube. Place the tubes in the centrifuge buckets SW40Ti (Beckman Coulter) and seal before leaving Class 2 hood. Weigh a set up balance tube to an identical weight (±0.02 g) before placing tubes in rotor and centrifuge in an ultracentrifuge for 2 h at 90,000 × *g* (22,500 rpm for the SW40 rotor). Remove tubes from the rotor in the Class 2 hood. Puncture side wall approximately 1 cm below the level of the band with a wide bore 21-G syringe needle. This will allow you to get a large enough angle to remove the entire viral band. The viral band is the lowest of the three bands that should be visible. The volume is *NOT* important at this step. Dilute the virus fraction with half a volume of TE pH 7.8. Layer the virus fraction on top of a second CsCl gradient prepared with 1 ml of 1.45 g/ml CsCl and 1.5 ml of 1.33 g/ml CsCl. Overlay mineral oil until the meniscus is 2 ml from the top of the tube and centrifuge in an ultracentrifuge for 18 h at 100,000 × *g* (23,800 rpm for the SW40 rotor). Recover the viral band as before (*see* **Note 11**). At this stage, the volume recovered is very important. Inject virus into a dialysis cassette and dialyze the banded virus twice against 1.5 l of Buffer A for 2 h and once against 1.5 l of Buffer B for 2 h at room temperature. Aliquot the virus and store at –70°C. Maintain sterility at all times.
10. Titration of HC-Ad vectors is usually performed by measuring the total viral particles using Optical Density (*see* **Note 12**).
11. In vitro and in vivo assessment of gene expression must be carried out before assessing the therapeutic efficacy of the vector in preclinical models (*see* **Note 13**).

3.5. Confirming Inducible Transgene Expression in Vitro

1. *In vitro* expression can be confirmed using ELISA, immunocytochemistry, or Western blotting (*see* **Note 14**). It is preferable to use more than one method to confirm expression in vitro before proceeding. In this example, we tested HC-Ad-mCMV-TetON-Flt3L for transgene expression in the presence and absence

of Dox by ELISA and by immunocytochemistry. We tested Ad-IL4-TRE-IL-13.PE for expression by immunofluorescence and bioactivity by the production of cytotoxic IL13.PE (*see* **Note 15**).

2. *ELISA:* Flt3L is a secretable protein; we measured Flt3L expression in the media of HC-Ad-infected glioma cells from mouse, rat, and human in the presence and absence of Dox. Cells were plated in 12-well plates at 50,000 cells (GL26; CNS-1; U251; U87 MG; IN859; IN2045) per well and allowed to adhere overnight. The following morning, media was replaced with 0.5 ml of media containing DOX (1 μg/ml) or without DOX and cells were then infected with HC-Ad-TetON-Flt3L (50,000 VP/CELL). After 72 h, Flt3L (transgene) was determined in the cell culture supernatant using ELISA (R&D Systems), exactly as outlined in the manufacturer's guide (*see* **Note 16**). Inducible expression of Flt3L was detected in the media by ELISA after six glioma cell lines were infected with HC-Ad-TetON-Flt3L and incubated with or without Dox (*see* **Fig. 4**).
3. *Immunocytochemistry:* GL26, CNS-1, U251, U87, IN859, and IN2045 cells were seeded in 12-well dishes (50,000 cells/well), on sterile coverslips, and allowed to adhere overnight. The following day, cells were incubated with 1 μg/ml DOX or

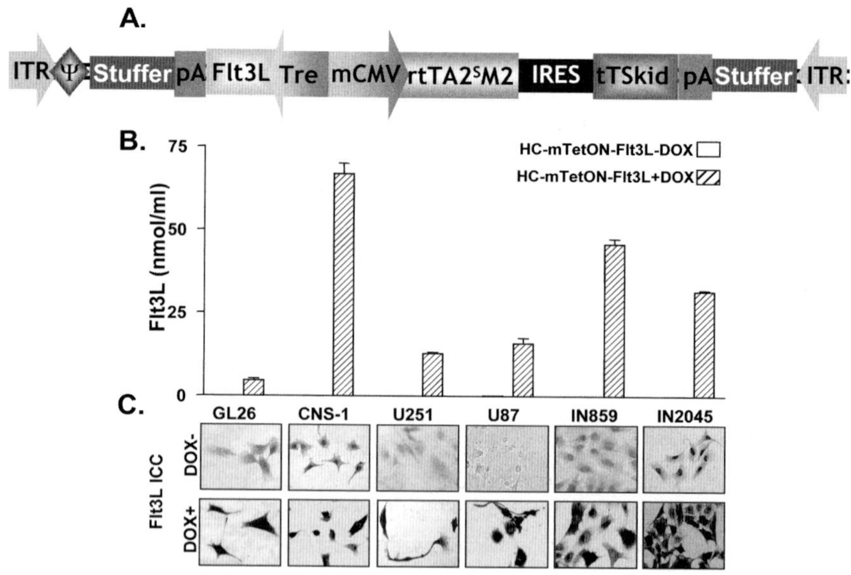

Fig. 4. Regulated expression of Flt3L from the high capacity adenoviral vetor (HC-Ad-mTetON-Flt3L) in human and murine glioma cells. Established human glioma cell lines (U251, U87), cultures from human glioma biopsies (IN859, IN 2045), rat (CNS-I), and mouse (GL26) glioma cells were infected with HC-Ad vector encoding Flt3L under the control of a TetON system driven by the murine CMV promoter (HC-Ad-mTetON-Flt3L) for 72 h in the presence or absence of the inducer doxycyline (Dox, 1 μg/ml). Transgene expression was determined by ELISA and immunocytochemistry (ICC).

without and were infected with HC-Ad-mCMV-TetON-Flt3L (50,000 VP/Cell). After 72 h, cells were washed cells two times with PBS and fixed for 10 min in 4% PFA at room temperature. Cells were then blocked (1 h, room temperature) with 10% Horse Serum in TBS + 0.5% Triton-X100. Stain 2 h with primary antibody (against transgene), use the dilution recommended by the manufacturer. Wash the cells and stain with secondary antibody (these can be either fluorescent tagged or Biotin tagged). If using biotinylated secondary Ab, incubate for 1 h, wash with PBS three times and add avidin-HRP (1:1000 in PBS, 1 h, room temperature). Wash in PBS, wash two times in 100 mM sodium acetate, and stain with DAB. Allow precipitate to develop (usually 1–5 min), wash with sodium acetate and then with PBS + 0.1% sodium azide. Mount on slides and visualize gene expression using visible light microscopy (*see* **Fig. 4**) (*see* **Note 17**).

4. *Western blotting:* Lyse cells in RIPA buffer and quantify protein concentration of cell lysates. For each sample dilute 30 μg of the cell lysate in 20 μl total volume in SDS–Tris buffer containing DTT. Boil samples for 10 min to denature proteins. Run 30 μg samples on an SDS–polyacrylamide gel (with protein standard markers) and transfer to nitrocellulose membrane (Amersham Biosciences). Block membranes and incubate with primary and secondary antibodies as recommended by the manufacturer. We use our custom Flt3L antibody at 1:1000 in blocking solution [TBS + 0.05% Tween20 (TBS-T) with 5% dry milk powder] overnight at 4°C on membranes blocked previously for 1 h with blocking solution. Membranes are then washed three times for 5 min each using TBS-T before addition of secondary antibody for 1 h at room temperature (goat anti-rabbit-HRP, 1:2000 in blocking solution, DAKO). Membranes are washed three times for 5 min each with TBS-T and developed using ECL Western Blotting Analysis kit (Amersham Biosciences) and visualize on X-OMAT LS film (Kodak).
5. *Immunofluorescence (Ads or HC-Ad)*: Cos7 cells were seeded at 50,000 cells per well in a 12-well dish, on coverslips, and allowed to adhere overnight. Cells were infected with Ad-IL4-TRE-IL-13.PE and Ad-TetON and were incubated with or without 1 μg/ml DOX for 48 h. After fixing as described above, transgenes were detected by immunofluorescence using primary antibodies against IL-4 or IL-13 (R&D Systems) in combination with a primary antibody against PE toxin developed in our laboratory. Wash cells in TBS+0.5% TritonX100 and incubate cells for 1 h with fluorescent labeled secondary antibody, protecting from light. Wash in PBS three times. On the final wash, add DAPI to visualize nuclei, wash once more with PBS and mount on slides with anti-Fade mounting media (ProFade) for analysis using an immunofluorescent microscope. Colocalization of the transgenes was observed by Confocal microscopy (*see* **Fig. 5**). Note that transgene expression was only observed when cells were incubated in the presence of Dox.
6. *Viability assay (MTS):* Cell viability was determined by the 3-(4,5-dimethylthiazol-2-yl)-5-(3-carboxymethoxyphenyl)-2-(4-sulfophenyl)-2H tetrazolium (MTS) assay. Cos7, IN859, and U251 cells were seeded at 5000 cells per well in a 96-well dish and infected with Ad-IL4-TRE-IL-13.PE and Ad-TetON. Cells were incubated with or without 1 μg/ml DOX for 48 h. Twenty micro-

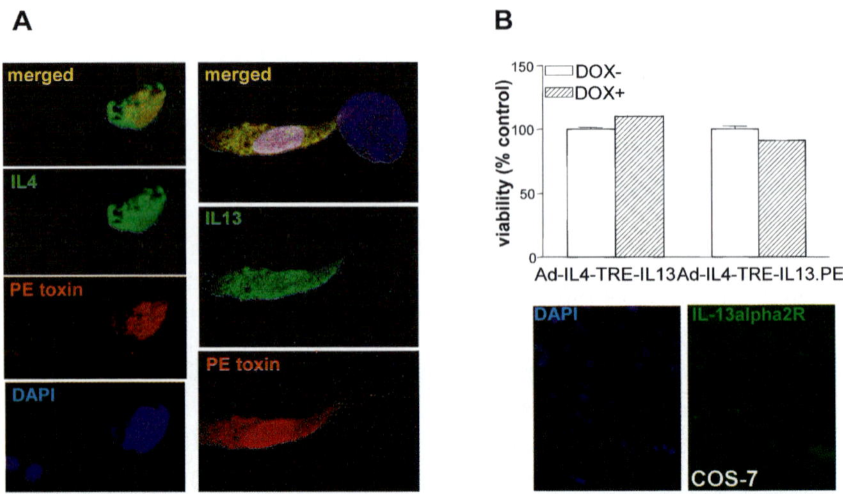

Fig. 5. mlL-13-PE toxin expression from the therapeutic (Ad-IL4-TRE-IL13.PE) adenoviral vector in COS-7 cells. **A.** COS-7 cells were infected with the therapeutic adenoviral vector (Ad-IL4-TRE-IL13. PE) that express mulL4 and mlL13 or the cytotoxin mulL-13-PE under the control of the bidirectional TRE promoter, which is activated by the transactivator (TetON) in the presence of the inducer Dox. Microphotographs show the regulated expression of the mlL-13-PE toxin, as determined by immunocytochemistry in the presence (Dox+) or absence (Dox–) of the inducer. **B.** COS-7 cells (were infected with control (Ad-IL4-TRE-IL13) and therapeutic (Ad-IL4-TRE-IL13.PE) adenoviral vectors. *Upper panels:* Cell viability was determined by MTS assay. Lower panels: IL-13alpha2R expression was determined by immunocytochemistry. Nuclei were stained with DAPI. Note that the therapeutic Ad-IL4-TRE-IL13.PE exert its cytotoxic effect only in cells expressing the glioma-restricted IL-I 3alpha2R.

liters of reaction solution containing MTS (final concentration 333 μg/ml) and an electron coupling reagent (phenazine ethosulfate, final concentration 25 μM) were added to each well containing 100 μl of culture medium. After 3 h at 37°C, the OD was read in a microplate spectrophotometer at a wavelength of 495 nm. The quantity of formazan product is directly proportional to the number of living cells in culture. Ad-IL4-TRE-IL13.PE was designed to produce PE toxin conjugated to a variant of IL-13 that binds specifically with the IL-13α2R receptor found almost exclusively on glioma cells. Cell viability was reduced 70% when the U251 and IN859 human glioma cells expressing the IL-13α2R (see **Fig. 6**) were incubated in the presence of Ad-IL4-TRE-IL13.PE in the "ON" state (Dox+) but remained unaffected in the "OFF" state, indicating that the expression of the chimeric toxin can be tightly regulated. Ad-IL4-TRE-IL13-PE did not affect the viability of COS-7 cells (African Green Monkey SV40-transfected kidney fibroblast cell

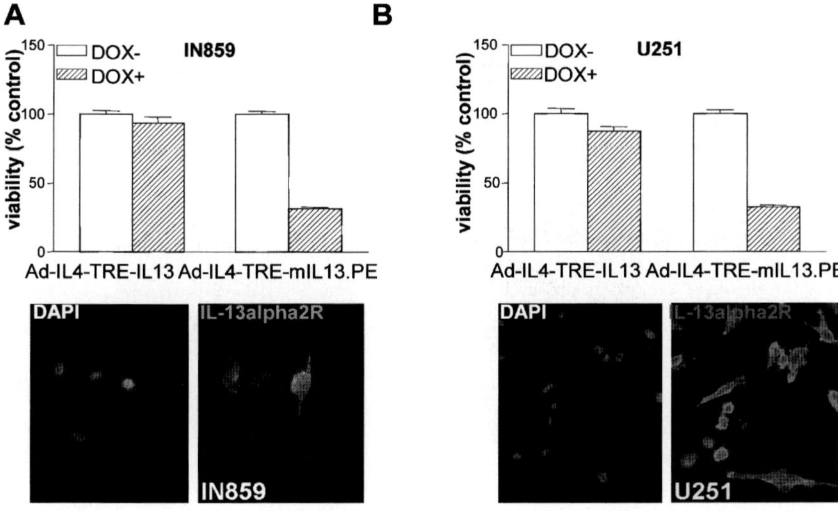

Fig. 6. mIL-13-PE toxin induces cell death in glioma cells. **A.** IN859 and **B.** U251 human glioma cells were infected with the therapeutic adenoviral vector (Ad-IL4-TRE-IL13.PE) or control vector Ad-IL4- TRE-IL13. *Upper panels:* Cell viability was determined by MTS assay. Lower panels: IL-13alpha2R expression was determined by immunocytochemistry. Nuclei were stained with DAPI. Note that the therapeutic Ad-IL4-TRE-IL13.PE exert its cytotoxic effect only in cells expressing the glioma-restricted IL- 13alpha2R.

line), which do not express the IL13alpha2 receptor (*see* **Fig. 5**). These results suggest that the cytotoxic effect of the chimeric is specific to glioma cells. The control vector Ad-IL4-TRE-IL13 did not affect the viability of human glioma cells or COS-7, neither in the presence nor in the absence of Dox.

3.6. Confirming Inducible Gene Expression in Vivo

Here we describe the detection of transgene expression *in vivo* using immunohistochemistry after stereotactic injection of inducible adenoviral vectors into the brain striatum. However, these methods can be applied to immunohistochemical analysis of any tissue of interest.

1. Male Fisher 344 rats of 200–250 g of body weight (Harlem Sprague Dawley Inc.) were used for in vivo HC-Ad-mediated gene delivery. Rats were given drinking water containing 2.0 mg/ml Dox (Sigma) and 1% sucrose or 1% sucrose alone 24 h before brain surgery and HC-Ad delivery.
2. On the day of surgery, the animal was anesthetized with ketamine and medatomadine, the head area was shaved, prepared with betadyne, and placed in

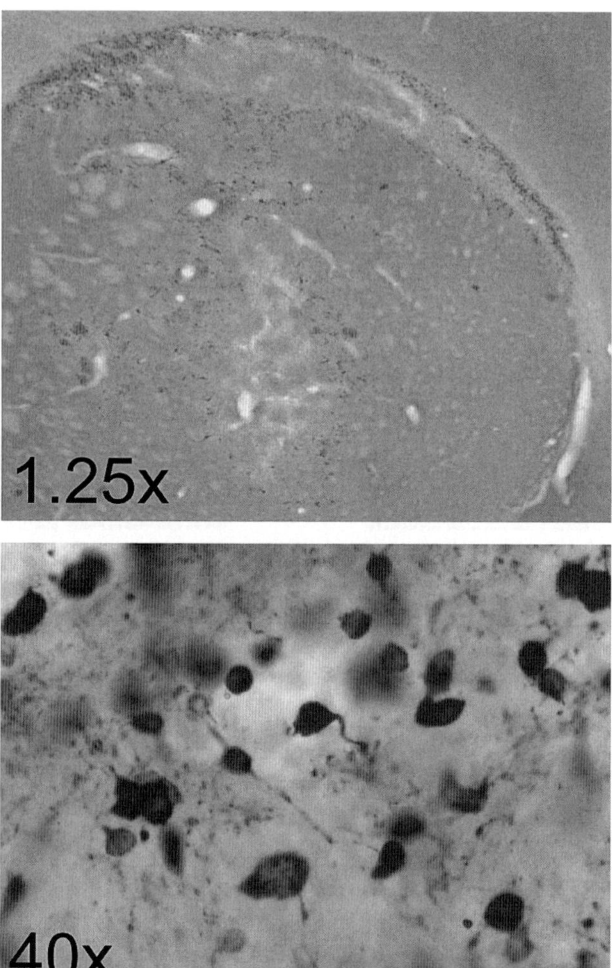

Fig. 7. Inducible Flt3L expression *in vivo*. Male Lewis rats fed Dox supplemented chow 24 h before injection of lx10^9 vp HC-Ad-mCMV-TetON-Flt3L into the brain striatum. After 7 days, rats were euthanized, fixed with PFA and coronal sections of the brain were taken using a vibratome (Leica). Brain sections were stained with rabbit anti-Flt3L as outlined in the materials and methods section and images were visualized on a Zeiss Axioplan 2 microscope under 1.25× n d 40× objectives. Expression of Flt3L was not observed in rats fed normal chow without Dox (not shown).

a stereotactic frame ready for surgery. A total of 1 × 10^9 vp of HC-Ad-mTetON-Flt3L, was injected in the rat striatum [coordinates from bregma (the contact between sagittal and coronal sutures) were the following: anterior, +1.0 mm; lateral, +3.0 mm; ventral, +4.0 mm] using a 10-μl Hamilton syringe (47). A total volume of 3 μl of HC-Ad vector diluted in 0.9% w/v saline per animal was injected in the striatum over a 5-min period. Subsequent to vector injection, the needle was left in place for a further 2 min before careful needle retraction. As a negative control, one group of rats received 3-μl saline injections.

3. After postoperative surgery, both untreated and treated animals received drinking water containing Dox and 1% sucrose, 1% sucrose alone, or water alone during the required times, and drinking water was changed daily.
4. After 6 days, animals were then sacrificed and brains were perfused with approximately 200 ml oxygenated Tyrode's solution by means of *trans*-cardial perfusion. Rats were next perfused with 250 ml of 4% paraformaldehyde, pH 7.4 (4% PFA), fixative, and brains were stored postfixed in 4% PFA for 3 days at 4°C followed by 3 days of washing in PBS containing 0.1% sodium azide.
5. Brains were serial sectioned using a vibratome to obtain 50-μm free-floating sections. Sections were stored in PBS containing 0.1% sodium azide at 4°C until ready for use.
6. Free-floating brain sections were washed with TBS + 0.5% Triton X-100 followed by 0.3% H_2O_2 incubation to inactivate endogenous peroxidase. Non-specific antibody sites and Fc receptors were blocked with 10% normal horse serum for 1 h.
7. Sections were incubated for 48 h at room temperature with rabbit polyclonal anti-Flt3L primary antibody (1:1000) (generated in our laboratory ***(15)***) diluted in TBS + 0.5% Triton + 1% horse serum + 0.1% sodium azide. Sections were washed three times with TBS plus 0.5% Triton and then incubated with goat anti-rabbit biotinylated secondary antibody (1:800) (Dako) for 4 h. The Avidin–biotinylated HRP complex was prepared and used for detection using a Vectastain ABC Elite kit (Vector Laboratories).
8. Following staining with diaminobenzidone (DAB) and glucose oxidase, sections were mounted on gelatin-coated glass slides, dehydrated through graded ethanol solutions, and carefully covered for microscopy (*see* **Fig. 7**).

4. Notes

1. TRE promoters can be either unidirectional or bidirectional. We have used both successfully to produce Ad vectors. Bidirectional promoters allow the simultaneous expression of two or more genes. In our experience, expression of two genes from a bidirectional promoter is more robust than expression of two genes, separated by an IRES, from a unidirectional promoter. We have also used bidirectional TRE promoter to simultaneously express four genes upon addition of DOX by using an IRES to separate two open reading frames (ORF) on either side of the bidirectional TRE promoter. However, if expression of only one transgene is required, then a unidirectional promoter is sufficient.
2. The transactivator used, $rtTA_2{}^{S}M2$, is a chimeric protein consisting of the DNA-binding domain of the Tet repressor rtTA fused with the transactivation domain

of VP-16. Point mutations changing the amino acid sequence of $rtTA_2{}^SM2$ have improved the induction of gene expression in the presence of DOX *(8)*.

3. We have determined that optimal transgene expression can be achieved by generating an HC-Ad genome where mCMV-TetON elements are in the opposite direction as Tre-transgene elements, and where mCMV and TRE are juxtaposed (*see* **Fig. 4**). The tTSkid ensures that basal activity of the TRE promoter remains low in the "OFF" state, whereas the close association of mCMV transcriptional enhancer elements in the mCMV promoter increases the activity of TRE promoter in the "ON" state.
4. Different RE require different buffer conditions for optimal activity. Product literature accompanying the RE will explain what conditions are required for optimal activity for each enzyme. Usually, two different RE can be used at the same time by selecting a buffer with optimal activity for both. However, sometimes, the RE buffers are not compatible and lead to suboptimal RE enzymatic activity. In this situation, sequential digestion of DNA with the RE may be performed, as long as DNA is precipitated from the first RE buffer before starting the second digestion.
5. Wear protective eye gear (mask or goggles) when cutting DNA bands out of the agarose gel. It is also important to note that UV light can cause point mutations, for example, by oxidizing Guanine to 8-Oxyguanine that binds with Adenine, not Cytosine, resulting in a GC-AT transition. Thus, it is important to limit the time that DNA is exposed to UV light. We use a low power UV box and limit DNA exposure to the UV light to less than 2 min.
6. Bubbles between the gel and the membrane can be removed by first overlaying one layer of filter paper on the membrane. Firmly but carefully a 4-ml tube or a 25-ml pipette can be rolled over the filter paper from one edge to the opposite edge 2–3 times. When doing this, it is important to take care not to break the gel or slide the membrane off the gel.
7. Cloning of HC-Ad vectors requires the manipulation of large plasmids, and recombination is often an issue. It is essential to use a recombinase deficient *Escherichia coli* bacterium, we use DH5α. Furthermore, we find that culture of bacteria transformed with pSTK120-transgene plasmids is more successful when bacteria are grown at 32°C and 160 g. After cloning, plasmids should be checked rigorously using REs to confirm that recombination events have not occurred. This can identify anomalies that otherwise would not be detected until the HC-Ad vectors are produced. In addition, although it has not been specifically mentioned here, DNA sequencing to confirm that the gene is present is advisable.
8. Considering that during viral replication transgenes are expressed in vector-producing cells, production of vectors encoding toxic transgenes is difficult and requires special strategies. We have produced adenoviral vectors encoding IL-13-PE toxin in 293 cells that are resistant to the toxin. During viral replication, IL-13-PE is expressed in 293 cells and can inhibit protein synthesis by ADP-ribosylation of EF-2, leading to low vector titers. Thus, we used 293 cells stably transfected with pHED-7 ***(21)***, which encodes the gene for ADP ribosylation-

resistant elongating factor-2 (EF-2) from CHO cells ***(22)***. In these cells, ADP-ribosylating toxins are incapable of EF-2 inactivation; thus, they can support the growth of adenoviruses expressing *Pseudomonas* or *Diphtheria* chimeric toxins, yielding high viral titers.

9. Another strategy to block the toxicity of these vectors during adenovirus production is to block the intracellular processing of the chimeric toxin. To inhibit protein synthesis, IL-13-PE toxin requires furin-mediated proteolytic cleavage that activates the exotoxin releasing the catalytic Domain III. Thus, adenoviral vectors encoding *Pseudomonas* or *Diphtheria* chimeric toxins can be produced in wild-type 293 cells in the presence of furin inhibitor Decanoyl-RVKR-CMK (5 μM, Calbiochem).
10. Transfection of pSTK120-transgene plasmids into 293Flpe cells can be achieved with high efficiency using the calcium phosphate precipitation method. It is necessary to make new transfection reagents and to avoid repeated freezing and then thawing of the reagents before use. Shorter times where DNA is added to 2× HBS (e.g., 10 s) work much better than longer times (e.g., 5 min). Polypropylene tubes improve transfection efficiency in our hands compared with polystyrene tubes. We have also tested various commercially available transfection reagents, we find that TransIT (Mirus) and HyFect (Denville) achieve similar high levels of transfection and high-quality viral preparations. Although useful for other applications, we did not have any success using either JetPEI (PolyPlus) or GeneJuice (Novagen) for transfecting HC-Ad vectors. Plasmid DNA must be of high quality to rescue HC-Ad vectors. The concentration of DNA must be at least 1 μg/ml and the OD_{260} : OD_{280} ratio must not be above 1.9.
11. Usually, only one viral band is evident during CsCl purification of HC-Ad viral vectors. This is desirable. Occasionally, we have observed a second viral band present; these preparations are usually of a lower quality when tested in vitro and *in vivo*.
12. Real-time qPCR can also be used to quantify the number of viral genomes present. We have found that the information obtained by qPCR is very useful in characterizing HC-Ad preparations. In particular, it allows a sensitive, quantitative measure of the presence of helper virus contamination in the sample. This protocol has already been described in detail (*see* Puntel et al., ***(23)***).
13. In our experience, a good-quality HC-Ad vector preparation should meet the following in vitro criteria:

    a. There should be less than 30-fold greater viral particles (measured with OD_{260nm}) than viral genomes as determined using qPCR.
    b. There should be less than 10 times as many viral genomes (qPCR) than transgene expressing units (TEU, determined by ICC on Flow cytometry).
    c. Contaminating helper viral particles (measured with plaque forming units assay) should be $<10^5$ pfu/ml.

14. Transgene expression in the presence of DOX can be confirmed using a myriad of techniques, the best of which is completely dependent on the transgene that

will be detected. We have outlined three such techniques in vitro, ELISA, ICC, and Western blotting that we most commonly use to titrate HC-Ad vectors in vitro. Several other techniques including reverse transcriptase PCR and flow cytometry may also be used and may be preferable depending on the availability of reagents to detect the transgene of interest.

15. Expression of many transgenes can be confirmed using immunoglobulins purchased from specific companies. However, it may be more desirable to "tag" the transgene (FLAG or Myc tags work well) before cloning it into pmCMV-TetON. Specific, high-affinity immunoglobulins have been developed against these tags improving the sensitivity of the assay for detecting the transgene.
16. For secretable proteins, e.g., human soluble Flt3L, much of the protein produced will be secreted. Thus, cell culture media from transfected cells can be used to determine expression of transgene in vitro.
17. Of the techniques we described to determine whether viral preps induce transgene expression in vitro (i.e., ELISA, ICC, and Western blot), only ICC can give an accurate assessment of the total number of transgene expressing particles in a viral preparation. We have recently attempted to overcome this problem by developing a method of titrating gene expression from Ad vectors using flow cytometry. We have found that this can give a very accurate and comparable assessment of the number of transgene expressing particles in a viral prep compared with ICC.

Acknowledgments

Work described in this paper was funded by the National Institute of Neurological Disorders and Strokes (NINDS), National Institutes of Health (NIH), grant number: 1U01 NS05246501 1RO1 NS44556.01; 1R21 NS054143.01; 1R21 NS05414301A2 1RO3 TW006273-01 to M.G.C.; NINDS, NIH grant number: 1 RO1 NS42893, 1R01 NS05419301A1, U54 4 NS04-5309, and R21 NS47298 to P.R.L.; F32 NS058156–01 to MC; by The Linda Tallen and David Paul Kane Foundation Annual Fellowship to M.G.C. and P.R.L.; the Bram and Elaine Goldsmith and the Medallions Group Endowed Chairs in Gene Therapeutics to PRL and MGC, respectively, and The Board of Governors at Cedars-Sinai Medical Center.

References

1. Wurm, F. M., Gwinn, K. A., and Kingston, R. E. (1986) Inducible overproduction of the mouse c-myc protein in mammalian cells, *Proc Natl Acad Sci USA* **83,** 5414–8.
2. Rivera, V. M., Clackson, T., Natesan, S., Pollock, R., Amara, J. F., Keenan, T., Magari, S. R., Phillips, T., Courage, N. L., Cerasoli, F., Jr., Holt, D. A., and Gilman, M. (1996) A humanized system for pharmacologic control of gene expression, *Nat Med* **2,** 1028–32.

3. Molin, M., Shoshan, M. C., Ohman-Forslund, K., Linder, S., and Akusjarvi, G. (1998) Two novel adenovirus vector systems permitting regulated protein expression in gene transfer experiments, *J Virol* **72,** 8358–61.
4. Goverdhana, S., Puntel, M., Xiong, W., Zirger, J. M., Barcia, C., Curtin, J. F., Soffer, E. B., Mondkar, S., King, G. D., Hu, J., Sciascia, S. A., Candolfi, M., Greengold, D. S., Lowenstein, P. R., and Castro, M. G. (2005) Regulatable gene expression systems for gene therapy applications: progress and future challenges, *Mol Ther* **12,** 189–211.
5. Xiong, W., Goverdhana, S., Sciascia, S. A., Candolfi, M., Zirger, J. M., Barcia, C., Curtin, J. F., King, G. D., Jaita, G., Liu, C., Kroeger, K., Agadjanian, H., Medina-Kauwe, L., Palmer, D., Ng, P., Lowenstein, P. R., and Castro, M. G. (2006) Regulatable gutless adenovirus vectors sustain inducible transgene expression in the brain in the presence of an immune response against adenoviruses, *J Virol* **80,** 27–37.
6. Knott, A., Garke, K., Urlinger, S., Guthmann, J., Muller, Y., Thellmann, M., and Hillen, W. (2002) Tetracycline-dependent gene regulation: combinations of transregulators yield a variety of expression windows, *Biotechniques* **32,** 796, 98, 800 passim.
7. Urlinger, S., Baron, U., Thellmann, M., Hasan, M. T., Bujard, H., and Hillen, W. (2000) Exploring the sequence space for tetracycline-dependent transcriptional activators: novel mutations yield expanded range and sensitivity, *Proc Natl Acad Sci U S A* **97,** 7963–8.
8. Freundlieb, S., Schirra-Muller, C., and Bujard, H. (1999) A tetracycline controlled activation/repression system with increased potential for gene transfer into mammalian cells, *J Gene Med* **1,** 4–12.
9. Lamartina, S., Silvi, L., Roscilli, G., Casimiro, D., Simon, A. J., Davies, M. E., Shiver, J. W., Rinaudo, C. D., Zampaglione, I., Fattori, E., Colloca, S., Gonzalez Paz, O., Laufer, R., Bujard, H., Cortese, R., Ciliberto, G., and Toniatti, C. (2003) Construction of an rtTA2(s)-m2/tts(kid)-based transcription regulatory switch that displays no basal activity, good inducibility, and high responsiveness to doxycycline in mice and non-human primates, *Mol Ther* **7,** 271–80.
10. Candolfi, M., Curtin, J. F., Xiong, W. D., Kroeger, K. M., Liu, C., Rentsendorj, A., Agadjanian, H., Medina-Kauwe, L., Palmer, D., Ng, P., Lowenstein, P. R., and Castro, M. G. (2006) Effective high-capacity gutless adenoviral vectors mediate transgene expression in human glioma cells, *Mol Ther* **14,** 371–81.
11. Candolfi, M., Pluhar, G. E., Kroeger, K., and Puntel, M. C. J., Barcia, C., Muhammad, A.G., Xiong, W., Liu, C., Mondkar, S., Kuoy, W., Kang, T., McNeil, EA., Freese, A. B., Ohlfest J. R., Moore, P., Palmer, D. Ng. P., Young, J.D., Lowenstein, P.R., Castro, M.G. (2006) Optimization of adenoviral vector mediated expression in the canine brain in vivo, and in canine glioma cells in vitro., *Neuro-oncology* **Under Review**.
12. Joseph Sambrook, and Russell, D. W. (2001) *Molecular Cloning: A Laboratory Manual*, Cold Spring Harbor Laboratory Press, Woodbury, NY, USA.
13. Umana, P., Gerdes, C. A., Stone, D., Davis, J. R., Ward, D., Castro, M. G., and Lowenstein, P. R. (2001) Efficient FLPe recombinase enables scalable production of helper-dependent adenoviral vectors with negligible helper-virus contamination, *Nat Biotechnol* **19,** 582–5.

14. Maleniak, T. C., Darling, J. L., Lowenstein, P. R., and Castro, M. G. (2001) Adenovirus-mediated expression of HSV1-TK or Fas ligand induces cell death in primary human glioma-derived cell cultures that are resistant to the chemotherapeutic agent CCNU, *Cancer Gene Ther* **8,** 589–98.
15. Ali, S., Curtin, J. F., Zirger, J. M., Xiong, W., King, G. D., Barcia, C., Liu, C., Puntel, M., Goverdhana, S., Lowenstein, P. R., and Castro, M. G. (2004) Inflammatory and anti-glioma effects of an adenovirus expressing human soluble Fms-like tyrosine kinase 3 ligand (hsFlt3L): treatment with hsFlt3L inhibits intracranial glioma progression, *Mol Ther* **10,** 1071–84.
16. Williams, J. C., Stone, D., Smith-Arica, J. R., Morris, I. D., Lowenstein, P. R., and Castro, M. G. (2001) Regulated, adenovirus-mediated delivery of tyrosine hydroxylase suppresses growth of estrogen-induced pituitary prolactinomas, *Mol Ther* **4,** 593–602.
17. Smith-Arica, J. R., Morelli, A. E., Larregina, A. T., Smith, J., Lowenstein, P. R., and Castro, M. G. (2000) Cell-type-specific and regulatable transgenesis in the adult brain: adenovirus-encoded combined transcriptional targeting and inducible transgene expression, *Mol Ther* **2,** 579–87.
18. Southgate, T. D., Kingston, P. A., and Castro, M. G. (2000) Gene Transfer into Neural Cells In Vitro using Adenoviral Vectors, John Wiley & Sons, Inc., New York.
19. Palmer, D., and Ng, P. (2003) Improved system for helper-dependent adenoviral vector production, *Mol Ther* **8,** 846–52.
20. Ng, P., Parks, R. J., Cummings, D. T., Evelegh, C. M., and Graham, F. L. (2000) An enhanced system for construction of adenoviral vectors by the two-plasmid rescue method, *Hum Gene Ther* **11,** 693–9.
21. Li, Y., McCadden, J., Ferrer, F., Kruszewski, M., Carducci, M., Simons, J., and Rodriguez, R. (2002) Prostate-specific expression of the diphtheria toxin A chain (DT-A): studies of inducibility and specificity of expression of prostate-specific antigen promoter-driven DT-A adenoviral-mediated gene transfer, *Cancer Res* **62,** 2576–82.
22. Kohno, K., and Uchida, T. (1987) Highly frequent single amino acid substitution in mammalian elongation factor 2 (EF-2) results in expression of resistance to EF-2-ADP-ribosylating toxins, *J Biol Chem* **262,** 12298–305.
23. Puntel, M., Curtin, J. F., Zirger, J. M., Muhammad, A. K., Xiong, W., Liu, C., Hu, J., Kroeger, K. M., Czer, P., Sciascia, S., Mondkar, S., Lowenstein, P. R., and Castro, M. G. (2006) Quantification of high-capacity helper-dependent adenoviral vector genomes in vitro and in vivo, using quantitative TaqMan real-time polymerase chain reaction, *Hum Gene Ther* **17,** 531–44.
24. Gerdes, C. A., Castro, M. G., and Lowenstein, P. R. (2000) Strong promoters are the key to highly efficient, noninflammatory and noncytotoxic adenoviral-mediated transgene delivery into the brain in vivo, *Mol Ther* **2,** 330–8.
25. Schiedner, G., Morral, N., Parks, R. J., Wu, Y., Koopmans, S. C., Langston, C., Graham, F. L., Beaudet, A. L., and Kochanek, S. (1998) Genomic DNA transfer with a high-capacity adenovirus vector results in improved in vivo gene expression and decreased toxicity, *Nat Genet* **18,** 180–3.

16

Liver-Directed Gene Therapy Using the *Sleeping Beauty* Transposon System

Lalitha R. Belur, R. Scott McIvor, and Andrew Wilber

Summary

Sleeping Beauty (SB) is a transposon system genetically reconstructed from teleost fish that mediates chromosomal integration of DNA sequences by a cut-and-paste mechanism. SB has been shown to mediate transposition in a variety of cells and tissues, has been used for the generation of transgenic animals and has been tested as a vector for gene therapy in several animal models of human disease. Here, we describe methods that we have developed for testing SB-mediated transposition, first in cultured mammalian cells, and then in vivo using a combination of rapid, high-volume tail vein injection for DNA delivery to the liver along with in vivo bioluminescence imaging to monitor sustained luciferase gene expression in individual animals.

Key Words: Gene therapy; non-viral; *Sleeping Beauty*; transposon; integration; liver; hydrodynamic delivery.

1. Introduction

DNA-mediated gene transfer presents an alternative to viral vectors for gene therapy. The primary limitation of DNA-mediated gene transfer, however, is the relatively short duration of gene expression. This is especially limiting in diseases such as genetic deficiencies, where extended gene expression of the therapeutic gene is required in order to achieve therapeutic benefit. This limitation can be circumvented, if persistence of gene expression can be achieved by transgene integration into host chromosomal DNA. Transposons are mobile genetic elements that integrate into host cell genomes and have been

From: *Methods in Molecular Biology, vol. 434: Volume 2: Design and Characterization of Gene Transfer Vectors*
Edited by: J. M. Le Doux © Humana Press, Totowa, NJ

used extensively as tools for insertional mutagenesis. The *Sleeping Beauty* (SB) transposon system, which was reconstructed from teleost fish sequences in the laboratory of Dr. Perry Hackett, consists of two components: (1) a transposon, which contains a sequence of interest flanked by transposon inverted repeats (IRs) and (2) SB transposase ***(1)***. During transposition, SB transposase recognizes the ends of the IRs, excises the transposon carrying the sequence of interest and then inserts the transposon into a host cell chromosome. For chromosomal integration, SB recognizes TA dinucleotides, which are widespread throughout the genome ***(1–4)***. Chromosomal abnormalities surrounding the target insertion site have not been reported, other than duplication of the target TA dinucleotide.

By introducing a therapeutic gene between the transposon IRs and supplying the transposase function either on the same plasmid or on a separate plasmid, SB can be adapted as a vector system for gene therapy. SB-mediated transposition has been shown to occur in a variety of cultured cell types, in zebrafish ***(5–7)*** and mouse embryos ***(8)***, in murine germ cells ***(9–12)***, in human primary blood lymphocytes ***(13)*** and in mouse somatic tissues, including the lung and the liver ***(14–21)***. SB thus provides a way to combine the advantages of non-viral plasmid-based gene transfer with an integrative capability, thus leading to long-term expression of the transgene through transposon integration.

There has been considerable interest in the liver as a target organ for gene therapy, not only because of its accessibility to the bloodstream for gene delivery and expression of gene products into the circulation, but also because of the central role of the liver in animal physiology and metabolism. Genes may be delivered to the liver by systemic administration or by infusion into the portal vein, hepatic artery or the bile duct. Liu et al. and Zhang et al. developed an effective method for non-viral gene delivery to mouse liver by systemic injection of unconjugated plasmid DNA ***(22,23)***. In this hydrodynamics-based technique, mice are injected rapidly (less than 10 s) through the tail vein with plasmid DNA suspended in a large volume (10% v/w) of injection fluid. This technique is based on the principle that the physical barriers protecting cells and tissues from unwanted invasion can be overcome by increasing luminal pressure to promote extravasation. This rapid, high-volume injection causes fluid to accumulate in the vena cava, a load that is too large for the heart to process, resulting in backup of fluid predominantly in the liver. Increased intra-hepatic pressure results in the release of fluid through fenestrated endothelium, thus leading to delivery and subsequent expression of DNA contained in the fluid when taken up by cells in hepatic tissue ***(24,25)***. Reports of transfection efficiency vary, with gene expression reported in 10% up to 40% of hepatocytes.

We have used this simple and effective hydrodynamics-based approach extensively for delivery and testing of novel transposon and transposase plasmid constructs in mouse liver. In this chapter, we first describe techniques to test for

SB-mediated transposition in cultured mammalian cells (*see* **Note 1**). We then describe the methods that we have assembled for non-viral DNA delivery to the liver using the SB transposon system in combination with in vivo bioluminescence imaging to monitor sustained expression of newly introduced genetic material (*see* **Note 2**).

2. Materials

2.1. Transfection of HT1080 Cells

1. Fugene transfection reagent is available from Roche Biochemicals (Indianapolis, IN, USA).
2. The transposon plasmid pT/SVNeo *(1)* contains the neomycin phosphotransferase gene (NEO) under transcriptional regulation of the simian virus 40 early promoter (*see* **Note 3**).
3. pCMVSB11 *(26)* encodes the SB transposase gene regulated by the cytomegalovirus (CMV) early promoter.
4. Geneticin (G418 sulfate) is available from Invitrogen Corporation (Carlsbad, CA, USA).
5. Tissue culture solutions: Dulbecco's modified Eagle medium (DMEM) is available from Invitrogen.
6. Dulbecco's phosphate-buffered saline (DPBS) is available from Invitrogen.
7. Penicillin–streptomycin–fungizone (PSF) solution is available from Invitrogen.
8. 0.05% trypsin with ethylenediaminetetraacetic acid (EDTA) is available from Invitrogen.
9. Crystal violet is available from Sigma Chemicals (St. Louis, MO, USA).
10. Tissue culture medium: DMEM supplemented with 10% fetal bovine serum (FBS) and 1% PSF.
11. Human HT-1080 fibrosarcoma cells are available from the American Type Culture Collection (Manassas, VA, USA) and are maintained in DMEM supplemented with FBS and PSF at 37°C in an atmosphere of 5% CO_2/air.

2.2. Rapid, High-Volume Delivery of Plasmid DNA

1. Ketamine HCl is available from Phoenix Pharmaceutical Incorporated (St. Joseph, MO, USA).
2. Acepromazine maleate is available from Phoenix Pharmaceutical Incorporated.
3. Butorphanol tartrate is obtained from Fort Dodge Animal Health (Fort Dodge, IA, USA).
4. Anesthetization cocktail: 8 mg/ml ketamine HCl, 0.1 mg/ml acepromazine maleate and 0.01 mg/ml butorphanol tartate. This drug cocktail is premixed and is stable for 4 weeks at room temperature. Protect drug from light.
5. The transposon pT2/CaL *(18)* encodes a firefly luciferase transgene under transcriptional regulation of the Cags promoter.

6. Plasmid pUbSB11 *(13)* encodes the SB transposase gene regulated by the ubiquitin C promoter.
7. Mice. We routinely use 8- to 10-week-old C57BL/6 mice, available from NCI (Frederick, MD, USA), although we have successfully used other strains (Balb/c, FVB/N, C57BL/6 NOD/SCID).

2.3. In Vivo luciferase Imaging

1. D-Luciferin firefly luciferase substrate, potassium salt (for in vivo imaging) is available from Caliper Life Sciences (formerly Xenogen Corporation, Hopkington, MA, USA). Store at –20°C.
2. Luciferin stock solution: Prepare a stock solution of luciferin (28.5 mg/ml) by dissolving 1 g of substrate in 35 ml DPBS. Filter sterilize solution through a 0.45-µm sterile filter. Store reconstituted substrate in 1 ml aliquots at –20°C. Protect substrate from light (*see* **Note 4**).

3. Methods

3.1. Transfection of HT1080 Cells

1. Plate cells in 6-cm plates at approximately 4–5 × 10^5 cells per well. Incubate the cultures overnight.
2. Approximately 16–24 h after cell plating, cells are co-transfected with plasmids pT/Neo and pCMV/SB10 (500 ng each) using Fugene 6.
3. To a clean Eppendorf tube add 100 µl of DMEM.
4. Add Fugene 6 reagent (4 µl per µg of total plasmid) directly into DMEM. Mix and incubate solution at room temperature for 5–15 min.
5. Add plasmid solution directly into DMEM/Fugene 6 solution. Mix and incubate at room temperature for an additional 5–15 min.
6. Add entire mixture of DMEM containing plasmid/Fugene 6 complex dropwise to cells. Mix by swirling plates before returning to incubator.
7. Two days post-transfection, viable cells (3 × 10^4) are plated into 10-cm dishes containing complete growth medium supplemented with G418 (800 µg/ml active drug). Change culture medium every 3–4 days, with complete growth medium supplemented with G418.
8. After 14 days of growth in selective medium, stain and fix drug-resistant colonies using a 95% ethanol solution containing 1% crystal violet (*see* **Note 5**).
9. Stained colonies are counted to determine the frequency of drug-resistant colony formation in the presence versus the absence of SB transposase.

3.2. Rapid, High-Volume Delivery of Plasmid DNA

1. Anesthetize mouse by intraperitoneal injection of 50 µl of anesthetization cocktail.
2. Dilute 5 µg of pT2/CaL (transposon plasmid) and 2.5 µg of pUbSB11 (transposase plasmid) into Ringer's solution at a volume of 1/10 the animal weight (ml/g) (*see* **Note 6**).

3. Dilate tail vein of mouse by immersing tail in a 55°C water bath for 2 min (*see* **Note 7**).
4. Inject mouse through the tail vein using a butterfly catheter consisting of a 27-G butterfly needle fitted to extension tubing and an adapter fitted onto a 3-ml syringe (*see* **Note 8**). Elapsed time from initiation to completion of injection is 4–8 s.
5. Monitor mouse respiration and return mouse to cage placed partially on a heating pad adjusted to medium heat for 2–3 h until recovery from anesthesia (*see* **Note 9**).

3.3. In Vivo Luciferase Imaging

1. Anesthetize mouse by intraperitoneal injection with a sedative dose (400 μl for a 25-g mouse) of the anesthetization cocktail.
2. Shave the underside (belly) of C57BL/6 mice with electric razor (*see* **Note 10**).
3. Inject 100 μl of stock luciferin substrate intraperitoneally 5 min before imaging (*see* **Note 11**).
4. Place the live, anesthetized mouse onto black construction paper in the center of the imaging chamber. Obtain exposure using an intensified, charge-coupled device (CCD) camera (Series 100, Caliper Life Sciences/Xenogen Corp., Alameda, CA, USA) for 1 s to 2 min. Raw values are recorded as photons of light emitted/s/cm^2(*see* **Note 12**).
5. Return mouse to cage placed partially on a heating pad set at medium heat for recovery (*see* **Note 9**).

4. Notes

1. *Transposon-mediated gene transfer into cultured cells.* A standard transposition assay has been devised to assess SB-mediated gene transfer in cultured somatic cells (*see* **Fig. 1**) *(1)*. In this assay, a transposon vector containing an SV40-regulated neomycin phosphotransferase (neo) gene is transfected with or without an expression vector encoding the CMV-regulated SB transposase into cultured cells (*see* **Fig. 1A**). Two days after transfection, the cells are subcultured into medium containing the neomycin analog G418 to determine the frequency of drug-resistant colony formation. The extent to which an increased frequency of drug-resistant colony formation is observed in cultures co-transfected with transposase-encoding plasmid relative to controls receiving an equal amount of control plasmid (pBlueScript) provides an assessment of transposition efficiency (*see* **Fig. 1B**). When expressed in this assay, the SB transposase mediates precise integration of a single copy of transposon sequence into chromosomes of cultured cells from fish, mice, and humans *(1)*. These data provide evidence for the effectiveness of the SB transposon system in mediating stable gene transfer in cultured vertebrate cells.
2. *Transposon-mediated gene transfer in the liver of adult mice.* The effect of SB-mediated transposition can be evaluated in vivo by comparing long-term expression levels after transposon delivery in the presence versus absence of

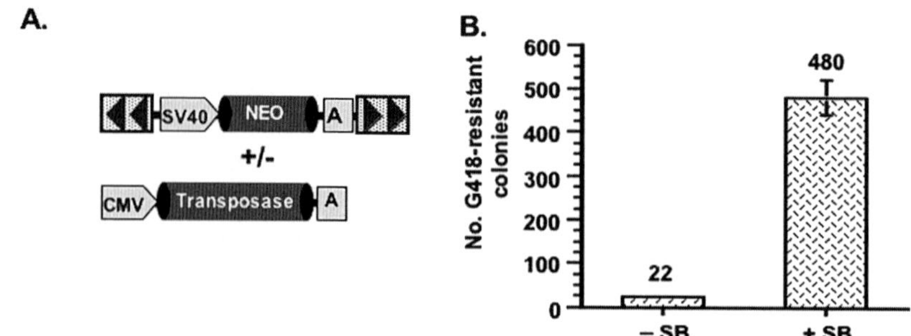

Fig. 1. *Sleeping Beauty*-mediated gene transfer in cultured cells. **(A)** Schematic diagram of transposon- and transposase-encoding plasmids used for colony-forming transposition assay. Boxes with triangles: Transposon inverted repeats. A: SV40 polyadenylation signal. Other sequences are described in the text. **(B)** Frequency of G418-resistant colony formation for HT1080 cells transfected with NEO-encoding transposon in the absence and presence of transposase. Results from three independent transfections showing average colony numbers designated above each bar +/– SE.

a source of transposase (*see* **Fig. 2**). We have used a non-invasive strategy to investigate stable expression of transgenes introduced by transposition. In this approach, adult mice are administered plasmid encoding the SB transposase (pUbSB11) along with a plasmid containing a firefly luciferase-encoding transposon (pT2CaL) (*see* **Fig. 2A**) using the rapid, high-volume injection technique described in **Subheading 3.2**. At various times after plasmid injection, animals are anesthetized and luciferase expression is assayed by in vivo bioluminescence imaging as described in **Subheading 3.3** *(28)*. After administration of luciferin substrate, luciferase enzyme activity is detected in the area of the liver as emitted light and visualized as a pseudocolor image superimposed over a black and white photograph of each animal (*see* **Fig. 2B**). Stable gene transfer resulting from transposition is evaluated by monitoring for persistent luciferase expression in individual animals over time. As shown in **Fig. 2C**, there was a 10-fold increase in the level of luciferase expression retained 15 weeks after co-injection of luciferase transposon along with transposase-encoding plasmid, relative to animals injected with luciferase transposon alone. These data provide evidence for the role of the transposase in mediating stable, long-term gene expression in experimental animals using the SB transposon system.

3. *Plasmid preparation.* Plasmids are purified by ion-exchange chromatography using the endofree plasmid maxi kit from Qiagen (Valencia, CA, USA) as per manufacturers instructions.
4. *Substrate stability.* Luciferin substrate is stable at room temperature for about 3 h in the dark. Unused substrate can be frozen and thawed 2–3 times without significant reduction in signal.

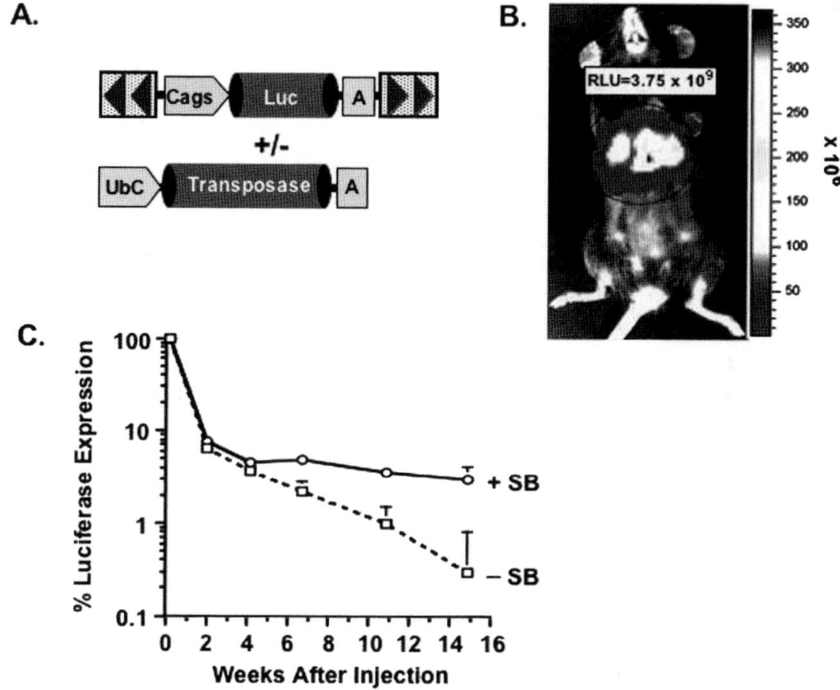

Fig. 2. Stable, *Sleeping Beauty*-mediated gene transfer and expression in mice. **(A)** Schematic diagram of transposon- and transposase-encoding plasmids used to achieve and monitor stable gene expression in living mice by in vivo bioluminescence imaging. **(B)** Image of mouse exhibiting luciferase-associated in vivo bioluminescence as photons emitted in the area of the liver. When converted to a grayscale image, the dark dot in the center of the luminescent image indicates the highest intensity of luciferase expression, followed by lesser intensity in the white area and the least amount of expression is depicted by the gray portion. **(C)** Mean % luciferase activity remaining at given time points relative to the level observed 1 day after infusion for each of the two groups of mice ($n = 5$; error bars: SE).

5. *Crystal violet staining*. Aspirate tissue culture medium from plate. Add 1 ml of crystal violet solution and swirl to coat plate. Fix and stain colonies for 1–2 min, then discard stain. Rinse plates once with 5–10 ml distilled water, then discard water. Invert plate and allow to dry. Count colonies.
6. *Mouse strain.* The amounts of transposon and transposase plasmids described here are applicable to the C57BL/6 strain. If using other strains of mice, a dose–response should be carried out in order to determine the optimal ratio of transposase to transposon plasmid that will facilitate long-term transgene expression. We *(18)* and others *(27)* have found that transposition is most effective when

expression of SB is regulated by the human ubiquitin or phosphoglycerate kinase (PGK) promoters.

7. *Tail vein dilation.* Ensure that the water bath is exactly 55ºC. Immersing the tail in water baths that are maintained at hotter temperatures or immersing the tail at 55ºC for too long will injure the tail.
8. *Injection timing.* Plasmids must be injected in 10 s or less, otherwise gene delivery to the liver will be very inefficient.
9. *Mouse recovery.* Mouse cage should be placed partially on a heating pad, so that the mice can move away from the heat if needed.
10. *Shaving mice.* Only mice with dark fur require shaving. Unshaved white mice can be imaged without a significant decrease in luminescence signal.
11. *Substrate injection.* Inject mice with luciferin substrate only after they are fully anesthetized, because the substrate sometimes counteracts the effects of the anesthetic.
12. *Imaging.* The limbs of the mouse can be taped to the construction paper if the animal starts to move during the imaging procedure.

References

1. Ivics, Z., Hackett, P. B., Plasterk, R. H., and Izsvak, Z. (1997) Molecular reconstruction of *Sleeping Beauty*, a Tc1-like transposon from fish, and its transposition in human cells. *Cell* **91**, 501–510.
2. Geurts, A. M., Hackett, C. S., Bell, J. B., Bergemann, T. L., Collier, L. S., Carlson, C. M., Largaespada, D. A., and Hackett, P. B. (2006) Structure-based prediction of insertion-site preferences of transposons into chromosomes. *Nucleic Acids Research* **34**, 2803–2811.
3. Carlson, C. M., Dupuy, A. J., Fritz, S., Roberg-Perez, K. J., Fletcher, C. F., and Largaespada, D. A. (2003) Transposon mutagenesis of the mouse germline. *Genetics* **165**, 243–256.
4. Yant, S. R., Wu, X., Huang, Y., Garrison, B., Burgess, S. M., and Kay, M. A. (2005) High-resolution genome-wide mapping of transposon integration in mammals *Molecular & Cellular Biology* **25**, 2085–2094.
5. Davidson, A. E., Balciunas, D., Mohn, D., Shaffer, J., Hermanson, S., Sivasubbu, S., Cliff, M. P., Hackett, P. B., and Ekker, S. C. (2003) Efficient gene delivery and gene expression in zebrafish using the *Sleeping Beauty* transposon. *Developmental Biology* **263**, 191–202.
6. Clark, K. J., Geurts, A. M., Bell, J. B., and Hackett, P. B. (2004) Transposon vectors for gene-trap insertional mutagenesis in vertebrates. *Genesis: the Journal of Genetics & Development* **39**, 225–233.
7. Wadman, S. A., Clark, K. J., and Hackett, P. B. (2005) Fishing for answers with transposons. *Marine Biotechnology* **7**, 135–141.
8. Dupuy, A. J., Clark, K., Carlson, C. M., Fritz, S., Davidson, A. E., Markley, K. M., Finley, K., Fletcher, C. F., Ekker, S. C., Hackett, P. B., Horn, S., and Largaespada, D. A. (2002) Mammalian germ-line transgenesis by transposition.

Proceedings of the National Academy of Sciences of the United States of America **99**, 44954–44959.

9. Dupuy, A. J., Fritz, S., and Largaespada, D. A. (2001) Transposition and gene disruption in the male germline of the mouse. *Genesis: the Journal of Genetics & Development* **30**, 82–88.
10. Fischer, S. E., Wienholds, E., and Plasterk, R. H. (2001) Regulated transposition of a fish transposon in the mouse germ line. *Proceedings of the National Academy of Sciences of the United States of America* **98**, 6759–6764.
11. Horie, K., Kuroiwa, A., Ikawa, M., Okabe, M., Kondoh, G., Matsuda, Y., and Takeda, J. (2001) Efficient chromosomal transposition of a Tc1/mariner- like transposon *Sleeping Beauty* in mice. *Proceedings of the National Academy of Sciences of the United States of America* **98**, 9191–9196.
12. Horie, K., Yusa, K., Yae, K., Odajima, J., Fischer, S. E., Keng, V. W., Hayakawa, T., Mizuno, S., Kondoh, G., Ijiri, T., Matsuda, Y., Plasterk, R. H., and Takeda, J. (2003) Characterization of *Sleeping Beauty* transposition and its application to genetic screening in mice. *Molecular & Cellular Biology* **23**, 9189–9207.
13. Huang, X., Wilber, A. C., Bao, L., Tuong, D., Tolar, J., Orchard, P. J., Levine, B. L., June, C. H., McIvor, R. S., Blazar, B. R., and Zhou, X. (2006) Stable gene transfer and expression in human primary T cells by the *Sleeping Beauty* transposon system. *Blood* **107**, 483–491.
14. Belur, L. R., Frandsen, J. L., Dupuy, A. J., Ingbar, D. H., Largaespada, D. A., Hackett, P. B., and McIvor, R.S. (2003) Gene insertion and long-term expression in lung mediated by the *Sleeping Beauty* transposon system. *Molecular Therapy: the Journal of the American Society of Gene Therapy* **8**, 501–507.
15. Liu, L., Sanz, S., Heggestad, A. D., Antharam, V., Notterpek, L., and Fletcher, B. S. (2004) Endothelial targeting of the *Sleeping Beauty* transposon within lung. *Molecular Therapy: the Journal of the American Society of Gene Therapy* **10**, 97–105.
16. Yant, S. R., Meuse, L., Chiu, W., Ivics, Z., Izsvak, Z., and Kay, M. A. (2000) Somatic integration and long-term transgene expression in normal and haemophilic mice using a DNA transposon system. *Nature Genetics* **25**, 35–41.
17. Yant, S. R., Ehrhardt, A., Mikkelsen, J. G., Meuse, L., Pham, T., and Kay, M. A. (2002) Transposition from a gutless adeno-transposon vector stabilizes transgene expression in vivo. *Nature Biotechnology* **20**, 999–1005.
18. Score, P. R., Belur, L. R., Frandsen, J. L., Guerts, J. L., Yamaguchi, T., Somia, N. V., Hackett, P. B., Largaespada, D. A., and McIvor, R. S. (2006) *Sleeping Beauty* -mediated transposition and long-term expression *in vivo:* use of the LoxP/Cre recombinase system to distinguish transposition-specific expression. *Molecular Therapy* **13**, 617–624.
19. Wilber, A., Frandsen, J. F., Geurts, J. L., Largaespada, D. A., Hackett, P. B., and McIvor, R. S. (2006) RNA as a source of transposase for *Sleeping Beauty*-mediated gene insertion and expression in somatic cells and tissues. *Molecular Therapy* **13**, 625–630.

20. Ohlfest, J. R., Frandsen, J. L., Fritz, S., Lobitz, P. D., Perkinson, S. G., Clark, K. J., Nelsestuen, G., Key, N. S., McIvor, R. S., Hackett, P. B., and Largaespada, D. A. Phenotypic correction and long-term expression of factor VIII in hemophilic mice by immunotolerization and nonviral gene transfer using the *Sleeping Beauty* transposon system. (2005) *Blood* **105**, 2691–2698.
21. Ohlfest, J. R., Demorest, Z. L., Motooka, Y., Vengco, I., Oh, S., Chen, E., Scappaticci, F. A., Saplis, R. J., Ekker, S. C., Low, W. C., Freese, A. B., and Largaespada, D. A. (2005) Combinatorial antiangiogenic gene therapy by nonviral gene transfer using the *Sleeping Beauty* transposon causes tumor regression and improves survival in mice bearing intracranial human glioblastoma. *Molecular Therapy: the Journal of the American Society of Gene Therapy* **12**, 778–788.
22. Liu, F., Song, Y., and Liu, D. Hydrodynamics-based transfection in animals by systemic administration of plasmid DNA. (1999) *Gene Therapy* **6**, 1258–1266.
23. Zhang, G., Budker, V., and Wolff, J. A. High levels of foreign gene expression in hepatocytes after tail vein injections of naked plasmid DNA. (1999) *Hum Gene Ther* **10**, 1735–1737.
24. Zhang, G., Gao, X., Song, Y. K., Vollmer, R., Stolz, D. B., Gasiorowski, J. Z., Dean, D. A., and Liu, D. (2004) Hydroporation as the mechanism of hydrodynamic delivery. *Gene Therapy* **11**, 675–682.
25. Suda, T., Gao, X., Stolz, D. B., and Liu, D. (2007) Structural impact of hydrodynamic injection on mouse liver. *Gene Therapy* **14**, 129–137.
26. Geurts, A. M., Yang, Y., Clark, K. J., Liu, G., Cui, Z., Dupuy, A. J., Bell, J. B., Largaespada, D. A., and Hackett, P. B. (2003) Gene transfer into genomes of human cells by the Sleeping Beauty transposon system. *Molecular Therapy: the Journal of the American Society of Gene Therapy* **8,** 108–117.
27. Mikkelsen, J. G., Yant, S. R., Meuse, L., Huang, Z., Xu, H., and Kay, M. A. (2003) Helper-Independent Sleeping Beauty transposon-transposase vectors for efficient nonviral gene delivery and persistent gene expression in vivo. *Molecular Therapy: the Journal of the American Society of Gene Therapy* **8**, 654–665.
28. Wilber, A., Frandsen, J. L., Wangensteen, K. J., Ekker, S. C., Wang, X., and McIvor, R. S. (2005) Dynamic gene expression after systemic delivery of plasmid DNA as determined by in vivo bioluminescence imaging. *Human Gene Therapy* **16**,1325–1332.

17

Generation and Functional Analysis of Zinc Finger Nucleases

Toni Cathomen, David J. Segal, Vincent Brondani, and Felix Müller-Lerch

Summary

The recent development of artificial endonucleases with tailored specificities has opened the door for a wide range of new applications, including the correction of mutated genes directly in the chromosome. This kind of gene therapy is based on homologous recombination, which can be stimulated by the creation of a targeted DNA double-strand break (DSB) near the site of the desired recombination event. Artificial nucleases containing zinc finger DNA-binding domains have provided important proofs of concept, showing that inserting a DSB in the target locus leads to gene correction frequencies of 1–18% in human cells. In this paper, we describe how zinc finger nucleases are assembled by polymerase chain reaction (PCR) and present two methods to assess these custom nucleases quickly in vitro and in a cell-based recombination assay.

Key Words: ZFN; artificial nuclease; custom nuclease; homologous recombination; HR; DNA double strand break; DSB.

1. Introduction

The targeted modification of complex genomes by homologous recombination (HR) is a powerful tool for genetic studies, biotechnology and gene therapy. Because HR is a rare event in mammalian cells *(1)*, the development of HR-based therapies has only become a possibility after the observation that the HR frequency can be dramatically enhanced by the creation of a targeted DNA double-strand break (DSB) near the site of the desired recombination

From: *Methods in Molecular Biology, vol. 434: Volume 2: Design and Characterization of Gene Transfer Vectors*
Edited by: J. M. Le Doux © Humana Press, Totowa, NJ

event. The recent development of artificial endonucleases with tailored specificities has thus opened the door for a wide range of new applications, including the correction of mutated genes directly in the chromosome. Because of their modularity in structure and function, zinc finger-based DNA-binding domains are highly suitable to direct an endonuclease domain to a specific genomic target site. Artificial nucleases containing a zinc finger DNA-binding domain have provided important proofs of concept. Several recent studies have reported gene correction frequencies of 1–18% in human cells when the targeted DSB was introduced by so-called zinc finger nucleases (ZFNs) ***(2–5)***. ZFNs have also been successfully used to stimulate HR in whole organisms such as drosophila ***(6)*** and plants ***(7)***.

ZFNs are composed of an engineered zinc finger DNA-binding domain tethered to the catalytic domain of the type II restriction enzyme *Fok*I ***(8,9)***. A single zinc finger motif consists of about 30 amino acids and includes an α-helix that generally contacts three bases in the major groove of DNA ***(10)***. Through a combination of rational design and selection by phage display, a database of zinc fingers motifs has been generated that includes DNA-binding domains able to recognize most of the 64 possible triples ***(11–19)***. Through modular assembly of such zinc finger motifs into tandem arrays, artificial DNA-binding domains can be engineered that recognize a wide variety of novel DNA target sites. Because the active ZFN is a dimer, two subunits are typically designed to recognize the target sequence in a tail-to-tail conformation ***(20,21)***. The C-terminal *Fok*I domains are positioned in proximity on the same face of the DNA helix by separating the two zinc finger-binding sites by a 6-bp spacer ***(22)***. The required combination of two ZFN subunits, each consisting of three fingers, thus provides enough specificity in any mammalian genome by defining a unique 18-bp site.

In this chapter, we describe how ZFNs are generated by using PCR and present two methods to assess them quickly in vitro or in a cell-based recombination assay.

2. Materials

2.1. Generation of ZFNs

1. Oligonucleotides for assembly and amplification PCR (*see* **Fig. 1A**): The oligonucleotides were ordered at MWG Biotech (High Purity Salt Free Purification; *see* **Note 1**):

   C1: 5′-cggggagaaaccctataagtgtccggagtgtggcaagtcgttctc
   C2: 5′-gcgtacccacacgggcgaaaagccgtacaaatgcccagaatgcggtaaatccttcagc
   C3: 5′-tcaacggacgcatacaggagagaagccatacaaatgtcccgaatgtgggaagagttttag

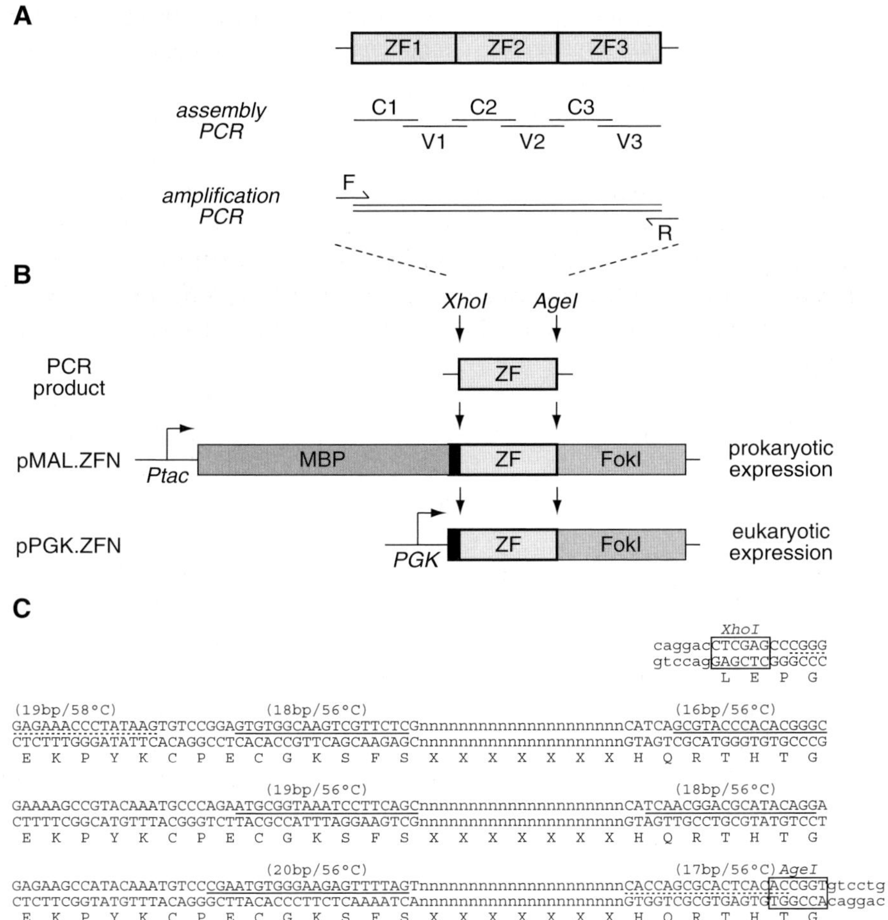

Fig. 1. Generation of zinc finger nucleases. (**A**) Polymerase chain reaction (PCR) assembly strategy. Six overlapping oligonucleotides coding for the constant (C1–C3) and variable regions (V1–V3) of the three zinc finger motifs are annealed and gaps filled in a first step (assembly PCR). The product is then amplified in a second step (amplification PCR) using two outside primers (F and R). The final PCR product encompassing the zinc finger DNA-binding domain is subsequently subcloned into appropriate expression vectors. (**B**) Schematic of a prokaryotic (pMAL.ZFN) and a eukaryotic (pPGK.ZFN) ZFN expression vector. Ptac: prokaryotic promoter, inducible by IPTG; PGK: human phosphoglycerate kinase promoter; MBP: maltose-binding protein; black box represents the HA tag; ZF: zinc finger DNA-binding domain consisting of three zinc finger motifs ZF1-ZF3; FokI: catalytic domain of the FokI endonuclease. (**C**) Coding sequence of an engineered zinc finger DNA-binding domain. Each line shows a single ZF motif comprising 28 residues. Xs designate the seven residues in the recognition helix that determine the DNA-binding specificity.

V1: 5′-gcccgtgtgggtacgctgatgnnnnnnnnnnnnnnnnnnnncgagaacgacttgccacac
V2: 5′-cctgtatgcgtccgttgatgnnnnnnnnnnnnnnnnnnngctgaaggatttaccgcat
V3: 5′-gtgtgagtgcgctggtgnnnnnnnnnnnnnnnnnnnnactaaaactcttcccacattcg
ZF-F: 5′-caggacctcgagcccggggagaaaccctataag
ZF-R: 5′-caggacaccggtgtgagtgcgctggtg

2. PCR: Vent DNA polymerase (New England BioLabs, Frankfurt, Germany) supplied with 10× ThermoPol-reaction buffer; dNTPs (10 mM, Roche, Mannheim, Germany).
3. TAE running buffer: 40 mM Tris–HCl, pH 8.0, 20 mM acetic acid, 1 mM ethylenediaminetetraacetic acid (EDTA); 10 mg/ml ethidium bromide; 6× loading dye [20% (w/v) Ficoll 400, 0.25% (w/v) bromophenol blue and 0.25 % (w/v) xylene cyanol FF, dissolved in water].
4. Agarose gel electrophoresis apparatus.
5. DNA extraction kit, such as Invisorb Spin DNA Extraction Kit (Invitek, Berlin, Germany).
6. T4 DNA ligase (400,000 U/ml) supplied with 10× ligase buffer and restriction endonucleases from New England BioLabs.
7. XL1-Blue cells (Stratagene, La Jolla, CA, USA).

2.2. *In Vitro Nuclease Assay*

1. The plasmid pMAL-c2X (New England BioLabs) is a prokaryotic expression vector that appends the maltose-binding protein (MBP) to the N-terminus of the expressed protein. The MBP fusion greatly improves the solubility of the ZFNs and does not appear to interfere with activity. The MBP fusion also enables purification to >90% in a single step over amylose resin (*see* **Note 2**).
2. Plasmid pMAL.ZFN (*see* **Fig. 1B**), a modified pMAL vector that places the N-terminal MBP in-frame with an engineered zinc finger peptide and a C-terminal *Fok*I cleavage domain, can be obtained from the authors.
3. DNA cleavage substrates: There are many ways to design target DNA for these experiments. The binding sites can be easily incorporated into a plasmid by constructing complementary oligonucleotides that contain the desired target site flanked by appropriate restriction sites (*see* **Fig. 2A–C**). We typically use the same target plasmids (TPs) that are used in the episomal recombination assay (*see* **Subheading 3.3.**). For the in vitro cleavage assay, the plasmid should be linearized in such a manner that cleavage by the ZFN will produce two fragments of different sizes. ZFN cleavage at the appropriate site will therefore indicate activity. Specificity of ZFNs can be assessed by using DNA substrates that include altered or non-target sites (*see* **Fig. 2B**).

Fig. 1. *(Continued)* Overlapping sequences between the oligonucleotides are underlined (dashed for the outside primers) and relevant cloning sites are indicated.

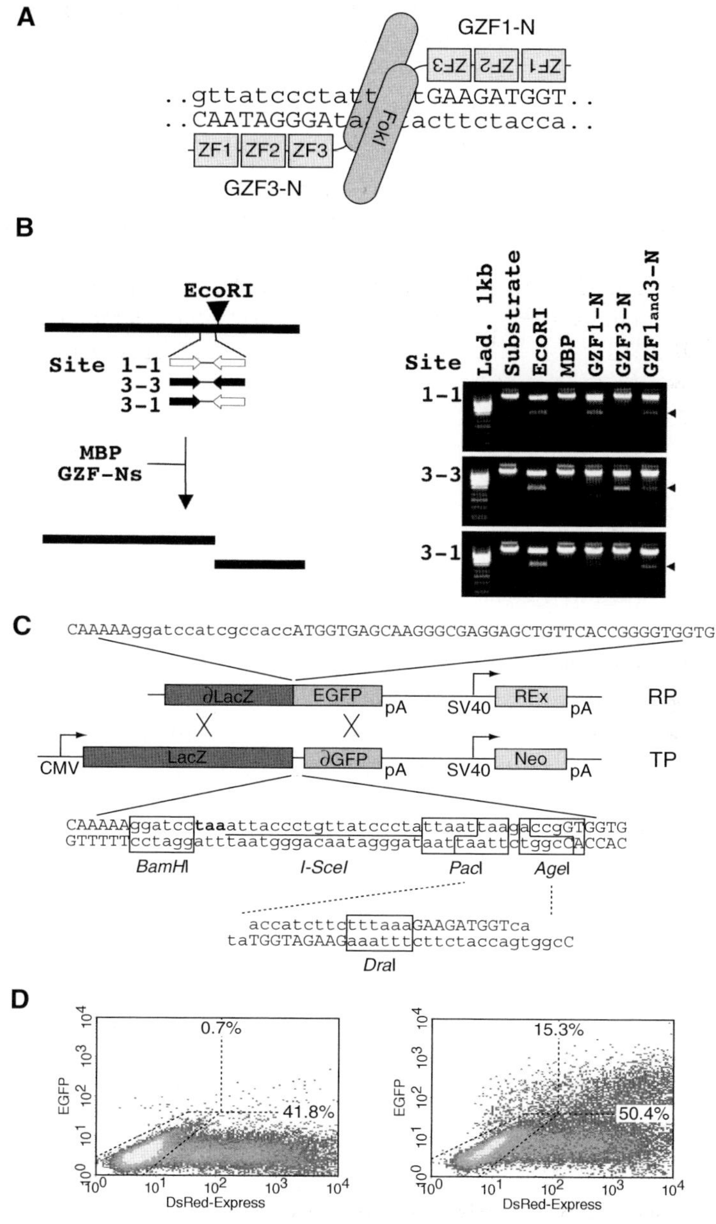

Fig. 2. Evaluation of zinc finger nucleases (ZFNs). (**A**) Schematic of a pair of ZFNs (GZF1-N and GZF3-N) bound to DNA. Individual ZF motifs (ZF1–ZF3) are shown as boxes. Each motif contacts three nucleotides of the target sequence (nucleotides contacted by the ZFs in capitals). The zinc fingers are fused to the catalytic domain of the *Fok*I endonuclease (FokI), which must dimerize in order to become active.

4. Cells: *Escherichia coli* ER2508 (New England BioLabs). These cells provide tight repression of protein expression before induction and high expression levels after induction.
5. Luria Broth (LB) powder (Sigma). Suspend 20 g in 1 l of distilled water; Autoclave. Carbenicillin (Omega Scientific, Tarzana, CA, USA), 100 mg/ml in water; Filter sterilize, and store at –20ºC. The antibiotic (a more stable analog of ampicillin) is added to LB media at 50 μg/ml just before use.
6. Isopropyl-beta-D-thiogalactopyranoside (IPTG) (Sigma), 0.5 M in water. Filter sterilize and store at –20ºC.
7. $ZnCl_2$, 100 mM in water.
8. Column buffer: 10 mM Tris–HCl (pH 8.0), 90 mM KCl, 0.1 mM $ZnCl_2$, 5 mM DTT, 0.2% Triton-X-100, 1× Complete protease inhibitor without EDTA (Roche). Prepare 50 ml for each ZFN to be purified.
9. Elution buffer: Dissolve 36 mg of maltose in 10 ml of column buffer (10 mM final).
10. Amylose resin (New England BioLabs).
11. Econo-Pac columns with 2-way stopcocks (Bio-Rad, Hercules, CA, USA), ring stand, clamps.
12. Sonicator, such as a Fisher Scientific Sonic Dismembrator, Model 100.

Fig. 2. *(Continued)* Consequently, the full target sequence is made up of two 9-bp half sites, which are separated by a 6-bp spacer. (**B**) In vitro cleavage assay. A schematic of the in vitro cleavage assay is shown on the left. A linear DNA substrate containing binding sites for GZF1-N (1-1), GZF3-N (3-3), or GZF3-N and GZF1-N (3-1) is cleaved in different size products. The right panel shows analysis of the cleavage reaction. Purified ZFNs were incubated with the linear DNA substrate and the extent of cleavage analyzed by agarose gel electrophoresis. Control reactions (substrate alone and cleavage with *Eco*RI) and protein control (MBP) are indicated. An arrow points to the cleavage products. (**C**) Experimental set up of cell-based recombination assay. Repair plasmid (RP) and target plasmid (TP) are shown schematically. The relevant sequences are given above or below, including LacZ- and EGFP-coding sequences (capital letters), the recognition site for *I-SceI* (underlined), and the relevant cloning sites (boxed). LacZ-EGFP expression from TP is prevented by a stop codon (bold), which terminates translation of LacZ, and a 33-bp truncation at the 5′-end of the EGFP open-reading frame. RP contains a 5′-deleted LacZ gene fused in-frame with a downstream EGFP open-reading frame. Expression of a LacZ-EGFP fusion protein is prevented by the 1197-bp deletion at the 5′-end and by omitting a promoter. To identify transfected cells, an expression cassette for DsRed-Express (Clontech, Mountain View, CA, USA) is located further downstream. Inserting a target site between the *Pac*I and *Age*I sites generates a customized TP for assessing novel ZFNs. An example is given for GZF1-N. (**D**) Analysis of gene repair by using flow cytometry. HEK293T cells transfected with RP, TP and a ZFN expression vector are analyzed by flow cytometry after 2 days. The right panel displays the extent of gene repair of a corresponding TP stimulated by the expression of GZF1-N, while the left panel depicts unstimulated recombination.

13. Sodium dodecyl sulfate–polyacrylamide gel electrophoresis (SDS–PAGE): 10–20% gradient Tris–HCl Ready-Gels, 50 μl, 10 wells (Bio-Rad), Broad Range SDS–PAGE Standards (Bio-Rad), and appropriate apparatus, buffers and stains.
14. BSA standards prepared to 50, 100, 200 and 400 μg/ml.
15. 10× Reaction buffer: 100 mM Tris–HCl (pH 8.0), 900 mM KCl, 1 mM $ZnCl_2$, 50 mM DTT. Prepare 1 ml.
16. $MgCl_2$, 200 mM in water.
17. 5× Loading Buffer: 4 mM Orange-G (Sigma), 30% glycerol.

2.3. Cell-Based Recombination Assay

1. Plasmids: Eukaryotic expression plasmid pPGK.ZFN, TP and repair plasmid (RP) are described in *(5)*. The plasmids and the corresponding plasmid maps can be obtained upon request.
2. Generation of TP: Oligonucleotides ZFBS-F: 5′-nnnnnnnnnntttaaannnnnnnnnca and ZFBS-R: 5′-ccggtgnnnnnnnnntttaaannnnnnnnnat (*see* **Note 3**); T4 DNA ligase (400,000 U/ml) supplied with 10× ligase buffer and restriction endonucleases from New England BioLabs; XL1-Blue cells (Stratagene).
3. HEK293T cells. These cells are a highly transfectable derivative of the 293 cell line into which the temperature sensitive gene for SV40 T-antigen was inserted (ATCC). The cells are maintained in Dulbecco's modified Eagle Medium (DMEM with high glucose; Gibco/Invitrogen, Karlsruhe, Germany) supplemented with 10% fetal bovine serum (FBS; Gibco/Invitrogen) and penicillin/streptomycin (100× solution; Gibco/Invitrogen).
4. Trypsin solution (0.05% Trypsin–EDTA; Gibco/Invitrogen).
5. CO_2 incubator with 5% CO_2 atmosphere, such as Heraeus HeraCell240.
6. 2× BES-buffered saline (BBS) for transfection: dissolve 4.28 g of BES (Sigma), 6.4 g of NaCl and 0.108 g of Na_2HPO_4 (or 0.161 g of Na_2HPO_4-$7H_2O$) in 360 ml of ddH_2O. Adjust pH to 6.96 with HCl at room temperature (RT) before adjusting volume to 400 ml with ddH_2O. Passage solution through a 0.2-μm filter and prepare 5 ml aliquots stored at –20°C.
7. 2.5 M $CaCl_2$: dissolve 13.5 g of $CaCl_2 \bullet 6H_2O$ (or 7.35 g of $CaCl_2 \bullet 2H_2O$) in 20 ml of ddH_2O. Filter sterilize (0.2 μm) and prepare 1 ml aliquots stored at –20°C.
8. Flow cytometer, such as FacsCalibur (BD Bioscience, Heidelberg, Germany), and software for analysis, such as CellQuest (BD Bioscience).

3. Methods

3.1. Generation of ZFNs

1. Design zinc finger sequence and the corresponding oligonucleotides: To determine the sequence that encodes a specific zinc finger DNA-binding domain recognizing your chosen target sequence, visit either one of two websites: http://www.zincfingertools.org [*(23)* Carlos Barbas' lab at The Scripps Research

Institute in La Jolla, CA, USA) or http://bindr.gdcb.iastate.edu/ZiFiT/ (Dobbs & Voytas laboratories at Iowa State University, Ames, IA, USA) (*see* **Note 4**).

2. Generate the sequence coding for the three zinc finger modules by assembly PCR, in which the overlapping oligonucleotides are annealed and the gaps between them closed (*see* **Fig. 1A**). Set up the PCR by mixing the following ingredients (keep all solutions on ice): 1 μl of each oligo (C1, C2, C3, V1, V2, and V3 at 25 pmol/μl; for sequence information *see* **Subheading 2.1.**; *see* **Note 5**), 0.5 μl of dNTPs (10 mM), 2.5 μl 10× ThermoPol-buffer, 0.125 μl of Vent polymerase (2 U/μl), and 15.9 μl of ddH_2O (end volume is 25 μl). Run PCR using the following cycling conditions: 1 cycle at 95°C for 2 min; 12 cycles at 95°C/56°C/72°C, for 30 s each segment; 1 cycle at 72°C for 5 min; keep at 4°C until further use.
3. Set up the PCR for the second engineering step, in which the product of **step 1** is used as a template for the amplification PCR (*see* **Fig. 1A**): 1 μl of the assembly PCR product, 0.75 μl of each primer (ZF-F and ZF-R at 10 pmol/μl; for sequence information *see* **Subheading 2.1.**; *see* **Note 6**), 0.5 μl of dNTPs (10 mM), 2.5 μl of 10× ThermoPol-buffer, 0.125 μl of Vent polymerase (2 U/μl) and 19.4 μl of ddH_2O (end volume is 25 μl). Run PCR using the following cycling conditions: 1 cycle at 95°C for 2 min; 25 cycles at 95°C/56°C/72°C, for 30 s each segment; 1 cycle at 72°C for 5 min; last cycle at 4°C until further use.
4. Pour a TAE-buffered 1.5% agarose gel containing 1 μg/ml ethidium bromide. Add 5 μl of 6× loading dye to the PCR product and load on gel. Run gel at 80 V for 40 min and isolate the 280-bp fragment from the gel. Extract the DNA using a DNA extraction kit and elute DNA in 30 μl of 10 mM Tris–HCl pH 8.0.
5. Digest 20 μl of the eluted DNA and a respective expression vector (pMAL.ZFN or pPGK.ZFN) with *Age*I and *Xho*I (*see* **Fig. 2B and 2C**) according standard procedures. Ligate the digested insert and backbone and use the ligation product to transform XL1-Blue cells (*see* **Note 7**).

3.2. In Vitro Nuclease Assay

1. Clone the assembled zinc fingers from **Subheading 3.1** into the pMAL.ZFN prokaryotic expression vector and place the zinc fingers in-frame between an N-terminal MBP and the C-terminal *Fok*I cleavage domain (*see* **Fig. 1B**). This plasmid is used to transform *E. coli* ER2508 cells.
2. Inoculate a 5-ml culture in LB/carbenicillin (50 μg/ml) and incubate overnight in a 37°C shaker.
3. Back dilute 2.5 ml of the overnight culture into 50 ml LB/carbenicillin (50 μg/ml) in a 250-ml flask. Incubate in a 37°C shaker until $A_{600} \approx 0.7$ OD (~2 h).
4. Induce ZFN expression by adding 30 μl of 0.5 M IPTG (0.3 mM final) and 50 μl of 100 mM $ZnCl_2$ (100 μM final). Continue incubation in a 37°C shaker for 2 h.
5. Transfer the culture to a 50-ml conical tube. Pellet the cells in a bench-top centrifuge at 1900 × *g* (Beckman CS-6R) for 20 min.

6. Discard the supernatant as media waste. If desired, the pellets can be frozen and stored at –80°C for several days. Thaw at RT.
7. Resuspend the pellet in 5 ml Column Buffer (*see* **Note 8**). The samples should be kept on ice during all subsequent steps.
8. Sonicate on ice 2 min at 50% duty cycle (10 s pulse, 10 s pause).
9. Pellet insoluble debris in high-speed centrifuge at 44,000 × *g* (Beckman J2-21M in JA20 rotor) for 30 min at 4°C. Alternatively, split sample into five 1.5-ml microcentrifuge tubes and centrifuge at 21,000 × *g* (Eppendorf Refrigerated Microcentrifuge 5417R) for 15 min at 4°C.
10. Filter the supernatant with a 0.2-μm syringe filter.
11. Prepare a chromatography column with 3 ml of amylose resin. This should produce a packed resin bed of approximately 1.5 ml. Do not let the column run dry at any step. Use of a flow cap or valve is recommended for this purpose. We typically perform **steps 11–15** at RT for convenience, but a cold room can be used if desired.
12. Equilibrate the resin with 10 vol (15 ml) of Column Buffer.
13. Apply the filtered supernatant from **step 10** to the column.
14. Wash with 10 vol (15 ml) of Column Buffer.
15. Elute in 10 × 0.5 ml of Column Buffer/10 mM maltose. Protein should elute in fractions 3–10. Add 0.25 ml glycerol (30% final) and store at –20°C.
16. Analyze 10 μl of the fractions by using SDS–PAGE. Samples should be compared to BSA standards of 0.5, 1, 2 and 4 μg (i.e., 10 μl of stocks) (*see* **Note 9**).
17. In a 30-μl reaction volume, prepare 1× Reaction Buffer, 1 μg (≈200 fmol) of linear DNA substrate and approximately 50 ng (≈500 fmol) of full-length (80 kDa) purified protein. It is often useful to examine a range of protein concentrations.
18. Incubate at RT (25°C) for 15 min.
19. Add 1.5 μl of 200 mM $MgCl_2$ (final = 10 mM) to initiate cleavage.
20. Incubate at 37°C for 30 min.
21. Add 5 μl of 5× Loading Buffer and analyze the reactions on a 1% agarose gel containing 0.5 μg/ml ethidium bromide (*see* **Fig. 2B**).

3.3. Cell-Based Recombination Assay

1. Clone the PCR product from **Subheading 3.1** into plasmid pPGK.ZFN, placing the zinc finger DNA-binding domain in-frame between an N-terminal HA tag and the C-terminal *Fok*I cleavage domain (*see* **Fig. 1B**).
2. Clone two antiparallel oligonucleotides encoding the zinc finger-binding site into the TP. To this end, digest 100 ng TP (*see* **Fig. 2C**) with 1 unit each of *Pac*I and *Age*I in a total volume of 20 μl for 2 h at 37°C (*see* **Note 10**). Then, heat inactivate the restriction enzymes for 20 min at 80°C. In the meantime, mix the two oligonucleotides: 1 μl each of ZFBS-F and ZFBS-R (100 pmol/μl each) and 98 μl of 150 mM NaCl. Heat the mix to 95°C for 5 min, then allow the heat block to slowly cool to RT (about 60 min), which permits efficient annealing of the two oligonucleotides.

3. Set up ligation by mixing 5 µl of the digested plasmid with 1 µl of double-stranded oligos, 4 µl ddH_2O, 1.2 µl 10× ligase buffer and 0.8 µl T4 DNA ligase (400 U/µl). Incubate for 15 min at RT and use 5 µl to transform 45 µl of chemically competent XL1-Blue cells. Successful cloning should be confirmed by *Dra*I digest (*see* **Note 3**), followed by sequencing.
4. For the cell-based recombination assay, seed 75,000 HEK293T cells in 0.5-ml culture medium into each well of a 24-well plate and incubate cells for 24 h.
5. Replace culture medium carefully with 0.5 ml fresh medium 1 h before transfection and prepare transfection mixture in Eppendorf tube. Mix 1.5 µg plasmid DNA (20 ng TP; 1.0 µg RP; 0.5 µg pPGK.ZFN; *see* **Note 11**) with 22.5 µl of H_2O, 2.5 µl of 2.5 M $CaCl_2$ and 25 µl of 2× BBS buffer. Mix transfection solution gently by pipetting up and down and incubate for 10 min at RT. Then, overlay cells carefully drop by drop with the mixture and incubate the culture overnight in the CO_2 incubator.
6. The next day, aspirate the transfection medium carefully from cells and overlay cells with 0.5 ml of fresh tissue culture medium. Place the cells in the CO_2 incubator for another 24 h.
7. The next day, aspirate medium and wash cells carefully with 1 ml PBS. Remove all PBS, add 80 µl of trypsin solution and leave in CO_2 incubator until cells detach. Add 400 µl of PBS supplemented with 20% FBS, resuspend the cells thoroughly by pipetting up and down several times, transfer cell suspension to a FACS tube (BD Bioscience) and keep tubes on ice until analysis.
8. Set up your flow cytometer (*see* **Note 12**).
9. Count 50,000 cells and evaluate the number of EGFP- and DsRedExpress-positive cells as shown in **Fig. 2D**. The fraction of transfected cells in which an HR event took place is calculated by dividing the number of EGFP-positive cells by the number of DsRedExpress-positive cells. An example for the calculation is given for the right panel: 15.3% (green cells)/50.4% (red cells) × 100 = 30.3% (corrected cells).

4. Notes

1. A relatively low degree of purification [high purity salt free (HPSF)] is sufficient to create zinc finger domains by the method described. We were very cautious when designing the overlapping sequences between the constant and the variable regions and between the different zinc finger modules in order to make sure that the PCR assembly strategy is as robust and as error-free as possible.
2. This detailed protocol is based on the general protocol provided in the New England BioLabs Protein Expression and Purification Kit manual.
3. Subcloning of the two annealed oligonucleotides into TP generates an additional *Dra*I site (*see* **Fig. 2C**), which can be used for screening mini-prep DNA.
4. The assumption that zinc finger motifs are functionally independent subunits is somewhat of an oversimplification. Cooperativity in zinc finger–DNA interaction and recognition of a fourth base in the target sequence (target-site-overlap)

can lead to the generation of DNA-binding domains with both low affinity and low specificity. Such issues can be addressed by elaborate selection strategies *(24–28)*, which are, however, beyond the scope of this chapter. Generally, zinc finger DNA-binding motifs recognizing 5′-GNNGNNGNN target sites tend to perform with sufficient affinity and specificity *(3,5)*. Further information on the zinc finger technology in general can be found on our website (http://www.charite.de/cathomen/) or the website of The Zinc Finger Consortium (http://www.zincfingers.org/), which has been established to promote the development of the engineered zinc fingers.

5. If possible, the variable region encoding the recognition helix (shown as "n" in **Subheading 2.1**) should be designed such that an analytic recognition site for a restriction endonuclease is introduced. This greatly facilitates restriction analysis and later discrimination between different zinc finger domains.
6. To recombine the PCR-amplified zinc finger cassette with a desired expression plasmid, primer ZF-F contains an *Xho*I recognition site whereas primer ZF-R comprises an *Age*I site.
7. Every newly engineered zinc finger cassette was scrutinized in the following order: After a thorough restriction analysis, expression of the ZFN was assessed by western blot analysis. Positive clones were then verified by sequencing. Using the described method, about 90% of mini-preps tested positive in the restriction analysis. About 75% of these minis were subsequently positively evaluated for ZFN expression by western blot analysis using an anti-HA antibody (Novus Biologicals, Littleton, CO, USA No. NB600-363) and about 2/3 of clones sequenced contained the correct sequence.
8. Mg^{2+} is required by the *Fok*I cleavage domain for activity. The purification is performed without Mg^{2+} so that the protein can be stored without loss of activity. Mg^{2+} is added in **step 19** before the cleavage reaction.
9. Often, the proteins are only partially purified by this protocol. However, this level of purification is typically sufficient to assess the binding and cleavage activity.
10. Alternatively, the oligonucleotides encoding the ZFN target site can be designed to be inserted into TP through *BamH*I and *Age*I. This, however, removes the binding site for *I-SceI*, which usually serves as a positive control in the cell-based recombination assay.
11. Plasmid pRK5.LHA-Sce1 *(29)*, which codes for the yeast homing endonuclease *I-SceI*, can be used instead of pPGK.ZFN as a positive control, whereas pCMV.Luc *(5)* encoding luciferase can be used as a negative control (*see* **Fig. 2C–D**).
12. For a FacsCalibur (BD Bioscience), we typically use the following settings for Voltage and AmpGain: FSC (E-1/4.40), SSC (270/6.23), FL1 (500/1.00), and FL2 (445/1.00). Compensations between the green and red channels are FL1–1.2% FL2 and FL2–34.5% FL1 (FSC, Forward Scatter; SSC, Side Scatter; FL1, green channel; FL2, red channel). You may have to adjust these numbers to some extent, depending on the age of your laser and/or the batch of HEK293T cells you are using in your laboratory.

Acknowledgments

The authors thank Shamim H. Rahman for discussions and careful reading of the manuscript. This chapter is based on work supported by grants CA311/1 from the German Research Foundation (T.C.) and CA103651 from the National Cancer Institute, NIH (D.J.S.).

References

1. Vasquez, K. M., Marburger, K., Intody, Z., and Wilson, J. H. (2001) Manipulating the mammalian genome by homologous recombination. *Proc Natl Acad Sci USA* **98:**8403–8410.
2. Urnov, F. D., Miller, J. C., Lee, Y. L., Beausejour, C. M., Rock, J. M., Augustus, S., Jamieson, A. C., Porteus, M. H., Gregory, P. D., and Holmes, M. C. (2005) Highly efficient endogenous human gene correction using designed zinc-finger nucleases. *Nature* **435:**646–651.
3. Porteus, M. H. (2006) Mammalian gene targeting with designed zinc finger nucleases. *Mol Ther* **13:**438–446.
4. Porteus, M. H. and Baltimore, D. (2003) Chimeric nucleases stimulate gene targeting in human cells. *Science* **300:**763.
5. Alwin, S., Gere, M. B., Guhl, E., Effertz, K., Barbas, C. F., 3rd, Segal, D. J., Weitzman, M. D., and Cathomen, T. (2005) Custom zinc-finger nucleases for use in human cells. *Mol Ther* **12:**610–617.
6. Beumer, K., Bhattacharyya, G., Bibikova, M., Trautman, J. K., and Carroll, D. (2006) Efficient gene targeting in Drosophila with zinc-finger nucleases. *Genetics* **172:**2391–2403.
7. Wright, D. A., Townsend, J. A., Winfrey, R. J., Jr., Irwin, P. A., Rajagopal, J., Lonosky, P. M., Hall, B. D., Jondle, M. D., and Voytas, D. F. (2005) High-frequency homologous recombination in plants mediated by zinc-finger nucleases. *Plant J* **44:**693–705.
8. Kim, Y. G., Cha, J., and Chandrasegaran, S. (1996) Hybrid restriction enzymes: zinc finger fusions to FokI cleavage domain. *Proc Natl Acad Sci USA* **93:** 1156–1160.
9. Durai, S., Mani, M., Kandavelou, K., Wu, J., Porteus, M. H., and Chandrasegaran, S. (2005) Zinc finger nucleases: custom-designed molecular scissors for genome engineering of plant and mammalian cells. *Nucleic Acids Res* **33:**5978–5990.
10. Pavletich, N. P. and Pabo, C. O. (1991) Zinc finger-DNA recognition: crystal structure of a Zif268-DNA complex at 2.1 A. *Science* **252:**809–817.
11. Rebar, E. J. and Pabo, C. O. (1994) Zinc finger phage: affinity selection of fingers with new DNA-binding specificities. *Science* **263:**671–673.
12. Choo, Y. and Klug, A. (1994) Selection of DNA binding sites for zinc fingers using rationally randomized DNA reveals coded interactions. *Proc Natl Acad Sci USA* **91:**11168–11172.

13. Jamieson, A. C., Wang, H., and Kim, S. H. (1996) A zinc finger directory for high-affinity DNA recognition. *Proc Natl Acad Sci USA* **93:**12834–12839.
14. Wu, H., Yang, W. P., and Barbas, C. F., 3rd. (1995) Building zinc fingers by selection: toward a therapeutic application. *Proc Natl Acad Sci USA* **92:** 344–348.
15. Liu, Q., Xia, Z., Zhong, X., and Case, C. C. (2002) Validated zinc finger protein designs for all 16 GNN DNA triplet targets. *J Biol Chem* **277:** 3850–3856.
16. Segal, D. J., Dreier, B., Beerli, R. R., and Barbas, C. F., 3rd. (1999) Toward controlling gene expression at will: selection and design of zinc finger domains recognizing each of the 5'-GNN-3' DNA target sequences. *Proc Natl Acad Sci USA* **96:**2758–2763.
17. Dreier, B., Beerli, R. R., Segal, D. J., Flippin, J. D., and Barbas, C. F., 3rd. (2001) Development of zinc finger domains for recognition of the 5'-ANN-3' family of DNA sequences and their use in the construction of artificial transcription factors. *J Biol Chem* **276:**29466–29478.
18. Dreier, B., Fuller, R. P., Segal, D. J., Lund, C. V., Blancafort, P., Huber, A., Koksch, B., and Barbas, C. F., 3rd. (2005) Development of zinc finger domains for recognition of the 5'-CNN-3' family DNA sequences and their use in the construction of artificial transcription factors. *J Biol Chem* **280:**35588–35597.
19. Blancafort, P., Magnenat, L., and Barbas, C. F., 3rd. (2003) Scanning the human genome with combinatorial transcription factor libraries. *Nat Biotechnol* **21:** 269–274.
20. Smith, J., Bibikova, M., Whitby, F. G., Reddy, A. R., Chandrasegaran, S., and Carroll, D. (2000) Requirements for double-strand cleavage by chimeric restriction enzymes with zinc finger DNA-recognition domains. *Nucleic Acids Res* **28:** 3361–3369.
21. Bitinaite, J., Wah, D. A., Aggarwal, A. K., and Schildkraut, I. (1998) FokI dimerization is required for DNA cleavage. *Proc Natl Acad Sci USA* **95:**10570–10575.
22. Bibikova, M., Carroll, D., Segal, D. J., Trautman, J. K., Smith, J., Kim, Y. G., and Chandrasegaran, S. (2001) Stimulation of homologous recombination through targeted cleavage by chimeric nucleases. *Mol Cell Biol* **21:**289–297.
23. Mandell, J. G. and Barbas, C. F., 3rd. (2006) Zinc Finger Tools: custom DNA-binding domains for transcription factors and nucleases. *Nucleic Acids Res* **34:**W516–523.
24. Greisman, H. A. and Pabo, C. O. (1997) A general strategy for selecting high-affinity zinc finger proteins for diverse DNA target sites. *Science* **275:**657–661.
25. Isalan, M., Klug, A., and Choo, Y. (2001) A rapid, generally applicable method to engineer zinc fingers illustrated by targeting the HIV-1 promoter. *Nat Biotechnol* **19:**656–660.
26. Cheng, X., Boyer, J. L., and Juliano, R. L. (1997) Selection of peptides that functionally replace a zinc finger in the Sp1 transcription factor by using a yeast combinatorial library. *Proc Natl Acad Sci USA* **94:**14120–14125.
27. Hughes, M. D., Zhang, Z. R., Sutherland, A. J., Santos, A. F., and Hine, A. V. (2005) Discovery of active proteins directly from combinatorial randomized protein

libraries without display, purification or sequencing: identification of novel zinc finger proteins. *Nucleic Acids Res* **33:**e32.

28. Hurt, J. A., Thibodeau, S. A., Hirsh, A. S., Pabo, C. O., and Joung, J. K. (2003) Highly specific zinc finger proteins obtained by directed domain shuffling and cell-based selection. *Proc Natl Acad Sci USA* **100:**12271–12276.
29. Radecke, F., Peter, I., Radecke, S., Gellhaus, K., Schwarz, K., and Cathomen, T. (2006) Targeted chromosomal gene modification in human cells by single-stranded oligodeoxynucleotides in the presence of a DNA double-strand break. *Mol Ther* **14:**798–808.

18

Conditional Gene Expression and Knockdown Using Lentivirus Vectors Encoding shRNA

Jolanta Szulc and Patrick Aebischer

Summary

Drug-inducible systems allowing the control of transgene expression and knockdown in mammalian cells are invaluable tools for genetic research, and could also play important roles in translational research or gene therapy. We and others have developed a lentivector-based, conditional gene expression system for drug-controllable expression of transgenes and small hairpin RNAs (shRNAs). This system is highly robust and versatile, governing tightly controlled expression of transgenes and endogenous cellular genes (through shRNAs) in various primary and established cell lines in vitro, as well as in vivo in the central nervous system or in human cancer cells xenotransplanted into nude mice. The goal of this article is to provide a concise methodology for construction and manipulation of this conditional lentiviral-based system, and quantitative analyses of drug-inducible transgene expression and gene knockdown both in vitro and in vivo.

Key Words: Conditional gene expression; drug-inducible gene knockdown; shRNA; lentiviral vectors; doxycycline.

1. Introduction

Drug-inducible systems for conditional transgene expression or knockdown of cellular genes are important tools in many areas of both basic and translational research. The drug-controllable knockdown allows for conditional expression of any cellular gene for which an effective shRNA can be designed. In principle, all systems designed for either conditional transgene expression or knockdown are based on activation of silent, minimal promoters or repression of active cellular promoters *(1)*. The activation-based systems

From: *Methods in Molecular Biology, vol. 434: Volume 2: Design and Characterization of Gene Transfer Vectors*
Edited by: J. M. Le Doux © Humana Press, Totowa, NJ

utilize chimeric transactivators that activate gene expression by binding to the minimal promoters in a drug-inducible fashion ***(2,3)***. On the contrary, the repression-based systems have been developed in which the transcriptional activity of cellular promoters is modulated by drug-controllable epigenetic repressors, usually containing a Krüppel-associated box (KRAB) domain. The KRAB domain found in many zinc finger proteins can silence both Pol II and Pol III promoters by triggering heterochromatin formation. When tethered to specific DNA regions within the context of chimeric proteins, KRAB can induce a general silencing of transcription within up to 3 kb from its binding site ***(4–6)***. When KRAB is fused to the tetR DNA-binding domain, the resulting tTRKRAB chimeric protein allows for the doxycycline-mediated control of any promoter placed nearby *tetO* sequences, in either "Tet-On" or "Tet-Off" configurations depending on the TetR version used. The main advantages of the repression-based systems are (1) highly reduced leakiness in comparison with the activation-based systems and (2) versatility; virtually any Pol II or Pol III promoter can be subjected to the epigenetic repression thus allowing for modulating strength or tissue-specificity of the transgene expression or knockdown of any cellular gene through small hairpin RNAs (shRNAs) or micro-hairpin RNAs (mihRNAs) ***(7–9)***. Taking advantage of these properties, the tTRKRAB-regulated, lentivector-based system was developed, which allowed for controllable transgene expression and knockdown of cellular genes both with a high degree of efficacy and without significant leakiness in vitro and in vivo ***(10–13)***. Because tTRKRAB can control both Pol II and Pol III promoters, internal monitoring devices can be built in, for instance by placing a GFP marker in the vector, as its expression will be inversely regulated with that of the gene targeted by RNA interference. Because of the promiscuous activity of KRAB and the flexibility of the vector design, the Tet-KRAB system can be applied for regulating mihRNA expressed from Pol II promoters, providing with the opportunity of conditional knockdown in a tissue-specific fashion. Additionally, the presented technology can be potentially applied for generating conditional shRNA libraries, modeling of human diseases, and preclinicalscreening of novel therapeutic compounds ***(14–17)***.

Here, we present the step-by-step methodology for drug-inducible transgene expression and knockdown, using the conditional lentiviral vector system based on doxycycline-controllable tTRKRAB repressor. In the first part, we guide the readers through the selection of potential shRNA oligonucleotides and cloning into the lentiviral vectors, generation of infectious recombinant viruses, their quantification, and methods for screening the effective shRNA sequences. Next, using GFP marker and TP53-specific shRNA, we outline the procedures for quantification of the conditional transgene expression and knockdown at mRNA level by reverse transcriptase-quantitative polymerase chain reaction

(RT-QPCR) and at the protein level by Western blot. Finally, we present an experimental approach to regulate GFP marker and TP53 in vivo, in human breast cancer cells xenotransplanted into nude mice.

2. Materials

2.1. shRNA Selection

iRNAi can be downloaded for free from http://mekentosj.com/irnai/

2.2. Annealing, Phosphorylation, and Cloning of shRNAs into the Conditional Lentiviral Vectors

1. Plasmid: Lentiviral vectors (LV) (pLVTHM, pLVETH-tTRKRAB, pLVCT-tTRKRAB, pLVUTHshp53-tTRKRAB), packaging plasmids (pCMV-dR8.74), envelope plasmids (pMD2G). The conditional tTRKRAB-based lentiviral vector-derived systems for dox-controllable knockdown, as well as vector maps, sequences, are available through Addgene (Cambridge, MA, USA; http://www.adgene.org).
2. shRNA Oligos (Sigma-Aldrich, St Louis, MO, USA), 0.05 μmole scale.
3. Annealing buffer: 100 mM potassium acetate, 30 mM HEPES pH 7.4, 2 mM magnesium acetate.
4. Restriction and modifying enzymes (NEB, Ipswich, MA, USA): T4 polynucleotide kinase (PNK), T4 Ligase, Cla I, Mlu I.
5. DH5α competent bacteria (Invitrogen, Carlsbad, CA, USA), Luria broth (LB).
6. QIAquick PCR purification Kit, QIAquick Gel Extraction Kit, QIAGEN Plasmid Mini Kit, QIAGEN Plasmid Maxi Kit (Qiagen, Valencia, CA, USA).
7. H1-F primer: GCATGTCGCTATGTGTTCTGGG (custom order, Sigma).

2.3. Vector Production and Quantification, Screening of shRNAs, Transduction of Target Cells, and Treatment with Doxycycline

1. DMEM supplemented with 10% fetal calf serum and 100 U/mL of penicillin and 100 mg/mL streptomycin, Phosphate-buffered saline (PBS), Trypsin 0.25% (Invitrogen)
2. Cell lines: HEK293FT cells (Invitrogen), MCF7 cells (ATCC, Manassas, VA, USA)
3. 2× HBS (for 500 ml): NaCl (8 g), KCl (0.38 g), Na_2HPO_4 (0.1 g), HEPES (5 g), glucose (1 g), adjust to pH 7.05, sterilize by filtration (0.22 μm).
4. 2.5 M $CaCl_2$.
5. H_2O bi-distilled.
6. SNET buffer: 10 mM Tris–HCl pH 8.0, 100 mM ethylenediaminetetraacetic acid (EDTA), 0.5% SDS. Just before use add: 100 μg/ml of proteinase K (Sigma) and 10 μg/ml of RNAase (Sigma).

2.4. Analysis of Conditional Transgene Expression and Knockdown by RT–QPCR

1. RNA isolation: Trizol (Invitrogen), isopropanol, 70% ethanol, DEPC-treated water; add 1:1000 vol of DEPC to double distilled water, shake at 37°C overnight and autoclave.
2. RT reaction: Superscript III reverse transcriptase (Invitrogen), Random Primers (Promega, Madison, WI, USA), RNAsin (Promega), dNTP mix (Promega)
3. QPCR probes and primers:
   a. human TP53 set (assay-on-demand target ID Hs00153349_m1) (Applied Biosystems, Foster City, CA, USA).
   b. 20× probe and primers mix for WPRE; 2 μM of each forward and reverse primer, 5 μM of FAM-labeled WPRE probe.
      i. forward primer: TGTGGATACGCTGCTTTAATG
      ii. reverse primer: CATAAAGAGACAGCAACCAGGA
      iii. probe: CTATTGCTTCCCGTATGGCTTTCATTTTC
   c. EF-1α endogenous controls (Applied Biosystems).
4. 2× Taqman Universal PCR MasterMix (Applied Biosystems).

2.5. Analysis of Conditional Transgene Expression and Gene Knockdown by Western Blot

1. RIPA buffer: 25 mM Tris–HCl pH 7.5, 1% Triton X-100, 0.5% sodium deoxycholate, 5 mM EDTA, 150 mM NaCl. Just before use add 1:100 cocktail of protease inhibitors (Sigma-Aldrich, St. Louis, MO, USA).
2. BCA Protein Assay kit (Pierce, Rockford, IL, USA).
3. 4–12% pre-cast Bis–Tris gels (Invitrogen).
4. 10× running buffer (Invitrogen).
5. 50× transfer buffer (Invitrogen).
6. 4× loading buffer (Invitrogen).
7. Nitrocellulose membrane.
8. Methanol.
9. Whatman 3-mm paper.
10. Antibodies: TP53 (Santa Cruz Biotechnology, Santa Cruz, CA, USA), GFP (Clontech, Mountain View, CA, USA), PCNA (Proliferating Cell Nuclear Antigen; Oncogene Research Products, San Diego, CA, USA) secondary antibodies conjugated with horseradish peroxidase (Amersham, Piscataway, NJ, USA).
11. Ponseau S solution (Sigma-Aldrich, USA).
12. Enhanced chemiluminescence kit ECL (Amersham), film (Kodak, Rochester, NJ, USA).

2.6. Conditional Transgene Expression and Gene Knockdown In Vivo

1. 17γ estradiol (0.72 mg/pellet, 60-day release time; Innovative Research of America, Sarasota, FL, USA).

2. Female NMRI *nu/nu* mice (6 weeks old, purchased from Janvier, Le Genest sur l'Isle, France).
3. Matrigel (Becton-Dickinson, Franklin Lakes, NJ, USA, 10 mg/ml).

3. Methods

3.1. Selection and Design of shRNA

1. Open iRNAi software and fetch the sequence from the Entrez server (from the top menu select: File and Fetch from Entrez).
2. Change the following in the settings menu (Window → settings) to generate Mlu I and Cla I sites at the ends of the hairpin: Forward Primer: 5′ Prefix: CGCGTCCCC; 3′ Suffix: TTTTTGGAAAT; Reverse Primer: 5′ Prefix: CGATTTCCAAAAA; 3′ Suffix: GGGGA. Set the search pattern to "AA-(N19)-" (*see* **Fig. 1**).

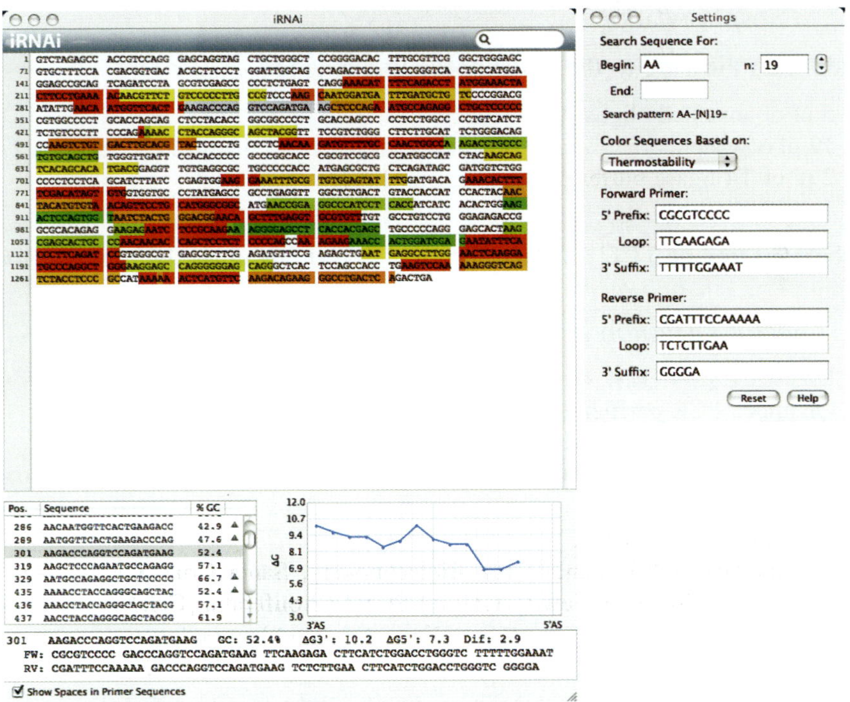

Fig. 1. Selection of small hairpin RNA (shRNA) oligos using iRNAi. Right: main window: mRNA of the target gene, the AA-(N)19-mers are highlighted, colors reflect thermostability of the 21-mer; middle-left: sequences of all AA-(N)19-mers; middle-right: thermostability curve; bottom: hairpin sequence of the selected oligo. Left: Program settings to generate AA-(N)19-mers and add MluI and ClaI restriction sites at the end of the hairpin.

3. Select 5–10 sequences based on the following rules: %GC 45-55; ΔG(3´´AS) > ΔG(5´AS), G or T at +1, the target sequence should be located 50–100 bp from the start codon (*see* **Note 1**).
4. Order the forward and reverse oligos from Sigma. The sequences can be directly copy-pasted from the bottom window (*see* **Fig. 1**).

3.2. Annealing, Phosphorylation, and Cloning of shRNAs into the Conditional LV

3.2.1. Annealing

1. Dissolve both oligos (forward and reverse) in water to the final concentration of 1 μM.
2. Mix 2 μl of each oligo with 46 μl of annealing buffer.
3. Denaturate 4 min at 95°C then incubate 10 min at 70°C. Cool down the solution to room temperature (RT) (can be stored at –20°C).

3.2.2. Phosphorylation

Add the following to the phosphorylation reaction.

a. 5 μl of annealed oligos.
b. 12 μl of water.
c. 2 μl of T4 ligase buffer containing 1 mM ATP.
d. 1 μl PNK.

Incubate 30 min at 37°C and heat inactivate PNK incubate 10 min at 70°C.

3.2.3. Vector Digestion and Purification

1. Digest the pLVTHM vector with Mlu I for 2 h at 37ºC, remove the salts using QIAquick PCR purification Kit.
2. Digest the vector opened with Mlu I with ClaI overnight at 37ºC.
3. Gel-purify the digested vector using QIAquick Gel Extraction Kit and quantify by gel electrophoresis.

3.2.4. Ligation

1. Combine the following:

   a. 5 μl of phosphorylated oligos.
   b. 1 μl of pLVTHM vector digested with MluI-ClaI (20–100 ng).
   c. 11 μl of water.
   d. 2 μl of T4 ligase buffer.
   e. 1 μl of T4 ligase (40 U).

2. Incubate 16 h at 16ºC and transform DH5α bacteria with 3 μl of ligation reaction by heat-shock using protocol provided by manufacturer.

3. Pick up 2–4 colonies per ligation, set up mini-prep cultures, and purify DNA using Qiagen Plasmid Mini Kit.
4. Check the clones by EcoR1-Cla1 digestion and electrophoresis. The positive clones should be: 10849 + 286 bp, the negative clones: 10849 + 236 bp.
5. Sequence the cloned hairpin in the positive clones using the H1-F primer.
6. Set up a midi-prep culture (250 ml) and purify DNA using QIAGEN Plasmid Maxi Kit.

3.3. Vector Production and Quantification

3.3.1. Transfection (see Note 2)

1. Plate 2–2.5 × 10^6 of 293T cells per 10-cm plate.
2. Prepare calcium-phosphate precipitate (1 ml for 10-cm plate):

    a. transfer vector (pLVTHM-shRNA)—20 μg.
    b. packaging plasmid (pCMV-dR8.74)—15 μg.
    c. envelope plasmid (pMD2G-VSVG)—6 μg.

3. Add water to 0.5 ml, add 0.5 ml 2× HBS and shake well.
4. Add 50 μl 2.5 M $CaCl_2$ and shake briefly, keep in RT for 20–25 min, add drop-wise on a plate and mix gently with a medium.
5. Change medium (6–8 h later); remove medium with precipitate and add 6 ml/plate of fresh medium.
6. After 36–40 h collect medium, spin at 3000 × *g* for 5 min at RT, pass through 0.45-μm filter. Aliquot, use fresh, or freeze at –70°C.

3.3.2. Titration

1. Plate 20,000 293T cells in 1 ml medium per well on 24-well plate.
2. 24 h later, before transduction count the cells in 1 well (expect 40–60,000).
3. Prepare 6, fourfold serial dilutions of virus in 250 μl total volume; take 100 μl for non-concentrated or 1 μl for concentrated vector stock as a first dilution. Add 1 ml of medium 2 h after infection.
4. 72 h later trypsinize the cells 1:5, collect 200,000 cells. Analyze fluorescence by using fluorescence-activated cell sorter (FACS) and read percentage from linear values (usually 5–10%). Titer is a number (percentage) of cells transduced by a given volume and counted at the day of transduction. Example below: ˜0.4 μl of virus transduced 6.52% of 50.000 cells (number of cells at the day of transduction). Calculate the number of transduce cells (0.0652 × 50,000) and multiple by dilution (3.260 × 2,500). The titer is 8.15 × 10^6 TU/ml (*see* **Table 1**) (*see* **Notes 3** and **4**).

3.4. Screening of shRNA

1. Plate the cells expressing the target gene on 24-well plate at density of 1–2 × 10^4 cells per well and culture overnight.

Table 1
Example of Lentiviral Vector Titration

Dilution	%GFP+ cells	MOI
1:1 (100 µl)	93.46	16.3
1:4 (25 µl)	87.51	4.1
1:16 (6.25 µl)	65.36	1.02
1:64 (1.5625 µl)	25.42	0.25
1:256 (0.391 µl)	6.52	0.0625
1:1024 (0.098 µl)	2.29	0.015

The 293T cells were transduced with 6 fourfold dilution of pLVTHM vector and percentage of GFP-positive cells was measured by fluorescence-activated cell sorter (FACS) 72 h later. The titer should be calculated form the most linear part of the dilution (between 1:64-1-1:1024).

2. Transduce the cells with MOI of 10 in a total volume of 250 µl for 1 h, add 1 ml of fresh medium, and culture for 5–7 days. Split the cells when necessary (*see* **Note 5**).
3. Analyze expression of the target gene by using RT-QPCR as described in **Subheading 3.7.3**. Efficient shRNA should result in >90% knockdown measured by using RT-QPCR (*see* **Note 6**).
4. Transfer the selected, most-efficient shRNA into a conditional vector by cloning MscI-FspI fragment from pLVTHMshRNA into pLVCT-tTRKRAB (*see* **Note 7**).
5. Produce and quantify the virus using pLVCTHshRNA-tTRKRAB transfer vector as described above in **Subheading 3.2** (*see* **Note 8**).

3.5. Transduction of Target Cells and Reversible Doxycycline Switches (see Note 9)

1. Plate MCF7 cells on 24-well plate at a density of 20,000 cells per well and culture overnight.
2. Count cells, remove cell medium, and add LVUTHshTP53-tTRKRAB supernatant at MOI 10 (5×10^5 TU) in a total volume of 250 µL for 2 h; add 2 mL of fresh medium and culture overnight. Perform transductions in duplicate (*see* **Note 10**).
3. Detach the cells using trypsin. Split the cells from 1 well into 2 wells and add doxycycline to one of the wells at 10 ng/µl and keep in culture for 7 days. Split the cells every 2–3 days and transfer to 6-well plate.
4. Detach the cells from each well and divide as follows:

   a. 100,000 cells for FACS analysis (*see* **Fig. 2**).
   b. 500,000 cells for RNA isolation and RT-QPCR analysis (*see* **Subheading 3.6.**).

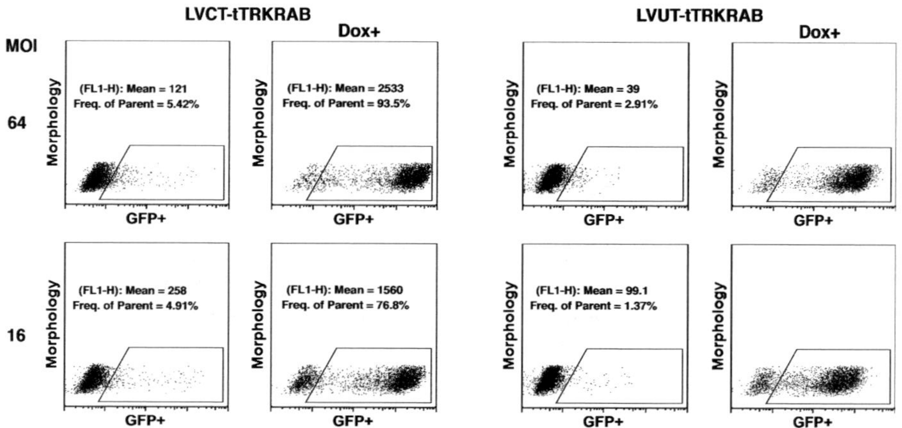

Fig. 2. Fluorescence-activated cell sorter (FACS) analysis of conditional GFP expression in vitro. 293T cells transduced at indicated MOI with LVCT-tTRKRAB (CAG promoter, GFP) and LVUT-tTRKRAB (human ubiquitin C promoter, GFP) were cultured in the presence or absence of dox for 5 days before FACS analysis ***(11)***.

c. 500,000 cells for protein isolation and Western blot analysis (see **Subheading 3.7.**).
d. 50,000 cells for Dox switches.

5. For "On to Off" Dox switch, wash the cells 3× in PBS and place in culture medium with Dox for 7 days. Split the cells every 2–3 days (*see* **Note 11**).
6. Prepare the cells for analyses as described in **step 4** above (*see* **Fig. 6**).

3.6. Analysis of Conditional Transgene Expression and Gene Knockdown by RT–QPCR

3.6.1. RNA Isolation (see Note 12)

1. Transfer the cells (*see* **Subheading 3.5., step 4b**) to 1.5-ml Eppendorf tubes, spin at 2000 × *g* for 5 min at 4°C, and remove the medium.
2. Resuspend the cell pellet completely in 1 ml of the Trizol reagent and incubate 5 min at RT.
3. Add 200 μl of chloroform, mix by shaking, and incubate for additional 5 min at RT.
4. Separate the phases by spinning at 13,000 × *g* for 10 min collect upper aqueous phase and transfer to a new tube.
5. Precipitate RNA by adding 500 μl of isopropanol, mix by inverting, and incubate 15–30 min at 4°C, spin at 13,000 × *g* for 10 min.
6. Wash RNA with 70% ethanol, dry briefly and resuspend in 200 μl DEPC-treated water. Measure RNA concentration using spectrometer and adjust all samples to 0.5 μg/ml.

3.6.2. Reverse Transcription

1. Combine in a tube:

   a. 2 μl RNA (1 μg).
   b. 1 μl Random Hexamers.
   c. 1 μl of 10 mM dNTPs mix (10 mM of each dATP, dCTP, dGTP, and dTTP).
   d. 9 μl water.

2. Heat the mixture to 65°C for 5 min and transfer quickly on ice for 1 min, spin briefly, and add:

   a. 4 μl of 5× reaction buffer.
   b. 1 μl 0.1M DTT.
   c. 1 μl RNAasin.
   d. 1 μl Superscript III RT.

3. Mix by pipetting up and down, and incubate the reaction at 50°C for 60 min.
4. Inactivate the reaction by heating to 70°C for 10 min and add 180 μl of water.

3.6.3. QPCR

1. Prepare reaction mix for each of the analyzed genes, i.e., TP53, WPRE (*see* **Note 13**), and EF-1α endogenous control (volumes for 1 reaction):

   a. 10 μl of 2× Taqman Reaction Mix.
   b. 1 μl of TP53 assay on demand or 20× probe and primers mix for WPRE.
   c. 4 μl of water.

2. Set up 4 tenfold serial dilutions of RT reaction from Wt MCF7 cells and LVUTHshp53-tTRKRAB-transduced MCF7 cells (Dox+) that will be used to calculate the standard curve for TP53 and GFP transgene (WPRE), respectively (*see* **Note 14**).
3. Aliquot 5 μl of each diluted RT reaction in duplicates on 96-well QPCR plate.
4. Add 15 μl of the reaction mix to each well and run the plates using real time PCR machine.
5. After the run, using the QPCR software provided with the machine calculate the Ct values and export the results to MS Excel; calculate mean values for each sample from the duplicates.
6. Generate the standard curves by plotting the Ct values of the standards (y) against logarithmic values of the standard dilutions (x) for TP53, WPRE and EF-1α; add trendline and display the equation as presented in **Fig. 3**.
7. Using obtained equations, calculate x parameter for each sample (xCt) from the mean Ct values, reverse the logarithmic value ($10^{\wedge(x\mathrm{Ct})}$) and normalize each sample to the endogenous control as presented in **Fig. 4** for GFP transgene (WPRE) and **Fig. 5** for TP53.

Dilution	Dilution log	Ct TP53	Ct WPRE	Ct EF-1
1	0	27.95	18.94	20.65
0.1	-1	31.26	22.62	24.34
0.01	-2	34.19	25.80	27.92
0.001	-3	37.25	28.18	31.11

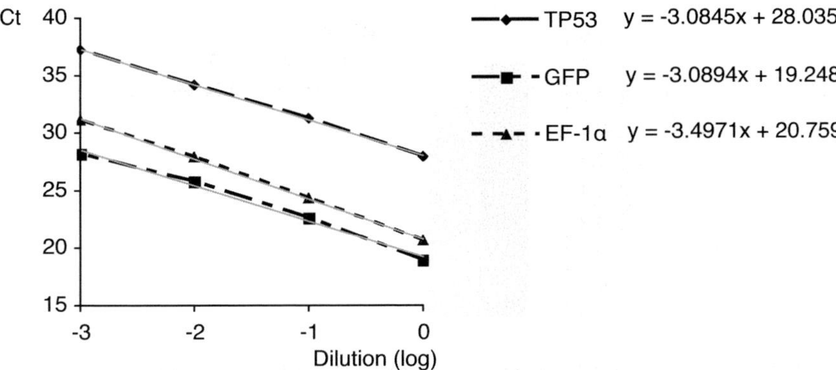

Fig. 3. The standard curves for the transgene (GFP), cellular gene (TP53), and endogenous control (EF-1α). The standard curves are generated by plotting the mean Ct values of the standards (*y*) against logarithmic values of the standard dilutions (*x*). The equations are calculated using trendlines of each curve.

3.7. Analysis of the Conditional Transgene Expression and Gene Knock Down by Western Blot

3.7.1. Protein Isolation

1. Transfer 10^6 cells (from **Subheading 3.4., step 4**) to 1.5-ml Eppendorf tubes, spin at 2000 × *g* for 5 min at 4°C, and remove the medium.
2. Resuspend the cell pellet completely in 0.5 ml of RIPA buffer and incubate 30 min on ice.
3. Spin at 13,000 × *g* at 4°C for 30 min and collect protein lysate (aqueous phase).

3.7.2. Protein Quantification Using BCA Protein Assay Kit

1. Prepare 7 2-fold dilution of the BSA standard in RIPA buffer.
2. Aliquot 10 μl of the protein on 96-well plate (round bottom). Use serial dilutions of BSA provided with a Kit as a standard.
3. Prepare the reaction mix by combining 200 μl of solution A with 4 μl of solution B from the BCA Protein Assay Kit and add 200 μl to each well.
4. Cover the plate with aluminum foil and incubate for 30 min at 37°C.
5. Read the absorbance using ELISA reader and calculate protein concentration for each sample.
6. Adjust protein concentration in each sample to 0.5 μg/μl.

	TP53			EF-1α			Normalized
	Ct	xCt	10^xCT	Ct	xCt	10^xCT	TP53/EF-1α
Wt	29.36	-0.430	0.37	21.96	-0.34	0.45304	0.82
Dox–	28.43	-0.13	0.75	20.61	0.04	1.1	0.68
Dox+	32.67	-1.5	0.03	21.28	-0.15	0.71	0.042

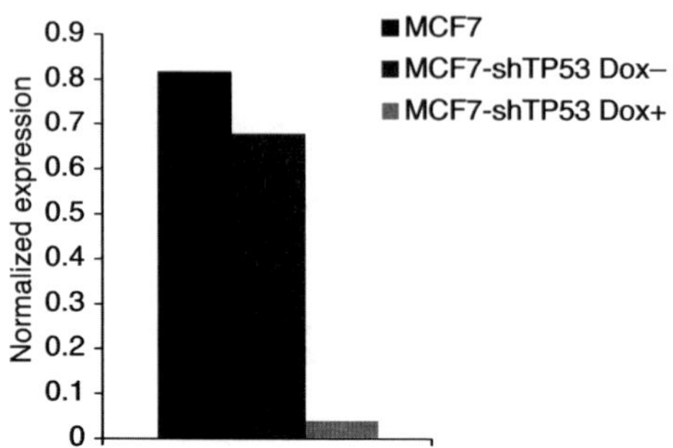

Fig. 4. RT-QPCR quantification of conditional GFP expression. The normalized expression values are obtained by calculating x parameter for each sample (*x*Ct) using the equations obtained from standard curves for GFP and the mean Ct values of the experimental samples obtained using set of probes and primers specific for WPRE, reversing the logarithmic value ($10^{\wedge(xCt)}$) and normalizing each sample to values calculated for EF-1α endogenous control.

3.7.3. Electrophoresis

1. Mix 9 µl (4.5 µg) of protein sample with 3 µl of 4× loading buffer.
2. Heat the samples to 100°C for 5 min and chill quickly on ice.
3. Load 5 µl of a control pre-stained protein ladder (Bio-Rad Laboratories, Hercules, CA, USA) to track sample migration in the first lane of the gel. Load the samples and 5 µl of a control pre-stained protein ladder to track sample migration in the first lane of the 4–12% pre-cast Tri–Bis gel and run at 100 V until the blue bromophenol dye has reached the bottom of the gel.

3.7.4. Transfer

1. Remove the gel from electrophoresis apparatus, cut off the stacking gel, and rinse the gel in transfer buffer.
2. Prepare a blotting sandwich and XCell SureLock™ Mini-Cell CE mark apparatus (Invitrogen). Place gel on a sheet of Whatman 3-mm paper soaked in transfer buffer. Apply a layer of nitrocellulose membrane (Bio-Rad Laboratories) to the gel and remove bubbles by smoothing with gloved fingers, being careful to keep the membrane wet. Cover membrane with another layer of wet Whatman 3-mm

	WPRE			EF-1α			Normalized
	Ct	xCt	10^xCT	Ct	xCt	10^xCT	WPRE/EF-1α
Wt	28.84	-3.105	0.001	21.96	-0.34	0.45	0.0017
Dox–	27.63	-2.712	0.0019	20.61	0.04	1.10146	0.0017
Dox+	23.38	-1.336	0.05	21.28	-0.15	0.71118	0.07

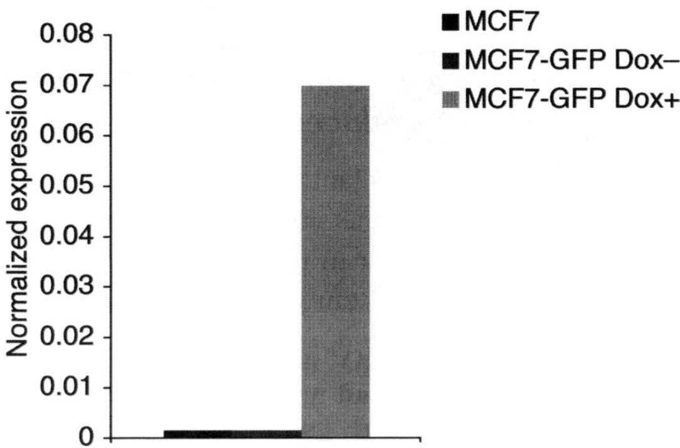

Fig. 5. Reverse transcriptase-quantitative polymerase chain reaction (RT-QPCR) quantification of conditional TP53 knockdown. The normalized expression values are obtained by calculating *x* parameter for each sample (*x*Ct) using the equations obtained from standard curves for TP53 and the mean Ct values of the experimental samples obtained using set of probes and primers specific for TP53, reversing the logarithmic value ($10^{\wedge(xCt)}$) and normalizing each sample to values calculated for EF-1α endogenous control.

paper and place in a blotting sandwich assembly covered with transfer buffer. Immerse the assembly in a blot cell filled with transfer buffer with the membrane side facing the positive electrode and the gel facing the negative electrode.

3. Apply power at a constant current of 475 mA and allow 90 min for the transfer. Perform electrophoresis on ice or on the cold room.
4. Check membrane for transfer of control protein ladder. Asses transfer of samples and relative concentration by staining the membrane with Ponseau S solution. Wash membrane in PBS for a few minutes to remove stain.

3.7.5. Detection

1. Remove the membrane from the sandwich and block in PBSTM (PBS-containing 0.2% Tween, 5% non-fat milk) for 30 min at RT on a rocking platform.
2. Add PBSTM containing primary antibodies: anti-human TP53 (1:1000), anti-GFP (1:1000), anti-PCNA (1:2000) and incubate 1 h at RT.
3. Remove PBSTM and wash three times with PBST; 10 min each wash.

4. Add PBSTM containing anti-mouse secondary antibody conjugated with HRP and incubate 1 h at RT with gentle rocking.
5. Remove PBSTM and wash three times with PBST; 10 min each wash.
6. Mix equal parts of ECL solutions 1 and 2 and add to the membrane for 2 min.
7. Drain the ECL solution mix, wrap the membrane with saran foil, place in the exposure cassette, and expose to the Kodak film (*see* **Fig. 6**).

3.8. Conditional Transgene Expression and Gene Knockdown In Vivo. Xenotransplantation of the Transduced Cells into Nude Mice

1. 48 h before injection of cells implant estrogen pellets subcutaneously into the mice. To implant the pellets, lift the skin on the lateral side of the mouse neck. Make an incision equal in diameter to that of the pellet. Using a pair of forceps, create a horizontal pocket about 2 cm beyond the incision site. Insert the pellet into the pocket with forceps.

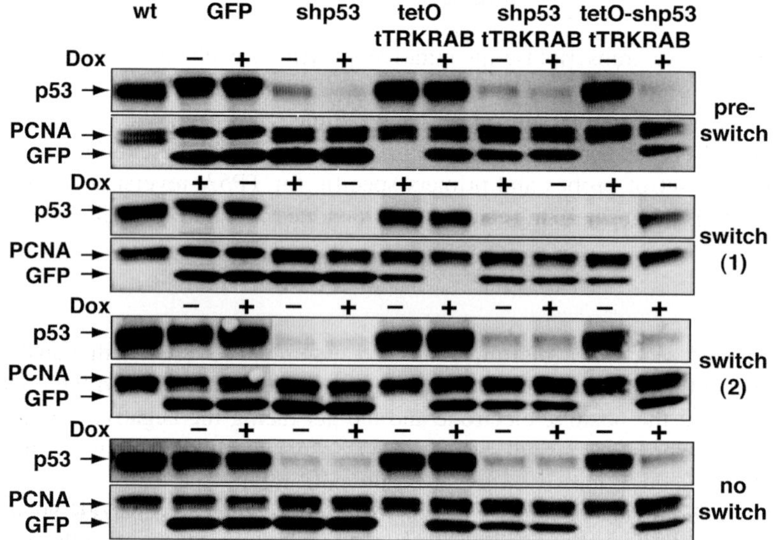

Fig. 6. Western blot analysis of conditional GFP marker expression and TP53 knockdown in human breast cancers cells. MCF7 human breast cancer cells were transduced with the lentiviral vectors at MOI 10 and cultured in the presence of dox for 5 days. Next, the GFP+ cells were sorted, cultured in the presence or absence of dox for 7 days (pre-switch), and kept in the same condition (no switch) or subjected to two subsequent switches at day 7 (switch 1) and 14 (switch 2); (– or +) represent absence or presence of dox after switch. Vector abbreviations: GFP (FUGW), shp53 (pLVUshp53), tetOtTRKRAB (pLVUT-tTRKRAB), shp53tTRKRAB (pLVUHshp53-tTRKRAB), tetOshp53tTRKRAB (pLVUTHshp53-tTRKRAB) *(11)*.

2. Expand the transduced cells in culture, detach from plastic using trypsin, wash once in PBS, count and resuspend at 10^8 cells/ml in PBS, and keep on ice to prevent aggregation.
3. Before injection mix the cell suspension 1:1 with Matrigel and inject 200 μl into the flank of the nude mice (10^7 cells pre mouse) (*see* **Note 16**).
4. After 2 weeks induce gene knockdown by doxycycline administration in the drinking water (2 g/L supplemented with 5% sucrose). Wrap the bottles with the aluminum foil to prevent antibiotic degradation, change Dox-containing water every 3–4 days.
5. After 7 days of induction kill the mice, dissect the subcutaneous tumors, and freeze the samples immediately on dry ice. The samples can be stored in –70ºC. For protein and RNA isolation, use about 5–8 mm^3 of the sample and homogenize quickly in RIPA buffer or Trizol, respectively. Proceed as described in **Subheadings 3.6** and **3.7.**
6. When GFP is used as a fluorescent marker, the status of the conditional knockdown can be evaluated using fluorescent stereomicroscope.

4. Notes

1. Select at least 5–10 sequences for screening; depending on a target gene, screening of 10 shRNAs usually result in 1–2 giving over 90% knockdown of a target gene expression measured at mRNA level. Always check shRNA specificity in a sequence database. The secondary structure of the target mRNA does not have significant effect on knockdown efficiency. The detailed algorithm incorporating eight criteria for selecting efficient siRNA was described ***(18)***. It has to be determined experimentally whether similar rules also apply for designing of functional shRNAs. Not every working siRNA sequence is equally effective when incorporated into shRNA. The discussion forum on the lentivectors, RNAi, and the conditional systems can be found at http://lentiweb.com.
2. Efficient transfection of 239T cells (>95% GFP+ cells) is critical for achieving good vector titers. The following factors are important: (1) quality of plasmid DNA preparation; most of the commercially availably kits provide good quality of plasmid DNA, (2) healthy 293T cells at low passage; the cells need to be split 1:10 every 2–3 days, and (3) exact pH 7.05 of the 2× HBS buffer; the buffer can be stored at –20ºC for months. The quality of the precipitate can be checked under the microscope. Particles should be fine, sandy, and abundant.
3. Vector titer is a number of infectious vector particles [transducing units (TU)] per volume (TU/ml). Its value may vary depending on the read-out, target cells used in the assay, and conditions of transduction. Titer can be measured by quantification of the integrated vector genomes (QPCR) or counting the cells expressing the marker gene: by FACS (fluorescence markers: GFP, RFP, YFP, etc.); by antibody staining and FACS (surface markers: NGFRdel, CD8del); by direct counting of single cells (cytochemical markers LacZ), or by antibiotic-resistant

colonies (resistance markers: puro, hygro, neo, etc.). Concentration of HIV p24 capsid protein (p24 ELISA) or activity of the reverse transcriptase (RT assay) can be used for quantification of vector physical particles (PP). Because the outcome of their outcome includes largely non-infectious particles, these assays cannot be regarded as vector titration (quantification of infectious particles, TU/ml). Relationship between the number of infectious particles (TU) and number of PP (pg of p24) can be used to determine a quality of concentrated vector preparation (infectivity): there are approximately 2000 molecules of p24 per physical particle (PP) of HIV, i.e., $2 \times 10^3 \times 24 \times 10^3$ kDa of p24 per PP. i.e., 48×10^6/Avogadro = $48 \times 10^6/6 \times 10^{23} = 8 \times 10^{-17}$ g of p24 per PP, i.e., approximately 1 PP per $10e^{-16}$ g of p24, i.e., 10^4 PP per pg of p24. A reasonably well-packaged, VSVG pseudotyped lentiviral vector should have an infectivity index in the range of 1 TU per 1000 PP to 1 TU per 100 PP (or less). Thus, the acceptable range is approximately 10–100 TU per 1 pg of p24. Below these values, your vector has experienced problems during packaging.

4. For titration by QPCR, the transduced cells have to be cultured for at least 5–7 days after transduction 5 to limit number of non-integrated particles. QPCR results can be calculated as described in **Subheading 3.7.3**. QPCR titration can be direct, when the result is calculated using DNA isolated from a clone of cells carrying known number of vector copies as a standard (the titer will be expressed as number of vector integrants per genome of target cells), or relative when a sample is analyzed side-by-side and compared with a sample of which functional titer was already measured by any other methods (the titer will be expressed as number of TU/ml). Protocol: Titration of the lentiviral vectors by QPCR.

   a. DNA isolation for QPCR titration: 7 days after transduction, remove the medium from the dilution 1:1, 1:4, and non-transduced cells; add 0.5 ml of SNET buffer complemented with proteinase K and RNAase, collect viscous lysate, and transfer to Eppendorf tubes and incubate at 55°C for 1 h.
   b. Precipitate DNA by adding 1 ml of 95% ethanol, mix by inverting, spin at 13,000 × *g* for 5 min, wash once with 0.5 ml of 70% ethanol, and resuspend in 1 ml of water.
   c. Prepare 5 serial 10-fold dilution of the genomic DNA isolated from the cells transduced with 1:1 vector dilution that will be used to calculate standard curve and follow the QPCR protocol as described in **Subheading 3.6.3**. Use 5 μl of each DNA sample; dilute if necessary.

5. Multiplicity of infection (MOI) is a number of infection particles per number of the target cells and can be only calculated from the titer.
6. Efficient transduction of the target cells (>90%) is crucial for accurate analysis of the knockdown. If necessary, the transduced cells can be sorted by FACS to achieve homogenous population. Alternatively, the GFP fluorescent marker can be changed to antibiotic resistance (e.g., puro, hygro, neo) gene of surface

marker (e.g., NGF-R, CD8, lacking the signaling domains). In case of short-lived proteins, knockdown efficiency can be also assayed at the protein level by Western blot.

7. The shRNA sequences can be directly cloned and screened in pLVETH-tTRKRAB vector. In this case, the titration and screening has to be performed in the presence of doxycycline.
8. For some applications (direct in vivo injection, stem cells transduction, LV transgenesis), the vector titer can be increased by concentration. Transfer 30 ml of virus to 33-ml Beckman conical tubes spin at 90,000 × *g* for 2 h at 4°C in Beckman SW28 swingle-bucket rotor. After spin, discard supernatant and resuspend the virus in a desired volume of serum-free medium (e.g., Optimem or Episerf) or PBS/1% BSA, aliquot, and store at –70°C. For transduction of fragile cells, the virus can be concentrated on a sucrose cushion. Place 4 ml of 20% sucrose on the bottom of the tube and overlay with 26 ml of viral supernatant.
9. Performance of the conditional system for the transgene expression as well as drug-inducible knockdown depends largely on the activity of the internal Pol II promoter in a given cell type. For optimal results, 3–4 various promoters should be tested using target cells. In most commonly used adherent cell lines CAG>UbiC>EF-1α>PGK and in most human cells hPGK>mPGK. Note that any Pol II promoter can be subjected to Dox-inducible regulation using the tTRKRAB system.
10. To achieve higher transduction efficiency, the target cells can be transduced in suspension. In this case, infect the cells at the time of plating in a total volume of 250 μl for 2 h as described in **step 2**.
11. In most cases, the kinetic of the Off to On switch upon Dox administration is faster than the reverse On to Off switch that lags because of drug accumulation in the transduced cells and additionally depends on target mRNA or protein stability. For the optimal results, keep the Dox concentration as low as possible. Titration of the drug-response using decreasing Dox concentrations may be helpful because drug-responsiveness of the system in vitro may differ between various cell lines. Some batches of FCS or FBS may contain traces of doxycycline that result in leakiness of the conditional system.
12. RNA is prone to degradation by RNAases. Always use sterile or autoclaved tips and tubes, DEPC-treated water. Use separate bench to avoid contamination by plasmid DNA or other samples that may interfere with QPCR read-out.
13. WPRE sequence is always a part of a transcript, and WPRE-specific primers and probe can be used to analyze expression of any transgene expressed from an internal Pol II promoter.
14. Standard curve has to be drawn using a RT reaction sample prepared from the cells expressing high levels of an analyzed gene; Wt cells can be used for analysis of TP53 and EF-1α and the vector transduced cells for analysis of GFP transgene expression.

15. The Licor Odyssey System Licor Biosciences, Lincoln, NE, USA) can be used for quantification of protein expression. Please follow the protocols and instructions provided by the manufacturer for details.
16. When transplanting 10^7 cells per mice, tumors appear at 10–14 days after subcutaneous injection. Reduced number of cells (about 10^6) is recommended to achieve slower kinetic of the tumor growth.

Acknowledgments

Authors thank Maciej Wiznerowicz for critical reading of this chapter.

References

1. Toniatti, C., Bujard, H., Cortese, R., and Ciliberto, G. (2004) Gene therapy progress and prospects: transcription regulatory systems. *Gene Ther* **11,** 649–57.
2. Gossen, M., and Bujard, H. (1992) Tight control of gene expression in mammalian cells by tetracycline-responsive promoters. *Proc Natl Acad Sci USA* **89,** 5547–51.
3. Gossen, M., Freundlieb, S., Bender, G., Muller, G., Hillen, W., and Bujard, H. (1995) Transcriptional activation by tetracyclines in mammalian cells. *Science* **268,** 1766–9.
4. Margolin, J. F., Friedman, J. R., Meyer, W. K., Vissing, H., Thiesen, H. J., and Rauscher, F. J., 3rd (1994) Kruppel-associated boxes are potent transcriptional repression domains. *Proc Natl Acad Sci USA* **91,** 4509–13.
5. Moosmann, P., Georgiev, O., Thiesen, H. J., Hagmann, M., and Schaffner, W. (1997) Silencing of RNA polymerases II and III-dependent transcription by the KRAB protein domain of KOX1, a Kruppel-type zinc finger factor. *Biol Chem* **378,** 669–77.
6. Urrutia, R. (2003) KRAB-containing zinc-finger repressor proteins. *Genome Biol* **4,** 231.
7. Brummelkamp, T. R., Bernards, R., and Agami, R. (2002) Stable suppression of tumorigenicity by virus-mediated RNA interference. *Cancer Cell* **2,** 243–7.
8. Stegmeier, F., Hu, G., Rickles, R. J., Hannon, G. J., and Elledge, S. J. (2005) A lentiviral microRNA-based system for single-copy polymerase II-regulated RNA interference in mammalian cells. *Proc Natl Acad Sci USA* **102,** 13212–7.
9. Dickins, R. A., Hemann, M. T., Zilfou, J. T., Simpson, D. R., Ibarra, I., Hannon, G. J., and Lowe, S. W. (2005) Probing tumor phenotypes using stable and regulated synthetic microRNA precursors. *Nat Genet* **37,** 1289–95.
10. Wiznerowicz, M. and Trono, D. (2005) Harnessing HIV for therapy, basic research and biotechnology. *Trends Biotechnol* **23,** 42–7.
11. Szulc, J., Wiznerowicz, M., Sauvain, M. O., Trono, D., and Aebischer, P. (2006) A versatile tool for conditional gene expression and knockdown. *Nat Methods* **3,** 109–16.
12. Miyake, K., Flygare, J., Kiefer, T., Utsugisawa, T., Richter, J., Ma, Z., Wiznerowicz, M., Trono, D., and Karlsson, S. (2005) Development of cellular

models for ribosomal protein S19 (RPS19)-deficient diamond-blackfan anemia using inducible expression of siRNA against RPS19. *Mol Ther* **11,** 627–37.

13. Arrighi, J. F., Pion, M., Wiznerowicz, M., Geijtenbeek, T. B., Garcia, E., Abraham, S., Leuba, F., Dutoit, V., Ducrey-Rundquist, O., van Kooyk, Y., Trono, D., and Piguet, V. (2004) Lentivirus-mediated RNA interference of DC-SIGN expression inhibits human immunodeficiency virus transmission from dendritic cells to T cells. *J Virol* **78,** 10848–55.
14. Moffat, J., Grueneberg, D. A., Yang, X., Kim, S. Y., Kloepfer, A. M., Hinkle, G., Piqani, B., Eisenhaure, T. M., Luo, B., Grenier, J. K., Carpenter, A. E., Foo, S. Y., Stewart, S. A., Stockwell, B. R., Hacohen, N., Hahn, W. C., Lander, E. S., Sabatini, D. M., and Root, D. E. (2006) A lentiviral RNAi library for human and mouse genes applied to an arrayed viral high-content screen. *Cell* **124,** 1283–98.
15. Berns, K., Hijmans, E. M., Mullenders, J., Brummelkamp, T. R., Velds, A., Heimerikx, M., Kerkhoven, R. M., Madiredjo, M., Nijkamp, W., Weigelt, B., Agami, R., Ge, W., Cavet, G., Linsley, P. S., Beijersbergen, R. L., and Bernards, R. (2004) A large-scale RNAi screen in human cells identifies new components of the p53 pathway. *Nature* **428,** 431–7.
16. Ngo, V. N., Davis, R. E., Lamy, L., Yu, X., Zhao, H., Lenz, G., Lam, L. T., Dave, S., Yang, L., Powell, J., and Staudt, L. M. (2006) A loss-of-function RNA interference screen for molecular targets in cancer. *Nature* **441,** 106–10.
17. Silva, J. M., Li, M. Z., Chang, K., Ge, W., Golding, M. C., Rickles, R. J., Siolas, D., Hu, G., Paddison, P. J., Schlabach, M. R., Sheth, N., Bradshaw, J., Burchard, J., Kulkarni, A., Cavet, G., Sachidanandam, R., McCombie, W. R., Cleary, M. A., Elledge, S. J., and Hannon, G. J. (2005) Second-generation shRNA libraries covering the mouse and human genomes. *Nat Genet* **37,** 1281–8.
18. Reynolds, A., Leake, D., Boese, Q., Scaringe, S., Marshall, W. S., and Khvorova, A. (2004) Rational siRNA design for RNA interference. *Nat Biotechnol* **22,** 326–30.

Index

Printed in the United States of America